DATE

VOLUME FIVE HUNDRED AND THIRTEEN

METHODS IN
ENZYMOLOGY

Nucleosomes, Histones & Chromatin
Part B

METHODS IN ENZYMOLOGY

Editors-in-Chief

JOHN N. ABELSON and MELVIN I. SIMON
Division of Biology
California Institute of Technology
Pasadena, California

Founding Editors

SIDNEY P. COLOWICK and NATHAN O. KAPLAN

VOLUME FIVE HUNDRED AND THIRTEEN

METHODS IN ENZYMOLOGY

Nucleosomes, Histones & Chromatin
Part B

Edited by

CARL WU
Howard Hughes Medical Institute
Janelia Farm Research Campus
Ashburn, VA, USA

C. DAVID ALLIS
Joy and Jack Fishman Professor
Head, Laboratory of Chromatin Biology and Epigenetics
The Rockefeller University
New York, NY, USA

AMSTERDAM • BOSTON • HEIDELBERG • LONDON
NEW YORK • OXFORD • PARIS • SAN DIEGO
SAN FRANCISCO • SINGAPORE • SYDNEY • TOKYO
Academic Press is an imprint of Elsevier

Academic Press is an imprint of Elsevier
525 B Street, Suite 1900, San Diego, CA 92101-4495, USA
225 Wyman Street, Waltham, MA 02451, USA
The Boulevard, Langford Lane, Kidlington, Oxford, OX51GB, UK
32, Jamestown Road, London NW1 7BY, UK
Radarweg 29, PO Box 211, 1000 AE Amsterdam, The Netherlands

First edition 2012

Copyright © 2012, Elsevier Inc. All Rights Reserved.

No part of this publication may be reproduced, stored in a retrieval system or transmitted in any form or by any means electronic, mechanical, photocopying, recording or otherwise without the prior written permission of the publisher

Permissions may be sought directly from Elsevier's Science & Technology Rights Department in Oxford, UK: phone (+44) (0) 1865 843830; fax (+44) (0) 1865 853333; email: permissions@elsevier.com. Alternatively you can submit your request online by visiting the Elsevier web site at http://elsevier.com/locate/permissions, and selecting *Obtaining permission to use Elsevier material*

Notice
No responsibility is assumed by the publisher for any injury and/or damage to persons or property as a matter of products liability, negligence or otherwise, or from any use or operation of any methods, products, instructions or ideas contained in the material herein. Because of rapid advances in the medical sciences, in particular, independent verification of diagnoses and drug dosages should be made

For information on all Academic Press publications visit our website at store.elsevier.com

ISBN: 978-0-12-391938-0
ISSN: 0076-6879

Printed and bound in United States of America
12 13 14   11 10 9 8 7 6 5 4 3 2 1

Working together to grow
libraries in developing countries

www.elsevier.com | www.bookaid.org | www.sabre.org

ELSEVIER   BOOK AID International   Sabre Foundation

CONTENTS

Contributors   xi
Preface   xv
Volumes in Series   xvii

Section 1
Single Molecule Studies of Nucleosomes

1. DNA Translocation of ATP-Dependent Chromatin Remodeling Factors Revealed by High-Resolution Optical Tweezers   3
Yongli Zhang, George Sirinakis, Greg Gundersen, Zhiqun Xi, and Ying Gao

    1. Introduction   4
    2. Instrument   6
    3. Remodeler Translocation on Bare DNA   13
    4. Nucleosome-Dependent Remodeler Translocation   18
    5. Data Analysis   25
    Acknowledgments   26
    References   26

2. Unzipping Single DNA Molecules to Study Nucleosome Structure and Dynamics   29
Ming Li and Michelle D. Wang

    1. Introduction   30
    2. Sample Preparation   33
    3. Instrumentation and Data Collection   39
    4. Data Processing   43
    5. Determination of Unzipping Accuracy and Precision   46
    6. Unzipping in Nucleosome Studies   48
    7. Conclusions   56
    Acknowledgments   56
    References   56

3. Monitoring Conformational Dynamics with Single-Molecule
   Fluorescence Energy Transfer: Applications in Nucleosome
   Remodeling   59
   Sebastian Deindl and Xiaowei Zhuang

   1. Introduction   60
   2. Preparation of Fluorescently Labeled Sample for Single-Molecule
      FRET Imaging   62
   3. Preparing PEG-Coated Slides with Sample Chamber   69
   4. Optical Setup and Single-Molecule FRET Data Acquisition   71
   5. Data Analysis   77
   6. Summary   83
   Acknowledgments   83
   References   83

Section 2
Higher Order Chromatin Interactions

4. 4C Technology: Protocols and Data Analysis   89
   Harmen J.G. van de Werken, Paula J.P. de Vree, Erik Splinter, Sjoerd J.B.
   Holwerda, Petra Klous, Elzo de Wit, and Wouter de Laat

   1. Introduction   90
   2. 4C Template Preparation   92
   3. 4C-Seq Primer Design and PCR   100
   4. High-Throughput Sequencing of 4C PCR Products   103
   5. Data Analysis   103
   Acknowledgments   110
   References   110

5. A Torrent of Data: Mapping Chromatin Organization Using
   5C and High-Throughput Sequencing   113
   James Fraser, Sylvain D. Ethier, Hisashi Miura, and Josée Dostie

   1. Introduction   114
   2. Preparing Torrent 5C Libraries   121
   3. Overview of the Ion Torrent Sequencing Protocol   129
   4. Torrent 5C Data Processing   131
   5. Conclusion   138
   Acknowledgments   138
   References   138

Section 3
Genome wide Analyses of Chromatin and Transcripts

6. Genome-Wide Mapping of Nucleosomes in Yeast Using Paired-End Sequencing — 145

Hope A. Cole, Bruce H. Howard, and David J. Clark

1. Introduction — 146
2. Preparation of Nucleosome Core Particles from Yeast — 148
3. Preparation of Core Particle DNA for Sequencing — 152
4. Paired-End Sequencing — 158
5. Bioinformatic Analysis of Nucleosome Sequences — 159
6. Some General Experimental Considerations — 165
Acknowledgments — 166
References — 167

7. Measuring Genome-Wide Nucleosome Turnover Using CATCH-IT — 169

Sheila S. Teves, Roger B. Deal, and Steven Henikoff

1. Introduction — 170
2. Covalent Attachment of Tagged Histones to Capture and Identify Turnover — 171
3. Modified Solexa Library Preparation — 178
References — 184

8. DNA Methyltransferase Accessibility Protocol for Individual Templates by Deep Sequencing — 185

Russell P. Darst, Nancy H. Nabilsi, Carolina E. Pardo, Alberto Riva, and Michael P. Kladde

1. Introduction — 186
2. Materials — 188
3. Protocols — 190
Acknowledgments — 202
References — 202

9. Genome-Wide *In Vitro* Reconstitution of Yeast Chromatin with *In Vivo*-Like Nucleosome Positioning — 205

Nils Krietenstein, Christian J. Wippo, Corinna Lieleg, and Philipp Korber

1. Introduction — 206
2. Preparation of Yeast WCE — 207

3. Preparation of Histones 211
4. Expansion of Genomic Plasmid Library 214
5. Chromatin Reconstitution by Salt Gradient Dialysis 217
6. Incubation of Salt Gradient Dialysis Chromatin with WCE or Purified Factors 227
7. Generation of Mononucleosomes by Limited MNase Digestion 228
8. Summary 229
Acknowledgments 230
References 230

10. Genome-Wide Mapping of Nucleosome Positions in Yeast Using High-Resolution MNase ChIP-Seq 233
Megha Wal and B. Franklin Pugh

1. Introduction 233
2. Methodology 235
Acknowledgments 249
References 249

11. Preparation of *Drosophila* Tissue Culture Cells from Different Stages of the Cell Cycle for Chromatin Immunoprecipitation Using Centrifugal Counterflow Elutriation and Fluorescence-Activated Cell Sorting 251
Nicole E. Follmer and Nicole J. Francis

1. Introduction 252
2. Centrifugal Elutriation of *Drosophila* Cells to Obtain G1, S, and G2 Populations 253
3. FACS Sorting of *Drosophila* Cells to Obtain Mitotic Cell Populations 261
4. Summary 268
Acknowledgments 268
References 268

12. Genome-Wide Polyadenylation Site Mapping 271
Vicent Pelechano, Stefan Wilkening, Aino Inkeri Järvelin, Manu M. Tekkedil, and Lars M. Steinmetz

1. Introduction 272
2. General Recommendations 275
3. Sample Preparation 276
4. Library Construction 282
5. Bioinformatic Analysis of Polyadenylation Site Sequencing Reads 289

6. Quality Control	291
7. Preparation of Double-Stranded Linkers	293
8. Summary	294
Acknowledgments	294
References	294

13. Genome-Wide Mapping of Nucleosome Occupancy, Histone Modifications, and Gene Expression Using Next-Generation Sequencing Technology   297

Gang Wei, Gangqing Hu, Kairong Cui, and Keji Zhao

1. Introduction	298
2. Mapping Genome-Wide Nucleosome Occupancy and Positions by High-Throughput Next-Generation Sequencing	299
3. Mapping Histone Modifications Using ChIP-Seq	302
4. Mapping Genome-Wide mRNA Profiles Using RNA-Seq	304
5. RNA-Seq Analysis with a Small Number of Cells	307
Acknowledgment	311
References	311

14. A Chemical Approach to Mapping Nucleosomes at Base Pair Resolution in Yeast   315

Kristin R. Brogaard, Liqun Xi, Ji-Ping Wang, and Jonathan Widom

1. Introduction	316
2. Construction of H4S47C *S. cerevisiae* Strain	317
3. Chemical Cleavage of Nucleosome Center Positions	318
4. Statistical Analysis of Chemical Mapping Data	324
5. Summary	332
Acknowledgments	333
References	333

Author Index	*335*
Subject Index	*349*

CONTRIBUTORS

Kristin R. Brogaard
Department of Molecular Biosciences, Northwestern University, Evanston, Illinois, USA

David J. Clark
Program in Genomics of Differentiation, *Eunice Kennedy Shriver* National Institute of Child Health and Human Development, National Institutes of Health, Bethesda, Maryland, USA

Hope A. Cole
Program in Genomics of Differentiation, *Eunice Kennedy Shriver* National Institute of Child Health and Human Development, National Institutes of Health, Bethesda, Maryland, USA

Kairong Cui
Systems Biology Center, NHLBI, NIH, Bethesda, Maryland, USA

Russell P. Darst
Department of Biochemistry and Molecular Biology, University of Florida Shands Cancer Center Program in Cancer Genetics, Epigenetics, and Tumor Virology, Gainesville, Florida, USA

Wouter de Laat
Hubrecht Institute-KNAW and University Medical Center Utrecht, Utrecht, The Netherlands

Paula J.P. de Vree
Hubrecht Institute-KNAW and University Medical Center Utrecht, Utrecht, The Netherlands

Elzo de Wit
Hubrecht Institute-KNAW and University Medical Center Utrecht, Utrecht, The Netherlands

Roger B. Deal
Division of Basic Sciences, Fred Hutchinson Cancer Research Center, Seattle, Washington, USA

Sebastian Deindl
Howard Hughes Medical Institute, and Department of Chemistry and Chemical Biology, Harvard University, Cambridge, Massachusetts, USA

Josée Dostie
Department of Biochemistry and Goodman Cancer Research Center, McGill University, Montréal, Québec, Canada

Sylvain D. Ethier
Department of Biochemistry and Goodman Cancer Research Center, McGill University, Montréal, Québec, Canada

Nicole E. Follmer
Department of Molecular and Cellular Biology, Harvard University, Cambridge, Massachusetts, USA

Nicole J. Francis
Department of Molecular and Cellular Biology, Harvard University, Cambridge, Massachusetts, USA

James Fraser
Department of Biochemistry and Goodman Cancer Research Center, McGill University, Montréal, Québec, Canada

Ying Gao
Department of Cell Biology, Yale University School of Medicine, New Haven, Connecticut, USA

Greg Gundersen
Department of Cell Biology, Yale University School of Medicine, New Haven, Connecticut, USA

Steven Henikoff
Division of Basic Sciences, Fred Hutchinson Cancer Research Center, and Howard Hughes Medical Institute, Seattle, Washington, USA

Sjoerd J.B. Holwerda
Hubrecht Institute-KNAW and University Medical Center Utrecht, Utrecht, The Netherlands

Bruce H. Howard
Program in Genomics of Differentiation, *Eunice Kennedy Shriver* National Institute of Child Health and Human Development, National Institutes of Health, Bethesda, Maryland, USA

Gangqing Hu
Systems Biology Center, NHLBI, NIH, Bethesda, Maryland, USA

Aino Inkeri Järvelin
Genome Biology Unit, European Molecular Biology Laboratory, Heidelberg, Germany

Michael P. Kladde
Department of Biochemistry and Molecular Biology, University of Florida Shands Cancer Center Program in Cancer Genetics, Epigenetics, and Tumor Virology, Gainesville, Florida, USA

Petra Klous
Hubrecht Institute-KNAW and University Medical Center Utrecht, Utrecht, The Netherlands

Philipp Korber
Adolf-Butenandt-Institut, Molecular Biology Unit, University of Munich, Munich, Germany

Nils Krietenstein
Adolf-Butenandt-Institut, Molecular Biology Unit, University of Munich, Munich, Germany

Ming Li
Department of Chemistry and Chemical Biology, Cornell University, Ithaca, New York, USA

Corinna Lieleg
Adolf-Butenandt-Institut, Molecular Biology Unit, University of Munich, Munich, Germany

Hisashi Miura
Department of Biochemistry and Goodman Cancer Research Center, McGill University, Montréal, Québec, Canada

Nancy H. Nabilsi
Department of Biochemistry and Molecular Biology, University of Florida Shands Cancer Center Program in Cancer Genetics, Epigenetics, and Tumor Virology, Gainesville, Florida, USA

Carolina E. Pardo
Department of Biochemistry and Molecular Biology, University of Florida Shands Cancer Center Program in Cancer Genetics, Epigenetics, and Tumor Virology, Gainesville, Florida, USA

Vicent Pelechano
Genome Biology Unit, European Molecular Biology Laboratory, Heidelberg, Germany

B. Franklin Pugh
Center for Eukaryotic Gene Regulation, Department of Biochemistry and Molecular Biology, The Pennsylvania State University, University Park, Pennsylvania, USA

Alberto Riva
Department of Molecular Genetics and Microbiology, University of Florida Genetics Institute, Gainesville, Florida, USA

George Sirinakis
Department of Cell Biology, Yale University School of Medicine, New Haven, Connecticut, USA

Erik Splinter
Hubrecht Institute-KNAW and University Medical Center Utrecht, Utrecht, The Netherlands

Lars M. Steinmetz
Genome Biology Unit, European Molecular Biology Laboratory, Heidelberg, Germany

Manu M. Tekkedil
Genome Biology Unit, European Molecular Biology Laboratory, Heidelberg, Germany

Sheila S. Teves
Division of Basic Sciences, Fred Hutchinson Cancer Research Center, and Molecular and Cellular Biology Program, University of Washington, Seattle, Washington, USA

Harmen J.G. van de Werken
Hubrecht Institute-KNAW and University Medical Center Utrecht, Utrecht, The Netherlands

Megha Wal
Center for Eukaryotic Gene Regulation, Department of Biochemistry and Molecular Biology, The Pennsylvania State University, University Park, Pennsylvania, USA

Ji-Ping Wang
Department of Statistics, Northwestern University, Evanston, Illinois, USA

Michelle D. Wang
Department of Physics, Laboratory of Atomic and Solid State Physics, and Howard Hughes Medical Institute, Cornell University, Ithaca, New York, USA

Gang Wei[1]
Systems Biology Center, NHLBI, NIH, Bethesda, Maryland, USA

Jonathan Widom
Department of Molecular Biosciences, Northwestern University, Evanston, Illinois, USA

Stefan Wilkening
Genome Biology Unit, European Molecular Biology Laboratory, Heidelberg, Germany

Christian J. Wippo
Adolf-Butenandt-Institut, Molecular Biology Unit, University of Munich, Munich, Germany

Liqun Xi
Department of Statistics, Northwestern University, Evanston, Illinois, USA

Zhiqun Xi
Department of Cell Biology, Yale University School of Medicine, New Haven, Connecticut, USA

Yongli Zhang
Department of Cell Biology, Yale University School of Medicine, New Haven, Connecticut, USA

Keji Zhao
Systems Biology Center, NHLBI, NIH, Bethesda, Maryland, USA

Xiaowei Zhuang
Howard Hughes Medical Institute; Department of Chemistry and Chemical Biology, and Department of Physics, Harvard University, Cambridge, Massachusetts, USA

---

[1] Present address: CAS-MPG Partner Institute for Computational Biology, SIBS, CAS, Shanghai, China

PREFACE

In 2004, when we edited the last three-volume series on *Chromatin and Chromatin Remodeling Enzymes*, remarkable progress had been made accessing the enzyme complexes that serve to remodel or modify histones and DNA, giving molecular insights into what might be considered "epigenetic landscapes" that lead to stable, and potentially heritable, states of gene expression. Clearly, that was only the tip of the iceberg. Over the past 8 years, interest in chromatin biology, and epigenetics in general, has sky-rocketed, in part due to numerous disease links to dysfunction in most, if not all, of the general classes of chromatin proteins, leading to the misactivation or missilencing of gene targets. Moreover, chromatin protein motifs such as bromodomains that recognize histone acetyl-lysine marks have now proven to be effective drug targets in several types of human disorders, lending support to the general view that modulating nonenzymatic, protein–protein interactions in chromatin-associated proteins can lead to therapeutically useful outcomes in humans. Researchers, in some cases, armed now with molecular insights gained from X-ray and NMR structures, are beginning to unravel the "rules" of effector recruitment to the chromatin template.

As well, the complexities of epigenetic landscapes have now been explored in the context of whole genomes, by combining RNA and chromatin immunoprecipitation with deep sequencing approaches. Importantly, this work is being done examining normal and disease states, stimulated by a flurry of recent findings of missense mutations in epigenetic-modifying activities, including the histone proteins themselves. Challenges remain such as determining "cause versus effect" in bringing about pathological disease states, even to the point of knowing whether histones and or methylated DNA are the physiologically relevant substrates for many of these activities. In addition, there remains a wide gap between reductionist, biochemical (vertical) approaches and the now-popular "omic" (horizontal) approaches, providing an excellent opportunity for clever new strategies to help "connect the dots" underlying these different scientific styles.

With the unquestioned importance of chromatin structure and function in human biology and disease, a wealth of new researchers are entering the field, bringing with them new expertise, methods, and approaches. This updated, two-volume series aims to bring many of these advances to the community, including exposure to genome-wide approaches, single-molecule microscopy methods, and peptide array-based assays. We are most grateful to our colleagues for their efforts in contributing to these volumes. The popularity of the field has also spawned numerous other sources of chromatin methods that the reader may refer to. If the speed of discovery and technology development in prior years is any guide, we anticipate that this volume set will soon require updating. However, we are confident that each of the series devoted to chromatin over the years continues to provide an enduring and evolving set of tools for many newcomers who seek to address the most fundamental problems in chromatin biology.

In closing, we wish to make a special note in appreciation of Jonathan Widom, who passed away in 2011. Jon was the foremost practitioner of chromatin biophysics in our era and contributed on many occasions to *Methods in Enzymology*. We owe him a debt of gratitude for leaving a legacy of key concepts in nucleosome dynamics, robust biophysical techniques, and invaluable reagents that he and his laboratory provided to the chromatin field for decades. It is fitting that the closing chapter of these volumes honors his most recent contribution—a new method for mapping nucleosome positions *in vivo* at base pair resolution. Jon will be greatly missed.

CARL WU
C. DAVID ALLIS

METHODS IN ENZYMOLOGY

VOLUME I. Preparation and Assay of Enzymes
Edited by SIDNEY P. COLOWICK AND NATHAN O. KAPLAN

VOLUME II. Preparation and Assay of Enzymes
Edited by SIDNEY P. COLOWICK AND NATHAN O. KAPLAN

VOLUME III. Preparation and Assay of Substrates
Edited by SIDNEY P. COLOWICK AND NATHAN O. KAPLAN

VOLUME IV. Special Techniques for the Enzymologist
Edited by SIDNEY P. COLOWICK AND NATHAN O. KAPLAN

VOLUME V. Preparation and Assay of Enzymes
Edited by SIDNEY P. COLOWICK AND NATHAN O. KAPLAN

VOLUME VI. Preparation and Assay of Enzymes *(Continued)*
Preparation and Assay of Substrates
Special Techniques
Edited by SIDNEY P. COLOWICK AND NATHAN O. KAPLAN

VOLUME VII. Cumulative Subject Index
Edited by SIDNEY P. COLOWICK AND NATHAN O. KAPLAN

VOLUME VIII. Complex Carbohydrates
Edited by ELIZABETH F. NEUFELD AND VICTOR GINSBURG

VOLUME IX. Carbohydrate Metabolism
Edited by WILLIS A. WOOD

VOLUME X. Oxidation and Phosphorylation
Edited by RONALD W. ESTABROOK AND MAYNARD E. PULLMAN

VOLUME XI. Enzyme Structure
Edited by C. H. W. HIRS

VOLUME XII. Nucleic Acids (Parts A and B)
Edited by LAWRENCE GROSSMAN AND KIVIE MOLDAVE

VOLUME XIII. Citric Acid Cycle
Edited by J. M. LOWENSTEIN

VOLUME XIV. Lipids
Edited by J. M. LOWENSTEIN

VOLUME XV. Steroids and Terpenoids
Edited by RAYMOND B. CLAYTON

VOLUME XVI. Fast Reactions
Edited by KENNETH KUSTIN

VOLUME XVII. Metabolism of Amino Acids and Amines (Parts A and B)
Edited by HERBERT TABOR AND CELIA WHITE TABOR

VOLUME XVIII. Vitamins and Coenzymes (Parts A, B, and C)
Edited by DONALD B. MCCORMICK AND LEMUEL D. WRIGHT

VOLUME XIX. Proteolytic Enzymes
Edited by GERTRUDE E. PERLMANN AND LASZLO LORAND

VOLUME XX. Nucleic Acids and Protein Synthesis (Part C)
Edited by KIVIE MOLDAVE AND LAWRENCE GROSSMAN

VOLUME XXI. Nucleic Acids (Part D)
Edited by LAWRENCE GROSSMAN AND KIVIE MOLDAVE

VOLUME XXII. Enzyme Purification and Related Techniques
Edited by WILLIAM B. JAKOBY

VOLUME XXIII. Photosynthesis (Part A)
Edited by ANTHONY SAN PIETRO

VOLUME XXIV. Photosynthesis and Nitrogen Fixation (Part B)
Edited by ANTHONY SAN PIETRO

VOLUME XXV. Enzyme Structure (Part B)
Edited by C. H. W. HIRS AND SERGE N. TIMASHEFF

VOLUME XXVI. Enzyme Structure (Part C)
Edited by C. H. W. HIRS AND SERGE N. TIMASHEFF

VOLUME XXVII. Enzyme Structure (Part D)
Edited by C. H. W. HIRS AND SERGE N. TIMASHEFF

VOLUME XXVIII. Complex Carbohydrates (Part B)
Edited by VICTOR GINSBURG

VOLUME XXIX. Nucleic Acids and Protein Synthesis (Part E)
Edited by LAWRENCE GROSSMAN AND KIVIE MOLDAVE

VOLUME XXX. Nucleic Acids and Protein Synthesis (Part F)
Edited by KIVIE MOLDAVE AND LAWRENCE GROSSMAN

VOLUME XXXI. Biomembranes (Part A)
Edited by SIDNEY FLEISCHER AND LESTER PACKER

VOLUME XXXII. Biomembranes (Part B)
Edited by SIDNEY FLEISCHER AND LESTER PACKER

VOLUME XXXIII. Cumulative Subject Index Volumes I-XXX
Edited by MARTHA G. DENNIS AND EDWARD A. DENNIS

VOLUME XXXIV. Affinity Techniques (Enzyme Purification: Part B)
Edited by WILLIAM B. JAKOBY AND MEIR WILCHEK

VOLUME XXXV. Lipids (Part B)
Edited by JOHN M. LOWENSTEIN

VOLUME XXXVI. Hormone Action (Part A: Steroid Hormones)
Edited by BERT W. O'MALLEY AND JOEL G. HARDMAN

VOLUME XXXVII. Hormone Action (Part B: Peptide Hormones)
Edited by BERT W. O'MALLEY AND JOEL G. HARDMAN

VOLUME XXXVIII. Hormone Action (Part C: Cyclic Nucleotides)
Edited by JOEL G. HARDMAN AND BERT W. O'MALLEY

VOLUME XXXIX. Hormone Action (Part D: Isolated Cells, Tissues, and Organ Systems)
Edited by JOEL G. HARDMAN AND BERT W. O'MALLEY

VOLUME XL. Hormone Action (Part E: Nuclear Structure and Function)
Edited by BERT W. O'MALLEY AND JOEL G. HARDMAN

VOLUME XLI. Carbohydrate Metabolism (Part B)
Edited by W. A. WOOD

VOLUME XLII. Carbohydrate Metabolism (Part C)
Edited by W. A. WOOD

VOLUME XLIII. Antibiotics
Edited by JOHN H. HASH

VOLUME XLIV. Immobilized Enzymes
Edited by KLAUS MOSBACH

VOLUME XLV. Proteolytic Enzymes (Part B)
Edited by LASZLO LORAND

VOLUME XLVI. Affinity Labeling
Edited by WILLIAM B. JAKOBY AND MEIR WILCHEK

VOLUME XLVII. Enzyme Structure (Part E)
Edited by C. H. W. HIRS AND SERGE N. TIMASHEFF

VOLUME XLVIII. Enzyme Structure (Part F)
Edited by C. H. W. HIRS AND SERGE N. TIMASHEFF

VOLUME XLIX. Enzyme Structure (Part G)
Edited by C. H. W. HIRS AND SERGE N. TIMASHEFF

VOLUME L. Complex Carbohydrates (Part C)
Edited by VICTOR GINSBURG

VOLUME LI. Purine and Pyrimidine Nucleotide Metabolism
Edited by PATRICIA A. HOFFEE AND MARY ELLEN JONES

VOLUME LII. Biomembranes (Part C: Biological Oxidations)
Edited by SIDNEY FLEISCHER AND LESTER PACKER

VOLUME LIII. Biomembranes (Part D: Biological Oxidations)
Edited by SIDNEY FLEISCHER AND LESTER PACKER

VOLUME LIV. Biomembranes (Part E: Biological Oxidations)
Edited by SIDNEY FLEISCHER AND LESTER PACKER

VOLUME LV. Biomembranes (Part F: Bioenergetics)
Edited by SIDNEY FLEISCHER AND LESTER PACKER

VOLUME LVI. Biomembranes (Part G: Bioenergetics)
Edited by SIDNEY FLEISCHER AND LESTER PACKER

VOLUME LVII. Bioluminescence and Chemiluminescence
Edited by MARLENE A. DELUCA

VOLUME LVIII. Cell Culture
Edited by WILLIAM B. JAKOBY AND IRA PASTAN

VOLUME LIX. Nucleic Acids and Protein Synthesis (Part G)
Edited by KIVIE MOLDAVE AND LAWRENCE GROSSMAN

VOLUME LX. Nucleic Acids and Protein Synthesis (Part H)
Edited by KIVIE MOLDAVE AND LAWRENCE GROSSMAN

VOLUME 61. Enzyme Structure (Part H)
Edited by C. H. W. HIRS AND SERGE N. TIMASHEFF

VOLUME 62. Vitamins and Coenzymes (Part D)
Edited by DONALD B. MCCORMICK AND LEMUEL D. WRIGHT

VOLUME 63. Enzyme Kinetics and Mechanism
(Part A: Initial Rate and Inhibitor Methods)
Edited by DANIEL L. PURICH

VOLUME 64. Enzyme Kinetics and Mechanism
(Part B: Isotopic Probes and Complex Enzyme Systems)
Edited by DANIEL L. PURICH

VOLUME 65. Nucleic Acids (Part I)
Edited by LAWRENCE GROSSMAN AND KIVIE MOLDAVE

VOLUME 66. Vitamins and Coenzymes (Part E)
Edited by DONALD B. MCCORMICK AND LEMUEL D. WRIGHT

VOLUME 67. Vitamins and Coenzymes (Part F)
Edited by DONALD B. MCCORMICK AND LEMUEL D. WRIGHT

VOLUME 68. Recombinant DNA
Edited by RAY WU

VOLUME 69. Photosynthesis and Nitrogen Fixation (Part C)
Edited by ANTHONY SAN PIETRO

VOLUME 70. Immunochemical Techniques (Part A)
Edited by HELEN VAN VUNAKIS AND JOHN J. LANGONE

VOLUME 71. Lipids (Part C)
Edited by JOHN M. LOWENSTEIN

VOLUME 72. Lipids (Part D)
Edited by JOHN M. LOWENSTEIN

VOLUME 73. Immunochemical Techniques (Part B)
Edited by JOHN J. LANGONE AND HELEN VAN VUNAKIS

VOLUME 74. Immunochemical Techniques (Part C)
Edited by JOHN J. LANGONE AND HELEN VAN VUNAKIS

VOLUME 75. Cumulative Subject Index Volumes XXXI, XXXII, XXXIV–LX
Edited by EDWARD A. DENNIS AND MARTHA G. DENNIS

VOLUME 76. Hemoglobins
Edited by ERALDO ANTONINI, LUIGI ROSSI-BERNARDI, AND EMILIA CHIANCONE

VOLUME 77. Detoxication and Drug Metabolism
Edited by WILLIAM B. JAKOBY

VOLUME 78. Interferons (Part A)
Edited by SIDNEY PESTKA

VOLUME 79. Interferons (Part B)
Edited by SIDNEY PESTKA

VOLUME 80. Proteolytic Enzymes (Part C)
Edited by LASZLO LORAND

VOLUME 81. Biomembranes (Part H: Visual Pigments and Purple Membranes, I)
Edited by LESTER PACKER

VOLUME 82. Structural and Contractile Proteins (Part A: Extracellular Matrix)
Edited by LEON W. CUNNINGHAM AND DIXIE W. FREDERIKSEN

VOLUME 83. Complex Carbohydrates (Part D)
Edited by VICTOR GINSBURG

VOLUME 84. Immunochemical Techniques (Part D: Selected Immunoassays)
Edited by JOHN J. LANGONE AND HELEN VAN VUNAKIS

VOLUME 85. Structural and Contractile Proteins (Part B: The Contractile Apparatus and the Cytoskeleton)
Edited by DIXIE W. FREDERIKSEN AND LEON W. CUNNINGHAM

VOLUME 86. Prostaglandins and Arachidonate Metabolites
Edited by WILLIAM E. M. LANDS AND WILLIAM L. SMITH

VOLUME 87. Enzyme Kinetics and Mechanism (Part C: Intermediates, Stereo-chemistry, and Rate Studies)
Edited by DANIEL L. PURICH

VOLUME 88. Biomembranes (Part I: Visual Pigments and Purple Membranes, II)
Edited by LESTER PACKER

VOLUME 89. Carbohydrate Metabolism (Part D)
Edited by WILLIS A. WOOD

VOLUME 90. Carbohydrate Metabolism (Part E)
Edited by WILLIS A. WOOD

VOLUME 91. Enzyme Structure (Part I)
Edited by C. H. W. HIRS AND SERGE N. TIMASHEFF

VOLUME 92. Immunochemical Techniques (Part E: Monoclonal Antibodies and General Immunoassay Methods)
Edited by JOHN J. LANGONE AND HELEN VAN VUNAKIS

VOLUME 93. Immunochemical Techniques (Part F: Conventional Antibodies, Fc Receptors, and Cytotoxicity)
Edited by JOHN J. LANGONE AND HELEN VAN VUNAKIS

VOLUME 94. Polyamines
Edited by HERBERT TABOR AND CELIA WHITE TABOR

VOLUME 95. Cumulative Subject Index Volumes 61–74, 76–80
Edited by EDWARD A. DENNIS AND MARTHA G. DENNIS

VOLUME 96. Biomembranes [Part J: Membrane Biogenesis: Assembly and Targeting (General Methods; Eukaryotes)]
Edited by SIDNEY FLEISCHER AND BECCA FLEISCHER

VOLUME 97. Biomembranes [Part K: Membrane Biogenesis: Assembly and Targeting (Prokaryotes, Mitochondria, and Chloroplasts)]
Edited by SIDNEY FLEISCHER AND BECCA FLEISCHER

VOLUME 98. Biomembranes (Part L: Membrane Biogenesis: Processing and Recycling)
Edited by SIDNEY FLEISCHER AND BECCA FLEISCHER

VOLUME 99. Hormone Action (Part F: Protein Kinases)
Edited by JACKIE D. CORBIN AND JOEL G. HARDMAN

VOLUME 100. Recombinant DNA (Part B)
Edited by RAY WU, LAWRENCE GROSSMAN, AND KIVIE MOLDAVE

VOLUME 101. Recombinant DNA (Part C)
Edited by RAY WU, LAWRENCE GROSSMAN, AND KIVIE MOLDAVE

VOLUME 102. Hormone Action (Part G: Calmodulin and Calcium-Binding Proteins)
Edited by ANTHONY R. MEANS AND BERT W. O'MALLEY

VOLUME 103. Hormone Action (Part H: Neuroendocrine Peptides)
Edited by P. MICHAEL CONN

VOLUME 104. Enzyme Purification and Related Techniques (Part C)
Edited by WILLIAM B. JAKOBY

VOLUME 105. Oxygen Radicals in Biological Systems
Edited by LESTER PACKER

VOLUME 106. Posttranslational Modifications (Part A)
Edited by FINN WOLD AND KIVIE MOLDAVE

VOLUME 107. Posttranslational Modifications (Part B)
Edited by FINN WOLD AND KIVIE MOLDAVE

VOLUME 108. Immunochemical Techniques (Part G: Separation and Characterization of Lymphoid Cells)
Edited by GIOVANNI DI SABATO, JOHN J. LANGONE, AND HELEN VAN VUNAKIS

VOLUME 109. Hormone Action (Part I: Peptide Hormones)
Edited by LUTZ BIRNBAUMER AND BERT W. O'MALLEY

VOLUME 110. Steroids and Isoprenoids (Part A)
Edited by JOHN H. LAW AND HANS C. RILLING

VOLUME 111. Steroids and Isoprenoids (Part B)
Edited by JOHN H. LAW AND HANS C. RILLING

VOLUME 112. Drug and Enzyme Targeting (Part A)
Edited by KENNETH J. WIDDER AND RALPH GREEN

VOLUME 113. Glutamate, Glutamine, Glutathione, and Related Compounds
Edited by ALTON MEISTER

VOLUME 114. Diffraction Methods for Biological Macromolecules (Part A)
Edited by HAROLD W. WYCKOFF, C. H. W. HIRS, AND SERGE N. TIMASHEFF

VOLUME 115. Diffraction Methods for Biological Macromolecules (Part B)
Edited by HAROLD W. WYCKOFF, C. H. W. HIRS, AND SERGE N. TIMASHEFF

VOLUME 116. Immunochemical Techniques
(Part H: Effectors and Mediators of Lymphoid Cell Functions)
Edited by GIOVANNI DI SABATO, JOHN J. LANGONE, AND HELEN VAN VUNAKIS

VOLUME 117. Enzyme Structure (Part J)
Edited by C. H. W. HIRS AND SERGE N. TIMASHEFF

VOLUME 118. Plant Molecular Biology
Edited by ARTHUR WEISSBACH AND HERBERT WEISSBACH

VOLUME 119. Interferons (Part C)
Edited by SIDNEY PESTKA

VOLUME 120. Cumulative Subject Index Volumes 81–94, 96–101

VOLUME 121. Immunochemical Techniques (Part I: Hybridoma Technology and Monoclonal Antibodies)
Edited by JOHN J. LANGONE AND HELEN VAN VUNAKIS

VOLUME 122. Vitamins and Coenzymes (Part G)
Edited by FRANK CHYTIL AND DONALD B. MCCORMICK

VOLUME 123. Vitamins and Coenzymes (Part H)
Edited by FRANK CHYTIL AND DONALD B. MCCORMICK

VOLUME 124. Hormone Action (Part J: Neuroendocrine Peptides)
Edited by P. MICHAEL CONN

VOLUME 125. Biomembranes (Part M: Transport in Bacteria, Mitochondria, and Chloroplasts: General Approaches and Transport Systems)
Edited by SIDNEY FLEISCHER AND BECCA FLEISCHER

VOLUME 126. Biomembranes (Part N: Transport in Bacteria, Mitochondria, and Chloroplasts: Protonmotive Force)
Edited by SIDNEY FLEISCHER AND BECCA FLEISCHER

VOLUME 127. Biomembranes (Part O: Protons and Water: Structure and Translocation)
Edited by LESTER PACKER

VOLUME 128. Plasma Lipoproteins (Part A: Preparation, Structure, and Molecular Biology)
Edited by JERE P. SEGREST AND JOHN J. ALBERS

VOLUME 129. Plasma Lipoproteins (Part B: Characterization, Cell Biology, and Metabolism)
Edited by JOHN J. ALBERS AND JERE P. SEGREST

VOLUME 130. Enzyme Structure (Part K)
Edited by C. H. W. HIRS AND SERGE N. TIMASHEFF

VOLUME 131. Enzyme Structure (Part L)
Edited by C. H. W. HIRS AND SERGE N. TIMASHEFF

VOLUME 132. Immunochemical Techniques (Part J: Phagocytosis and Cell-Mediated Cytotoxicity)
Edited by GIOVANNI DI SABATO AND JOHANNES EVERSE

VOLUME 133. Bioluminescence and Chemiluminescence (Part B)
Edited by MARLENE DELUCA AND WILLIAM D. MCELROY

VOLUME 134. Structural and Contractile Proteins (Part C: The Contractile Apparatus and the Cytoskeleton)
Edited by RICHARD B. VALLEE

VOLUME 135. Immobilized Enzymes and Cells (Part B)
Edited by KLAUS MOSBACH

VOLUME 136. Immobilized Enzymes and Cells (Part C)
Edited by KLAUS MOSBACH

VOLUME 137. Immobilized Enzymes and Cells (Part D)
Edited by KLAUS MOSBACH

VOLUME 138. Complex Carbohydrates (Part E)
Edited by VICTOR GINSBURG

VOLUME 139. Cellular Regulators (Part A: Calcium- and Calmodulin-Binding Proteins)
Edited by ANTHONY R. MEANS AND P. MICHAEL CONN

VOLUME 140. Cumulative Subject Index Volumes 102–119, 121–134

VOLUME 141. Cellular Regulators (Part B: Calcium and Lipids)
Edited by P. MICHAEL CONN AND ANTHONY R. MEANS

VOLUME 142. Metabolism of Aromatic Amino Acids and Amines
Edited by SEYMOUR KAUFMAN

VOLUME 143. Sulfur and Sulfur Amino Acids
Edited by WILLIAM B. JAKOBY AND OWEN GRIFFITH

VOLUME 144. Structural and Contractile Proteins (Part D: Extracellular Matrix)
Edited by LEON W. CUNNINGHAM

VOLUME 145. Structural and Contractile Proteins (Part E: Extracellular Matrix)
Edited by LEON W. CUNNINGHAM

VOLUME 146. Peptide Growth Factors (Part A)
Edited by DAVID BARNES AND DAVID A. SIRBASKU

VOLUME 147. Peptide Growth Factors (Part B)
Edited by DAVID BARNES AND DAVID A. SIRBASKU

VOLUME 148. Plant Cell Membranes
Edited by LESTER PACKER AND ROLAND DOUCE

VOLUME 149. Drug and Enzyme Targeting (Part B)
Edited by RALPH GREEN AND KENNETH J. WIDDER

VOLUME 150. Immunochemical Techniques (Part K: *In Vitro* Models of B and T Cell Functions and Lymphoid Cell Receptors)
Edited by GIOVANNI DI SABATO

VOLUME 151. Molecular Genetics of Mammalian Cells
Edited by MICHAEL M. GOTTESMAN

VOLUME 152. Guide to Molecular Cloning Techniques
Edited by SHELBY L. BERGER AND ALAN R. KIMMEL

VOLUME 153. Recombinant DNA (Part D)
Edited by RAY WU AND LAWRENCE GROSSMAN

VOLUME 154. Recombinant DNA (Part E)
Edited by RAY WU AND LAWRENCE GROSSMAN

VOLUME 155. Recombinant DNA (Part F)
Edited by RAY WU

VOLUME 156. Biomembranes (Part P: ATP-Driven Pumps and Related Transport: The Na, K-Pump)
Edited by SIDNEY FLEISCHER AND BECCA FLEISCHER

VOLUME 157. Biomembranes (Part Q: ATP-Driven Pumps and Related Transport: Calcium, Proton, and Potassium Pumps)
Edited by SIDNEY FLEISCHER AND BECCA FLEISCHER

VOLUME 158. Metalloproteins (Part A)
Edited by JAMES F. RIORDAN AND BERT L. VALLEE

VOLUME 159. Initiation and Termination of Cyclic Nucleotide Action
Edited by JACKIE D. CORBIN AND ROGER A. JOHNSON

VOLUME 160. Biomass (Part A: Cellulose and Hemicellulose)
Edited by WILLIS A. WOOD AND SCOTT T. KELLOGG

VOLUME 161. Biomass (Part B: Lignin, Pectin, and Chitin)
Edited by WILLIS A. WOOD AND SCOTT T. KELLOGG

VOLUME 162. Immunochemical Techniques (Part L: Chemotaxis and Inflammation)
Edited by GIOVANNI DI SABATO

VOLUME 163. Immunochemical Techniques (Part M: Chemotaxis and Inflammation)
Edited by GIOVANNI DI SABATO

VOLUME 164. Ribosomes
Edited by HARRY F. NOLLER, JR., AND KIVIE MOLDAVE

VOLUME 165. Microbial Toxins: Tools for Enzymology
Edited by SIDNEY HARSHMAN

VOLUME 166. Branched-Chain Amino Acids
Edited by ROBERT HARRIS AND JOHN R. SOKATCH

VOLUME 167. Cyanobacteria
Edited by LESTER PACKER AND ALEXANDER N. GLAZER

VOLUME 168. Hormone Action (Part K: Neuroendocrine Peptides)
Edited by P. MICHAEL CONN

VOLUME 169. Platelets: Receptors, Adhesion, Secretion (Part A)
Edited by JACEK HAWIGER

VOLUME 170. Nucleosomes
Edited by PAUL M. WASSARMAN AND ROGER D. KORNBERG

VOLUME 171. Biomembranes (Part R: Transport Theory: Cells and Model Membranes)
Edited by SIDNEY FLEISCHER AND BECCA FLEISCHER

VOLUME 172. Biomembranes (Part S: Transport: Membrane Isolation and Characterization)
Edited by SIDNEY FLEISCHER AND BECCA FLEISCHER

VOLUME 173. Biomembranes [Part T: Cellular and Subcellular Transport: Eukaryotic (Nonepithelial) Cells]
Edited by SIDNEY FLEISCHER AND BECCA FLEISCHER

VOLUME 174. Biomembranes [Part U: Cellular and Subcellular Transport: Eukaryotic (Nonepithelial) Cells]
Edited by SIDNEY FLEISCHER AND BECCA FLEISCHER

VOLUME 175. Cumulative Subject Index Volumes 135–139, 141–167

VOLUME 176. Nuclear Magnetic Resonance (Part A: Spectral Techniques and Dynamics)
Edited by NORMAN J. OPPENHEIMER AND THOMAS L. JAMES

VOLUME 177. Nuclear Magnetic Resonance (Part B: Structure and Mechanism)
Edited by NORMAN J. OPPENHEIMER AND THOMAS L. JAMES

VOLUME 178. Antibodies, Antigens, and Molecular Mimicry
Edited by JOHN J. LANGONE

VOLUME 179. Complex Carbohydrates (Part F)
Edited by VICTOR GINSBURG

VOLUME 180. RNA Processing (Part A: General Methods)
Edited by JAMES E. DAHLBERG AND JOHN N. ABELSON

VOLUME 181. RNA Processing (Part B: Specific Methods)
Edited by JAMES E. DAHLBERG AND JOHN N. ABELSON

VOLUME 182. Guide to Protein Purification
Edited by MURRAY P. DEUTSCHER

VOLUME 183. Molecular Evolution: Computer Analysis of Protein and Nucleic Acid Sequences
Edited by RUSSELL F. DOOLITTLE

VOLUME 184. Avidin-Biotin Technology
Edited by MEIR WILCHEK AND EDWARD A. BAYER

VOLUME 185. Gene Expression Technology
Edited by DAVID V. GOEDDEL

VOLUME 186. Oxygen Radicals in Biological Systems (Part B: Oxygen Radicals and Antioxidants)
Edited by LESTER PACKER AND ALEXANDER N. GLAZER

VOLUME 187. Arachidonate Related Lipid Mediators
Edited by ROBERT C. MURPHY AND FRANK A. FITZPATRICK

VOLUME 188. Hydrocarbons and Methylotrophy
Edited by MARY E. LIDSTROM

VOLUME 189. Retinoids (Part A: Molecular and Metabolic Aspects)
Edited by LESTER PACKER

VOLUME 190. Retinoids (Part B: Cell Differentiation and Clinical Applications)
Edited by LESTER PACKER

VOLUME 191. Biomembranes (Part V: Cellular and Subcellular Transport: Epithelial Cells)
Edited by SIDNEY FLEISCHER AND BECCA FLEISCHER

VOLUME 192. Biomembranes (Part W: Cellular and Subcellular Transport: Epithelial Cells)
Edited by SIDNEY FLEISCHER AND BECCA FLEISCHER

VOLUME 193. Mass Spectrometry
Edited by JAMES A. MCCLOSKEY

VOLUME 194. Guide to Yeast Genetics and Molecular Biology
Edited by CHRISTINE GUTHRIE AND GERALD R. FINK

VOLUME 195. Adenylyl Cyclase, G Proteins, and Guanylyl Cyclase
Edited by ROGER A. JOHNSON AND JACKIE D. CORBIN

VOLUME 196. Molecular Motors and the Cytoskeleton
Edited by RICHARD B. VALLEE

VOLUME 197. Phospholipases
Edited by EDWARD A. DENNIS

VOLUME 198. Peptide Growth Factors (Part C)
Edited by DAVID BARNES, J. P. MATHER, AND GORDON H. SATO

VOLUME 199. Cumulative Subject Index Volumes 168–174, 176–194

VOLUME 200. Protein Phosphorylation (Part A: Protein Kinases: Assays, Purification, Antibodies, Functional Analysis, Cloning, and Expression)
Edited by TONY HUNTER AND BARTHOLOMEW M. SEFTON

VOLUME 201. Protein Phosphorylation (Part B: Analysis of Protein Phosphorylation, Protein Kinase Inhibitors, and Protein Phosphatases)
Edited by TONY HUNTER AND BARTHOLOMEW M. SEFTON

VOLUME 202. Molecular Design and Modeling: Concepts and Applications (Part A: Proteins, Peptides, and Enzymes)
Edited by JOHN J. LANGONE

VOLUME 203. Molecular Design and Modeling: Concepts and Applications (Part B: Antibodies and Antigens, Nucleic Acids, Polysaccharides, and Drugs)
Edited by JOHN J. LANGONE

VOLUME 204. Bacterial Genetic Systems
Edited by JEFFREY H. MILLER

VOLUME 205. Metallobiochemistry (Part B: Metallothionein and Related Molecules)
Edited by JAMES F. RIORDAN AND BERT L. VALLEE

VOLUME 206. Cytochrome P450
Edited by MICHAEL R. WATERMAN AND ERIC F. JOHNSON

VOLUME 207. Ion Channels
Edited by BERNARDO RUDY AND LINDA E. IVERSON

VOLUME 208. Protein–DNA Interactions
Edited by ROBERT T. SAUER

VOLUME 209. Phospholipid Biosynthesis
Edited by EDWARD A. DENNIS AND DENNIS E. VANCE

VOLUME 210. Numerical Computer Methods
Edited by LUDWIG BRAND AND MICHAEL L. JOHNSON

VOLUME 211. DNA Structures (Part A: Synthesis and Physical Analysis of DNA)
Edited by DAVID M. J. LILLEY AND JAMES E. DAHLBERG

VOLUME 212. DNA Structures (Part B: Chemical and Electrophoretic Analysis of DNA)
Edited by DAVID M. J. LILLEY AND JAMES E. DAHLBERG

VOLUME 213. Carotenoids (Part A: Chemistry, Separation, Quantitation, and Antioxidation)
Edited by LESTER PACKER

VOLUME 214. Carotenoids (Part B: Metabolism, Genetics, and Biosynthesis)
Edited by LESTER PACKER

VOLUME 215. Platelets: Receptors, Adhesion, Secretion (Part B)
Edited by JACEK J. HAWIGER

VOLUME 216. Recombinant DNA (Part G)
Edited by RAY WU

VOLUME 217. Recombinant DNA (Part H)
Edited by RAY WU

VOLUME 218. Recombinant DNA (Part I)
Edited by RAY WU

VOLUME 219. Reconstitution of Intracellular Transport
Edited by JAMES E. ROTHMAN

VOLUME 220. Membrane Fusion Techniques (Part A)
Edited by NEJAT DÜZGÜNEŞ

VOLUME 221. Membrane Fusion Techniques (Part B)
Edited by NEJAT DÜZGÜNEŞ

VOLUME 222. Proteolytic Enzymes in Coagulation, Fibrinolysis, and Complement Activation (Part A: Mammalian Blood Coagulation

Factors and Inhibitors)
Edited by LASZLO LORAND AND KENNETH G. MANN

VOLUME 223. Proteolytic Enzymes in Coagulation, Fibrinolysis, and Complement Activation (Part B: Complement Activation, Fibrinolysis, and Nonmammalian Blood Coagulation Factors)
Edited by LASZLO LORAND AND KENNETH G. MANN

VOLUME 224. Molecular Evolution: Producing the Biochemical Data
Edited by ELIZABETH ANNE ZIMMER, THOMAS J. WHITE, REBECCA L. CANN, AND ALLAN C. WILSON

VOLUME 225. Guide to Techniques in Mouse Development
Edited by PAUL M. WASSARMAN AND MELVIN L. DEPAMPHILIS

VOLUME 226. Metallobiochemistry (Part C: Spectroscopic and Physical Methods for Probing Metal Ion Environments in Metalloenzymes and Metalloproteins)
Edited by JAMES F. RIORDAN AND BERT L. VALLEE

VOLUME 227. Metallobiochemistry (Part D: Physical and Spectroscopic Methods for Probing Metal Ion Environments in Metalloproteins)
Edited by JAMES F. RIORDAN AND BERT L. VALLEE

VOLUME 228. Aqueous Two-Phase Systems
Edited by HARRY WALTER AND GÖTE JOHANSSON

VOLUME 229. Cumulative Subject Index Volumes 195–198, 200–227

VOLUME 230. Guide to Techniques in Glycobiology
Edited by WILLIAM J. LENNARZ AND GERALD W. HART

VOLUME 231. Hemoglobins (Part B: Biochemical and Analytical Methods)
Edited by JOHANNES EVERSE, KIM D. VANDEGRIFF, AND ROBERT M. WINSLOW

VOLUME 232. Hemoglobins (Part C: Biophysical Methods)
Edited by JOHANNES EVERSE, KIM D. VANDEGRIFF, AND ROBERT M. WINSLOW

VOLUME 233. Oxygen Radicals in Biological Systems (Part C)
Edited by LESTER PACKER

VOLUME 234. Oxygen Radicals in Biological Systems (Part D)
Edited by LESTER PACKER

VOLUME 235. Bacterial Pathogenesis (Part A: Identification and Regulation of Virulence Factors)
Edited by VIRGINIA L. CLARK AND PATRIK M. BAVOIL

VOLUME 236. Bacterial Pathogenesis (Part B: Integration of Pathogenic Bacteria with Host Cells)
Edited by VIRGINIA L. CLARK AND PATRIK M. BAVOIL

VOLUME 237. Heterotrimeric G Proteins
Edited by RAVI IYENGAR

VOLUME 238. Heterotrimeric G-Protein Effectors
Edited by RAVI IYENGAR

VOLUME 239. Nuclear Magnetic Resonance (Part C)
Edited by THOMAS L. JAMES AND NORMAN J. OPPENHEIMER

VOLUME 240. Numerical Computer Methods (Part B)
Edited by MICHAEL L. JOHNSON AND LUDWIG BRAND

VOLUME 241. Retroviral Proteases
Edited by LAWRENCE C. KUO AND JULES A. SHAFER

VOLUME 242. Neoglycoconjugates (Part A)
Edited by Y. C. LEE AND REIKO T. LEE

VOLUME 243. Inorganic Microbial Sulfur Metabolism
Edited by HARRY D. PECK, JR., AND JEAN LEGALL

VOLUME 244. Proteolytic Enzymes: Serine and Cysteine Peptidases
Edited by ALAN J. BARRETT

VOLUME 245. Extracellular Matrix Components
Edited by E. RUOSLAHTI AND E. ENGVALL

VOLUME 246. Biochemical Spectroscopy
Edited by KENNETH SAUER

VOLUME 247. Neoglycoconjugates (Part B: Biomedical Applications)
Edited by Y. C. LEE AND REIKO T. LEE

VOLUME 248. Proteolytic Enzymes: Aspartic and Metallo Peptidases
Edited by ALAN J. BARRETT

VOLUME 249. Enzyme Kinetics and Mechanism (Part D: Developments in Enzyme Dynamics)
Edited by DANIEL L. PURICH

VOLUME 250. Lipid Modifications of Proteins
Edited by PATRICK J. CASEY AND JANICE E. BUSS

VOLUME 251. Biothiols (Part A: Monothiols and Dithiols, Protein Thiols, and Thiyl Radicals)
Edited by LESTER PACKER

VOLUME 252. Biothiols (Part B: Glutathione and Thioredoxin; Thiols in Signal Transduction and Gene Regulation)
Edited by LESTER PACKER

VOLUME 253. Adhesion of Microbial Pathogens
Edited by RON J. DOYLE AND ITZHAK OFEK

VOLUME 254. Oncogene Techniques
Edited by PETER K. VOGT AND INDER M. VERMA

VOLUME 255. Small GTPases and Their Regulators (Part A: Ras Family)
Edited by W. E. BALCH, CHANNING J. DER, AND ALAN HALL

VOLUME 256. Small GTPases and Their Regulators (Part B: Rho Family)
Edited by W. E. BALCH, CHANNING J. DER, AND ALAN HALL

VOLUME 257. Small GTPases and Their Regulators (Part C: Proteins Involved in Transport)
Edited by W. E. BALCH, CHANNING J. DER, AND ALAN HALL

VOLUME 258. Redox-Active Amino Acids in Biology
Edited by JUDITH P. KLINMAN

VOLUME 259. Energetics of Biological Macromolecules
Edited by MICHAEL L. JOHNSON AND GARY K. ACKERS

VOLUME 260. Mitochondrial Biogenesis and Genetics (Part A)
Edited by GIUSEPPE M. ATTARDI AND ANNE CHOMYN

VOLUME 261. Nuclear Magnetic Resonance and Nucleic Acids
Edited by THOMAS L. JAMES

VOLUME 262. DNA Replication
Edited by JUDITH L. CAMPBELL

VOLUME 263. Plasma Lipoproteins (Part C: Quantitation)
Edited by WILLIAM A. BRADLEY, SANDRA H. GIANTURCO, AND JERE P. SEGREST

VOLUME 264. Mitochondrial Biogenesis and Genetics (Part B)
Edited by GIUSEPPE M. ATTARDI AND ANNE CHOMYN

VOLUME 265. Cumulative Subject Index Volumes 228, 230–262

VOLUME 266. Computer Methods for Macromolecular Sequence Analysis
Edited by RUSSELL F. DOOLITTLE

VOLUME 267. Combinatorial Chemistry
Edited by JOHN N. ABELSON

VOLUME 268. Nitric Oxide (Part A: Sources and Detection of NO; NO Synthase)
Edited by LESTER PACKER

VOLUME 269. Nitric Oxide (Part B: Physiological and Pathological Processes)
Edited by LESTER PACKER

VOLUME 270. High Resolution Separation and Analysis of Biological Macromolecules (Part A: Fundamentals)
Edited by BARRY L. KARGER AND WILLIAM S. HANCOCK

VOLUME 271. High Resolution Separation and Analysis of Biological Macromolecules (Part B: Applications)
Edited by BARRY L. KARGER AND WILLIAM S. HANCOCK

VOLUME 272. Cytochrome P450 (Part B)
Edited by ERIC F. JOHNSON AND MICHAEL R. WATERMAN

VOLUME 273. RNA Polymerase and Associated Factors (Part A)
Edited by SANKAR ADHYA

VOLUME 274. RNA Polymerase and Associated Factors (Part B)
Edited by SANKAR ADHYA

VOLUME 275. Viral Polymerases and Related Proteins
Edited by LAWRENCE C. KUO, DAVID B. OLSEN, AND STEVEN S. CARROLL

VOLUME 276. Macromolecular Crystallography (Part A)
Edited by CHARLES W. CARTER, JR., AND ROBERT M. SWEET

VOLUME 277. Macromolecular Crystallography (Part B)
Edited by CHARLES W. CARTER, JR., AND ROBERT M. SWEET

VOLUME 278. Fluorescence Spectroscopy
Edited by LUDWIG BRAND AND MICHAEL L. JOHNSON

VOLUME 279. Vitamins and Coenzymes (Part I)
Edited by DONALD B. MCCORMICK, JOHN W. SUTTIE, AND CONRAD WAGNER

VOLUME 280. Vitamins and Coenzymes (Part J)
Edited by DONALD B. MCCORMICK, JOHN W. SUTTIE, AND CONRAD WAGNER

VOLUME 281. Vitamins and Coenzymes (Part K)
Edited by DONALD B. MCCORMICK, JOHN W. SUTTIE, AND CONRAD WAGNER

VOLUME 282. Vitamins and Coenzymes (Part L)
Edited by DONALD B. MCCORMICK, JOHN W. SUTTIE, AND CONRAD WAGNER

VOLUME 283. Cell Cycle Control
Edited by WILLIAM G. DUNPHY

VOLUME 284. Lipases (Part A: Biotechnology)
Edited by BYRON RUBIN AND EDWARD A. DENNIS

VOLUME 285. Cumulative Subject Index Volumes 263, 264, 266–284, 286–289

VOLUME 286. Lipases (Part B: Enzyme Characterization and Utilization)
Edited by BYRON RUBIN AND EDWARD A. DENNIS

VOLUME 287. Chemokines
Edited by RICHARD HORUK

VOLUME 288. Chemokine Receptors
Edited by RICHARD HORUK

VOLUME 289. Solid Phase Peptide Synthesis
Edited by GREGG B. FIELDS

VOLUME 290. Molecular Chaperones
Edited by GEORGE H. LORIMER AND THOMAS BALDWIN

VOLUME 291. Caged Compounds
Edited by GERARD MARRIOTT

VOLUME 292. ABC Transporters: Biochemical, Cellular, and Molecular Aspects
Edited by SURESH V. AMBUDKAR AND MICHAEL M. GOTTESMAN

VOLUME 293. Ion Channels (Part B)
Edited by P. MICHAEL CONN

VOLUME 294. Ion Channels (Part C)
Edited by P. MICHAEL CONN

VOLUME 295. Energetics of Biological Macromolecules (Part B)
Edited by GARY K. ACKERS AND MICHAEL L. JOHNSON

VOLUME 296. Neurotransmitter Transporters
Edited by SUSAN G. AMARA

VOLUME 297. Photosynthesis: Molecular Biology of Energy Capture
Edited by LEE MCINTOSH

VOLUME 298. Molecular Motors and the Cytoskeleton (Part B)
Edited by RICHARD B. VALLEE

VOLUME 299. Oxidants and Antioxidants (Part A)
Edited by LESTER PACKER

VOLUME 300. Oxidants and Antioxidants (Part B)
Edited by LESTER PACKER

VOLUME 301. Nitric Oxide: Biological and Antioxidant Activities (Part C)
Edited by LESTER PACKER

VOLUME 302. Green Fluorescent Protein
Edited by P. MICHAEL CONN

VOLUME 303. cDNA Preparation and Display
Edited by SHERMAN M. WEISSMAN

VOLUME 304. Chromatin
Edited by PAUL M. WASSARMAN AND ALAN P. WOLFFE

VOLUME 305. Bioluminescence and Chemiluminescence (Part C)
Edited by THOMAS O. BALDWIN AND MIRIAM M. ZIEGLER

VOLUME 306. Expression of Recombinant Genes in Eukaryotic Systems
Edited by JOSEPH C. GLORIOSO AND MARTIN C. SCHMIDT

VOLUME 307. Confocal Microscopy
Edited by P. MICHAEL CONN

VOLUME 308. Enzyme Kinetics and Mechanism (Part E: Energetics of Enzyme Catalysis)
Edited by DANIEL L. PURICH AND VERN L. SCHRAMM

VOLUME 309. Amyloid, Prions, and Other Protein Aggregates
Edited by RONALD WETZEL

VOLUME 310. Biofilms
Edited by RON J. DOYLE

VOLUME 311. Sphingolipid Metabolism and Cell Signaling (Part A)
Edited by ALFRED H. MERRILL, JR., AND YUSUF A. HANNUN

VOLUME 312. Sphingolipid Metabolism and Cell Signaling (Part B)
Edited by ALFRED H. MERRILL, JR., AND YUSUF A. HANNUN

VOLUME 313. Antisense Technology
(Part A: General Methods, Methods of Delivery, and RNA Studies)
Edited by M. IAN PHILLIPS

VOLUME 314. Antisense Technology (Part B: Applications)
Edited by M. IAN PHILLIPS

VOLUME 315. Vertebrate Phototransduction and the Visual Cycle (Part A)
Edited by KRZYSZTOF PALCZEWSKI

VOLUME 316. Vertebrate Phototransduction and the Visual Cycle (Part B)
Edited by KRZYSZTOF PALCZEWSKI

VOLUME 317. RNA–Ligand Interactions (Part A: Structural Biology Methods)
Edited by DANIEL W. CELANDER AND JOHN N. ABELSON

VOLUME 318. RNA–Ligand Interactions (Part B: Molecular Biology Methods)
Edited by DANIEL W. CELANDER AND JOHN N. ABELSON

VOLUME 319. Singlet Oxygen, UV-A, and Ozone
Edited by LESTER PACKER AND HELMUT SIES

VOLUME 320. Cumulative Subject Index Volumes 290–319

VOLUME 321. Numerical Computer Methods (Part C)
Edited by MICHAEL L. JOHNSON AND LUDWIG BRAND

VOLUME 322. Apoptosis
Edited by JOHN C. REED

VOLUME 323. Energetics of Biological Macromolecules (Part C)
Edited by MICHAEL L. JOHNSON AND GARY K. ACKERS

VOLUME 324. Branched-Chain Amino Acids (Part B)
Edited by ROBERT A. HARRIS AND JOHN R. SOKATCH

VOLUME 325. Regulators and Effectors of Small GTPases
(Part D: Rho Family)
Edited by W. E. BALCH, CHANNING J. DER, AND ALAN HALL

VOLUME 326. Applications of Chimeric Genes and Hybrid Proteins
(Part A: Gene Expression and Protein Purification)
Edited by JEREMY THORNER, SCOTT D. EMR, AND JOHN N. ABELSON

VOLUME 327. Applications of Chimeric Genes and Hybrid Proteins
(Part B: Cell Biology and Physiology)
Edited by JEREMY THORNER, SCOTT D. EMR, AND JOHN N. ABELSON

VOLUME 328. Applications of Chimeric Genes and Hybrid Proteins (Part C:
Protein–Protein Interactions and Genomics)
Edited by JEREMY THORNER, SCOTT D. EMR, AND JOHN N. ABELSON

VOLUME 329. Regulators and Effectors of Small GTPases (Part E: GTPases
Involved in Vesicular Traffic)
Edited by W. E. BALCH, CHANNING J. DER, AND ALAN HALL

VOLUME 330. Hyperthermophilic Enzymes (Part A)
Edited by MICHAEL W. W. ADAMS AND ROBERT M. KELLY

VOLUME 331. Hyperthermophilic Enzymes (Part B)
Edited by MICHAEL W. W. ADAMS AND ROBERT M. KELLY

VOLUME 332. Regulators and Effectors of Small GTPases (Part F:
Ras Family I)
Edited by W. E. BALCH, CHANNING J. DER, AND ALAN HALL

VOLUME 333. Regulators and Effectors of Small GTPases (Part G: Ras
Family II)
Edited by W. E. BALCH, CHANNING J. DER, AND ALAN HALL

VOLUME 334. Hyperthermophilic Enzymes (Part C)
Edited by MICHAEL W. W. ADAMS AND ROBERT M. KELLY

VOLUME 335. Flavonoids and Other Polyphenols
Edited by LESTER PACKER

VOLUME 336. Microbial Growth in Biofilms (Part A: Developmental and
Molecular Biological Aspects)
Edited by RON J. DOYLE

VOLUME 337. Microbial Growth in Biofilms (Part B: Special Environments
and Physicochemical Aspects)
Edited by RON J. DOYLE

VOLUME 338. Nuclear Magnetic Resonance of Biological Macromolecules
(Part A)
Edited by THOMAS L. JAMES, VOLKER DÖTSCH, AND ULI SCHMITZ

VOLUME 339. Nuclear Magnetic Resonance of Biological Macromolecules
(Part B)
Edited by THOMAS L. JAMES, VOLKER DÖTSCH, AND ULI SCHMITZ

VOLUME 340. Drug–Nucleic Acid Interactions
Edited by JONATHAN B. CHAIRES AND MICHAEL J. WARING

VOLUME 341. Ribonucleases (Part A)
Edited by ALLEN W. NICHOLSON

VOLUME 342. Ribonucleases (Part B)
Edited by ALLEN W. NICHOLSON

VOLUME 343. G Protein Pathways (Part A: Receptors)
Edited by RAVI IYENGAR AND JOHN D. HILDEBRANDT

VOLUME 344. G Protein Pathways (Part B: G Proteins and Their Regulators)
Edited by RAVI IYENGAR AND JOHN D. HILDEBRANDT

VOLUME 345. G Protein Pathways (Part C: Effector Mechanisms)
Edited by RAVI IYENGAR AND JOHN D. HILDEBRANDT

VOLUME 346. Gene Therapy Methods
Edited by M. IAN PHILLIPS

VOLUME 347. Protein Sensors and Reactive Oxygen Species (Part A: Selenoproteins and Thioredoxin)
Edited by HELMUT SIES AND LESTER PACKER

VOLUME 348. Protein Sensors and Reactive Oxygen Species (Part B: Thiol Enzymes and Proteins)
Edited by HELMUT SIES AND LESTER PACKER

VOLUME 349. Superoxide Dismutase
Edited by LESTER PACKER

VOLUME 350. Guide to Yeast Genetics and Molecular and Cell Biology (Part B)
Edited by CHRISTINE GUTHRIE AND GERALD R. FINK

VOLUME 351. Guide to Yeast Genetics and Molecular and Cell Biology (Part C)
Edited by CHRISTINE GUTHRIE AND GERALD R. FINK

VOLUME 352. Redox Cell Biology and Genetics (Part A)
Edited by CHANDAN K. SEN AND LESTER PACKER

VOLUME 353. Redox Cell Biology and Genetics (Part B)
Edited by CHANDAN K. SEN AND LESTER PACKER

VOLUME 354. Enzyme Kinetics and Mechanisms (Part F: Detection and Characterization of Enzyme Reaction Intermediates)
Edited by DANIEL L. PURICH

VOLUME 355. Cumulative Subject Index Volumes 321–354

VOLUME 356. Laser Capture Microscopy and Microdissection
Edited by P. MICHAEL CONN

VOLUME 357. Cytochrome P450, Part C
Edited by ERIC F. JOHNSON AND MICHAEL R. WATERMAN

VOLUME 358. Bacterial Pathogenesis (Part C: Identification, Regulation, and Function of Virulence Factors)
Edited by VIRGINIA L. CLARK AND PATRIK M. BAVOIL

VOLUME 359. Nitric Oxide (Part D)
Edited by ENRIQUE CADENAS AND LESTER PACKER

VOLUME 360. Biophotonics (Part A)
Edited by GERARD MARRIOTT AND IAN PARKER

VOLUME 361. Biophotonics (Part B)
Edited by GERARD MARRIOTT AND IAN PARKER

VOLUME 362. Recognition of Carbohydrates in Biological Systems (Part A)
Edited by YUAN C. LEE AND REIKO T. LEE

VOLUME 363. Recognition of Carbohydrates in Biological Systems (Part B)
Edited by YUAN C. LEE AND REIKO T. LEE

VOLUME 364. Nuclear Receptors
Edited by DAVID W. RUSSELL AND DAVID J. MANGELSDORF

VOLUME 365. Differentiation of Embryonic Stem Cells
Edited by PAUL M. WASSAUMAN AND GORDON M. KELLER

VOLUME 366. Protein Phosphatases
Edited by SUSANNE KLUMPP AND JOSEF KRIEGLSTEIN

VOLUME 367. Liposomes (Part A)
Edited by NEJAT DÜZGÜNEŞ

VOLUME 368. Macromolecular Crystallography (Part C)
Edited by CHARLES W. CARTER, JR., AND ROBERT M. SWEET

VOLUME 369. Combinational Chemistry (Part B)
Edited by GUILLERMO A. MORALES AND BARRY A. BUNIN

VOLUME 370. RNA Polymerases and Associated Factors (Part C)
Edited by SANKAR L. ADHYA AND SUSAN GARGES

VOLUME 371. RNA Polymerases and Associated Factors (Part D)
Edited by SANKAR L. ADHYA AND SUSAN GARGES

VOLUME 372. Liposomes (Part B)
Edited by NEJAT DÜZGÜNEŞ

VOLUME 373. Liposomes (Part C)
Edited by NEJAT DÜZGÜNEŞ

VOLUME 374. Macromolecular Crystallography (Part D)
Edited by CHARLES W. CARTER, JR., AND ROBERT W. SWEET

VOLUME 375. Chromatin and Chromatin Remodeling Enzymes (Part A)
Edited by C. DAVID ALLIS AND CARL WU

VOLUME 376. Chromatin and Chromatin Remodeling Enzymes (Part B)
Edited by C. DAVID ALLIS AND CARL WU

VOLUME 377. Chromatin and Chromatin Remodeling Enzymes (Part C)
Edited by C. DAVID ALLIS AND CARL WU

VOLUME 378. Quinones and Quinone Enzymes (Part A)
Edited by HELMUT SIES AND LESTER PACKER

VOLUME 379. Energetics of Biological Macromolecules (Part D)
Edited by JO M. HOLT, MICHAEL L. JOHNSON, AND GARY K. ACKERS

VOLUME 380. Energetics of Biological Macromolecules (Part E)
Edited by JO M. HOLT, MICHAEL L. JOHNSON, AND GARY K. ACKERS

VOLUME 381. Oxygen Sensing
Edited by CHANDAN K. SEN AND GREGG L. SEMENZA

VOLUME 382. Quinones and Quinone Enzymes (Part B)
Edited by HELMUT SIES AND LESTER PACKER

VOLUME 383. Numerical Computer Methods (Part D)
Edited by LUDWIG BRAND AND MICHAEL L. JOHNSON

VOLUME 384. Numerical Computer Methods (Part E)
Edited by LUDWIG BRAND AND MICHAEL L. JOHNSON

VOLUME 385. Imaging in Biological Research (Part A)
Edited by P. MICHAEL CONN

VOLUME 386. Imaging in Biological Research (Part B)
Edited by P. MICHAEL CONN

VOLUME 387. Liposomes (Part D)
Edited by NEJAT DÜZGÜNEŞ

VOLUME 388. Protein Engineering
Edited by DAN E. ROBERTSON AND JOSEPH P. NOEL

VOLUME 389. Regulators of G-Protein Signaling (Part A)
Edited by DAVID P. SIDEROVSKI

VOLUME 390. Regulators of G-Protein Signaling (Part B)
Edited by DAVID P. SIDEROVSKI

VOLUME 391. Liposomes (Part E)
Edited by NEJAT DÜZGÜNEŞ

VOLUME 392. RNA Interference
Edited by ENGELKE ROSSI

VOLUME 393. Circadian Rhythms
Edited by MICHAEL W. YOUNG

VOLUME 394. Nuclear Magnetic Resonance of Biological Macromolecules (Part C)
Edited by THOMAS L. JAMES

VOLUME 395. Producing the Biochemical Data (Part B)
Edited by ELIZABETH A. ZIMMER AND ERIC H. ROALSON

VOLUME 396. Nitric Oxide (Part E)
Edited by LESTER PACKER AND ENRIQUE CADENAS

VOLUME 397. Environmental Microbiology
Edited by JARED R. LEADBETTER

VOLUME 398. Ubiquitin and Protein Degradation (Part A)
Edited by RAYMOND J. DESHAIES

VOLUME 399. Ubiquitin and Protein Degradation (Part B)
Edited by RAYMOND J. DESHAIES

VOLUME 400. Phase II Conjugation Enzymes and Transport Systems
Edited by HELMUT SIES AND LESTER PACKER

VOLUME 401. Glutathione Transferases and Gamma Glutamyl Transpeptidases
Edited by HELMUT SIES AND LESTER PACKER

VOLUME 402. Biological Mass Spectrometry
Edited by A. L. BURLINGAME

VOLUME 403. GTPases Regulating Membrane Targeting and Fusion
Edited by WILLIAM E. BALCH, CHANNING J. DER, AND ALAN HALL

VOLUME 404. GTPases Regulating Membrane Dynamics
Edited by WILLIAM E. BALCH, CHANNING J. DER, AND ALAN HALL

VOLUME 405. Mass Spectrometry: Modified Proteins and Glycoconjugates
Edited by A. L. BURLINGAME

VOLUME 406. Regulators and Effectors of Small GTPases: Rho Family
Edited by WILLIAM E. BALCH, CHANNING J. DER, AND ALAN HALL

VOLUME 407. Regulators and Effectors of Small GTPases: Ras Family
Edited by WILLIAM E. BALCH, CHANNING J. DER, AND ALAN HALL

VOLUME 408. DNA Repair (Part A)
Edited by JUDITH L. CAMPBELL AND PAUL MODRICH

VOLUME 409. DNA Repair (Part B)
Edited by JUDITH L. CAMPBELL AND PAUL MODRICH

VOLUME 410. DNA Microarrays (Part A: Array Platforms and Web-Bench Protocols)
Edited by ALAN KIMMEL AND BRIAN OLIVER

VOLUME 411. DNA Microarrays (Part B: Databases and Statistics)
Edited by ALAN KIMMEL AND BRIAN OLIVER

VOLUME 412. Amyloid, Prions, and Other Protein Aggregates (Part B)
Edited by INDU KHETERPAL AND RONALD WETZEL

VOLUME 413. Amyloid, Prions, and Other Protein Aggregates (Part C)
Edited by INDU KHETERPAL AND RONALD WETZEL

VOLUME 414. Measuring Biological Responses with Automated Microscopy
Edited by JAMES INGLESE

VOLUME 415. Glycobiology
Edited by MINORU FUKUDA

VOLUME 416. Glycomics
Edited by MINORU FUKUDA

VOLUME 417. Functional Glycomics
Edited by MINORU FUKUDA

VOLUME 418. Embryonic Stem Cells
Edited by IRINA KLIMANSKAYA AND ROBERT LANZA

VOLUME 419. Adult Stem Cells
Edited by IRINA KLIMANSKAYA AND ROBERT LANZA

VOLUME 420. Stem Cell Tools and Other Experimental Protocols
Edited by IRINA KLIMANSKAYA AND ROBERT LANZA

VOLUME 421. Advanced Bacterial Genetics: Use of Transposons and Phage for Genomic Engineering
Edited by KELLY T. HUGHES

VOLUME 422. Two-Component Signaling Systems, Part A
Edited by MELVIN I. SIMON, BRIAN R. CRANE, AND ALEXANDRINE CRANE

VOLUME 423. Two-Component Signaling Systems, Part B
Edited by MELVIN I. SIMON, BRIAN R. CRANE, AND ALEXANDRINE CRANE

VOLUME 424. RNA Editing
Edited by JONATHA M. GOTT

VOLUME 425. RNA Modification
Edited by JONATHA M. GOTT

VOLUME 426. Integrins
Edited by DAVID CHERESH

VOLUME 427. MicroRNA Methods
Edited by JOHN J. ROSSI

VOLUME 428. Osmosensing and Osmosignaling
Edited by HELMUT SIES AND DIETER HAUSSINGER

VOLUME 429. Translation Initiation: Extract Systems and Molecular Genetics
Edited by JON LORSCH

VOLUME 430. Translation Initiation: Reconstituted Systems and Biophysical Methods
Edited by JON LORSCH

VOLUME 431. Translation Initiation: Cell Biology, High-Throughput and Chemical-Based Approaches
Edited by JON LORSCH

VOLUME 432. Lipidomics and Bioactive Lipids: Mass-Spectrometry–Based Lipid Analysis
Edited by H. ALEX BROWN

VOLUME 433. Lipidomics and Bioactive Lipids: Specialized Analytical Methods and Lipids in Disease
Edited by H. ALEX BROWN

VOLUME 434. Lipidomics and Bioactive Lipids: Lipids and Cell Signaling
Edited by H. ALEX BROWN

VOLUME 435. Oxygen Biology and Hypoxia
Edited by HELMUT SIES AND BERNHARD BRÜNE

VOLUME 436. Globins and Other Nitric Oxide-Reactive Protiens (Part A)
Edited by ROBERT K. POOLE

VOLUME 437. Globins and Other Nitric Oxide-Reactive Protiens (Part B)
Edited by ROBERT K. POOLE

VOLUME 438. Small GTPases in Disease (Part A)
Edited by WILLIAM E. BALCH, CHANNING J. DER, AND ALAN HALL

VOLUME 439. Small GTPases in Disease (Part B)
Edited by WILLIAM E. BALCH, CHANNING J. DER, AND ALAN HALL

VOLUME 440. Nitric Oxide, Part F Oxidative and Nitrosative Stress in Redox Regulation of Cell Signaling
Edited by ENRIQUE CADENAS AND LESTER PACKER

VOLUME 441. Nitric Oxide, Part G Oxidative and Nitrosative Stress in Redox Regulation of Cell Signaling
Edited by ENRIQUE CADENAS AND LESTER PACKER

VOLUME 442. Programmed Cell Death, General Principles for Studying Cell Death (Part A)
Edited by ROYA KHOSRAVI-FAR, ZAHRA ZAKERI, RICHARD A. LOCKSHIN, AND MAURO PIACENTINI

VOLUME 443. Angiogenesis: *In Vitro* Systems
Edited by DAVID A. CHERESH

VOLUME 444. Angiogenesis: *In Vivo* Systems (Part A)
Edited by DAVID A. CHERESH

VOLUME 445. Angiogenesis: *In Vivo* Systems (Part B)
Edited by DAVID A. CHERESH

VOLUME 446. Programmed Cell Death, The Biology and Therapeutic Implications of Cell Death (Part B)
Edited by ROYA KHOSRAVI-FAR, ZAHRA ZAKERI, RICHARD A. LOCKSHIN, AND MAURO PIACENTINI

VOLUME 447. RNA Turnover in Bacteria, Archaea and Organelles
Edited by LYNNE E. MAQUAT AND CECILIA M. ARRAIANO

VOLUME 448. RNA Turnover in Eukaryotes: Nucleases, Pathways and Analysis of mRNA Decay
Edited by LYNNE E. MAQUAT AND MEGERDITCH KILEDJIAN

VOLUME 449. RNA Turnover in Eukaryotes: Analysis of Specialized and Quality Control RNA Decay Pathways
Edited by LYNNE E. MAQUAT AND MEGERDITCH KILEDJIAN

VOLUME 450. Fluorescence Spectroscopy
Edited by LUDWIG BRAND AND MICHAEL L. JOHNSON

VOLUME 451. Autophagy: Lower Eukaryotes and Non-Mammalian Systems (Part A)
Edited by DANIEL J. KLIONSKY

VOLUME 452. Autophagy in Mammalian Systems (Part B)
Edited by DANIEL J. KLIONSKY

VOLUME 453. Autophagy in Disease and Clinical Applications (Part C)
Edited by DANIEL J. KLIONSKY

VOLUME 454. Computer Methods (Part A)
Edited by MICHAEL L. JOHNSON AND LUDWIG BRAND

VOLUME 455. Biothermodynamics (Part A)
Edited by MICHAEL L. JOHNSON, JO M. HOLT, AND GARY K. ACKERS (RETIRED)

VOLUME 456. Mitochondrial Function, Part A: Mitochondrial Electron Transport Complexes and Reactive Oxygen Species
Edited by WILLIAM S. ALLISON AND IMMO E. SCHEFFLER

VOLUME 457. Mitochondrial Function, Part B: Mitochondrial Protein Kinases, Protein Phosphatases and Mitochondrial Diseases
Edited by WILLIAM S. ALLISON AND ANNE N. MURPHY

VOLUME 458. Complex Enzymes in Microbial Natural Product Biosynthesis, Part A: Overview Articles and Peptides
Edited by DAVID A. HOPWOOD

VOLUME 459. Complex Enzymes in Microbial Natural Product Biosynthesis, Part B: Polyketides, Aminocoumarins and Carbohydrates
Edited by DAVID A. HOPWOOD

VOLUME 460. Chemokines, Part A
Edited by TRACY M. HANDEL AND DAMON J. HAMEL

VOLUME 461. Chemokines, Part B
Edited by TRACY M. HANDEL AND DAMON J. HAMEL

VOLUME 462. Non-Natural Amino Acids
Edited by TOM W. MUIR AND JOHN N. ABELSON

VOLUME 463. Guide to Protein Purification, 2nd Edition
Edited by RICHARD R. BURGESS AND MURRAY P. DEUTSCHER

VOLUME 464. Liposomes, Part F
Edited by NEJAT DÜZGÜNEŞ

VOLUME 465. Liposomes, Part G
Edited by NEJAT DÜZGÜNEŞ

VOLUME 466. Biothermodynamics, Part B
Edited by MICHAEL L. JOHNSON, GARY K. ACKERS, AND JO M. HOLT

VOLUME 467. Computer Methods Part B
Edited by MICHAEL L. JOHNSON AND LUDWIG BRAND

VOLUME 468. Biophysical, Chemical, and Functional Probes of RNA Structure, Interactions and Folding: Part A
Edited by DANIEL HERSCHLAG

VOLUME 469. Biophysical, Chemical, and Functional Probes of RNA Structure, Interactions and Folding: Part B
Edited by DANIEL HERSCHLAG

VOLUME 470. Guide to Yeast Genetics: Functional Genomics, Proteomics, and Other Systems Analysis, 2nd Edition
Edited by GERALD FINK, JONATHAN WEISSMAN, AND CHRISTINE GUTHRIE

VOLUME 471. Two-Component Signaling Systems, Part C
Edited by MELVIN I. SIMON, BRIAN R. CRANE, AND ALEXANDRINE CRANE

VOLUME 472. Single Molecule Tools, Part A: Fluorescence Based Approaches
Edited by NILS G. WALTER

VOLUME 473. Thiol Redox Transitions in Cell Signaling, Part A Chemistry and Biochemistry of Low Molecular Weight and Protein Thiols
Edited by ENRIQUE CADENAS AND LESTER PACKER

VOLUME 474. Thiol Redox Transitions in Cell Signaling, Part B Cellular Localization and Signaling
Edited by ENRIQUE CADENAS AND LESTER PACKER

VOLUME 475. Single Molecule Tools, Part B: Super-Resolution, Particle Tracking, Multiparameter, and Force Based Methods
Edited by NILS G. WALTER

VOLUME 476. Guide to Techniques in Mouse Development, Part A Mice, Embryos, and Cells, 2nd Edition
Edited by PAUL M. WASSARMAN AND PHILIPPE M. SORIANO

VOLUME 477. Guide to Techniques in Mouse Development, Part B Mouse Molecular Genetics, 2nd Edition
Edited by PAUL M. WASSARMAN AND PHILIPPE M. SORIANO

VOLUME 478. Glycomics
Edited by MINORU FUKUDA

VOLUME 479. Functional Glycomics
Edited by MINORU FUKUDA

VOLUME 480. Glycobiology
Edited by MINORU FUKUDA

VOLUME 481. Cryo-EM, Part A: Sample Preparation and Data Collection
Edited by GRANT J. JENSEN

VOLUME 482. Cryo-EM, Part B: 3-D Reconstruction
Edited by GRANT J. JENSEN

VOLUME 483. Cryo-EM, Part C: Analyses, Interpretation, and Case Studies
Edited by GRANT J. JENSEN

VOLUME 484. Constitutive Activity in Receptors and Other Proteins, Part A
Edited by P. MICHAEL CONN

VOLUME 485. Constitutive Activity in Receptors and Other Proteins, Part B
Edited by P. MICHAEL CONN

VOLUME 486. Research on Nitrification and Related Processes, Part A
Edited by MARTIN G. KLOTZ

VOLUME 487. Computer Methods, Part C
Edited by MICHAEL L. JOHNSON AND LUDWIG BRAND

VOLUME 488. Biothermodynamics, Part C
Edited by MICHAEL L. JOHNSON, JO M. HOLT, AND GARY K. ACKERS

VOLUME 489. The Unfolded Protein Response and Cellular Stress, Part A
Edited by P. MICHAEL CONN

VOLUME 490. The Unfolded Protein Response and Cellular Stress, Part B
Edited by P. MICHAEL CONN

VOLUME 491. The Unfolded Protein Response and Cellular Stress, Part C
Edited by P. MICHAEL CONN

VOLUME 492. Biothermodynamics, Part D
Edited by MICHAEL L. JOHNSON, JO M. HOLT, AND GARY K. ACKERS

VOLUME 493. Fragment-Based Drug Design
Tools, Practical Approaches, and Examples
Edited by LAWRENCE C. KUO

VOLUME 494. Methods in Methane Metabolism, Part A
Methanogenesis
Edited by AMY C. ROSENZWEIG AND STEPHEN W. RAGSDALE

VOLUME 495. Methods in Methane Metabolism, Part B
Methanotrophy
Edited by AMY C. ROSENZWEIG AND STEPHEN W. RAGSDALE

VOLUME 496. Research on Nitrification and Related Processes, Part B
Edited by MARTIN G. KLOTZ AND LISA Y. STEIN

VOLUME 497. Synthetic Biology, Part A
Methods for Part/Device Characterization and Chassis Engineering
Edited by CHRISTOPHER VOIGT

VOLUME 498. Synthetic Biology, Part B
Computer Aided Design and DNA Assembly
Edited by CHRISTOPHER VOIGT

VOLUME 499. Biology of Serpins
Edited by JAMES C. WHISSTOCK AND PHILLIP I. BIRD

VOLUME 500. Methods in Systems Biology
Edited by DANIEL JAMESON, MALKHEY VERMA, AND HANS V. WESTERHOFF

VOLUME 501. Serpin Structure and Evolution
Edited by JAMES C. WHISSTOCK AND PHILLIP I. BIRD

VOLUME 502. Protein Engineering for Therapeutics, Part A
Edited by K. DANE WITTRUP AND GREGORY L. VERDINE

VOLUME 503. Protein Engineering for Therapeutics, Part B
Edited by K. DANE WITTRUP AND GREGORY L. VERDINE

VOLUME 504. Imaging and Spectroscopic Analysis of Living Cells
Optical and Spectroscopic Techniques
Edited by P. MICHAEL CONN

VOLUME 505. Imaging and Spectroscopic Analysis of Living Cells
Live Cell Imaging of Cellular Elements and Functions
Edited by P. MICHAEL CONN

VOLUME 506. Imaging and Spectroscopic Analysis of Living Cells
Imaging Live Cells in Health and Disease
Edited by P. MICHAEL CONN

VOLUME 507. Gene Transfer Vectors for Clinical Application
Edited by THEODORE FRIEDMANN

VOLUME 508. Nanomedicine
Cancer, Diabetes, and Cardiovascular, Central Nervous System, Pulmonary and Inflammatory Diseases
Edited by NEJAT DÜZGÜNEŞ

VOLUME 509. Nanomedicine
Infectious Diseases, Immunotherapy, Diagnostics, Antifibrotics, Toxicology and Gene Medicine
Edited by NEJAT DÜZGÜNEŞ

VOLUME 510. Cellulases
Edited by HARRY J. GILBERT

VOLUME 511. RNA Helicases
Edited by ECKHARD JANKOWSKY

VOLUME 512. Nucleosomes, Histones & Chromatin, Part A
Edited by CARL WU AND C. DAVID ALLIS

VOLUME 513. Nucleosomes, Histones & Chromatin, Part B
Edited by CARL WU AND C. DAVID ALLIS

SECTION 1

Single Molecule Studies of Nucleosomes

CHAPTER ONE

DNA Translocation of ATP-Dependent Chromatin Remodeling Factors Revealed by High-Resolution Optical Tweezers

Yongli Zhang[1], George Sirinakis, Greg Gundersen, Zhiqun Xi, Ying Gao

Department of Cell Biology, Yale University School of Medicine, New Haven, Connecticut, USA
[1]Corresponding author: e-mail address: yongli.zhang@yale.edu

Contents

1. Introduction — 4
2. Instrument — 6
   2.1 Instrumentation of high-resolution dual-trap optical tweezers — 6
   2.2 Flow cell assembly — 11
   2.3 Tweezer calibration — 12
3. Remodeler Translocation on Bare DNA — 13
   3.1 Experimental setup — 13
   3.2 Preparation of the DNA substrate — 13
   3.3 Preparation of the tethered remodeler system — 14
   3.4 Procedure of single-molecule experiments on optical tweezers — 15
4. Nucleosome-Dependent Remodeler Translocation — 18
   4.1 Experimental design — 18
   4.2 Purification of SWI/SNF and RSC — 20
   4.3 Preparation of the DNA containing NPSs — 20
   4.4 DNA labeling with biotin and digoxigenin by DNA polymerase extension — 22
   4.5 Nucleosome reconstitution by salt-dialysis method — 22
   4.6 AFM imaging of nucleosomal arrays — 23
   4.7 Pulling nucleosomal arrays — 23
   4.8 Nucleosome-dependent SWI/SNF and RSC translocation — 23
5. Data Analysis — 25
Acknowledgments — 26
References — 26

Abstract

ATP-dependent chromatin remodeling complexes (remodelers) use the energy of ATP hydrolysis to regulate chromatin structures by repositioning and reconfiguring nucleosomes. Ensemble experiments have suggested that remodeler ATPases are DNA translocases, molecular motors capable of processively moving along DNA. This concept of DNA translocation has become a foundation for understanding the molecular mechanisms of ATP-dependent chromatin remodeling and its biological functions. However, quantitative characterizations of DNA translocation by representative remodelers are rare. Furthermore, it is unclear how a unified theory of chromatin remodeling is built upon this foundation. To address these problems, high-resolution optical tweezers have been applied to investigate remodeler translocation on bare DNA and nucleosomal DNA substrates at a single-molecule level. Our strategy is to hold two ends of a single DNA molecule and measure remodeler translocation by detecting the end-to-end extension and tension changes of the DNA molecule in response to chromatin remodeling. These single-molecule assays can reveal detailed kinetics of remodeler translocation, including velocity, processivity, stall force, pauses, direction changes, and even step size. Here we describe instruments, reagents, sample preparations, and detailed protocols for the single-molecule experiments. We show that optical tweezer force microscopy is a powerful and friendly tool for studies of chromatin structures and remodeling.

1. INTRODUCTION

Remodelers are a large family of protein complexes involved in all DNA-related transactions in eukaryotes, including gene transcription, replication, recombination, and repair (Bowman, 2010; Clapier & Cairns, 2009; Hargreaves & Crabtree, 2011; Smith & Peterson, 2005). They contain evolutionarily conserved and specialized ATPase motors and perform common and diverse catalytic activities and biological functions (Flaus, Martin, Barton, & Owen-Hughes, 2006). Based on the sequence homology of their ATPases, remodelers belong to the SWI2/SNF2 family of the SF2 helicase/translocase superfamily and can be further divided into the SWI/SNF-like, ISWI, Mi-2/CHD, and INO80 subfamilies (Clapier & Cairns, 2009). Besides the catalytic ATPase domains, remodelers contain various accessory protein domains or subunits that help remodelers bind onto nucleosomes, interact with transcription factors, recognize histone modifications or variants, and regulate remodeler activities. This modular structure allows remodelers to have both common and diverse functions. All remodelers are capable of mobilizing nucleosomes *in vitro* (Hamiche, Sandaltzopoulos, Gdula, & Wu, 1999; Langst, Bonte, Corona, & Becker, 1999). *In vivo*, remodelers often have overlapping activities (Boeger, Griesenbeck, Strattan, & Kornberg, 2004; Gkikopoulos et al., 2011). However, remodelers can have important and distinct biochemical functions. Remodelers in the SWI/SNF-like subfamily can disassemble nucleosomes and disrupt folded chromatin structures (Sinha, Watanabe, Johnson, Moazed, & Peterson, 2009), leading to gene activation and recombination. In contrast, remodelers in the ISWI and CHD1 subfamilies can assemble nucleosomes and generally help chromatin folding, resulting in gene repression (Gkikopoulos et al., 2011). Specifically, CHD1 can assemble the histone variant H3.3 into chromatin in a genome-wide, replication-independent manner (Konev et al., 2007). Finally, INO80 and SWR1 can exchange histone H2A in nucleosomes with its variant H2A.Z (Mizuguchi et al., 2004; Papamichos-Chronakis & Peterson, 2008). Thus, remodelers are conserved and specialized molecular machines with important and diverse biological functions (Cairns, 2005).

The molecular mechanisms of ATP-dependent chromatin remodeling are poorly understood. Growing evidence indicates that remodeler ATPases are DNA translocases (Saha, Wittmeyer, & Cairns, 2002). The evidence includes their homology with other translocases or helicases and DNA length-dependent ATPase activities. However, this evidence is rather indirect and obtained only for a few remodelers (Whitehouse, Stockdale, Flaus, Szczelkun, & Owen-Hughes, 2003). Especially, some motors in the SF2 superfamily do not seem to be DNA translocases (Pyle, 2008). Therefore, more direct tests of remodeler translocation are required. We have provided some of the first direct evidence for remodeler translocation using optical tweezers (Zhang et al., 2006). We measured the translocation processivities, velocities, step sizes, and forces of SWI/SNF and RSC on nucleosomal DNA, and the corresponding parameters of a minimal RSC complex on bare DNA (Sirinakis et al., 2011). Interestingly, these remodelers show distinct translocation properties, especially compared to Rad54, a motor protein in the SWI2/SNF2 family involved in DNA recombination and repair (Amitani, Baskin, & Kowalczykowski, 2006). For example, the minimal RSC complex and Rad54 move on DNA with dramatically different average velocities (25 vs. 301 bp/s) and processivities (35 vs. $\sim$11,500 bp). Therefore, despite their conserved structures, SWI2/SNF2 motors have different translocation properties, a fact that is important to understanding their functional diversities.

Optical tweezers generally use one or two optical traps to hold polystyrene or silica beads as force and displacement sensors (Moffitt, Chemla, Smith, & Bustamante, 2008). Optical traps are diffraction-limited light spots formed by focusing collimated laser beams using objectives with high numerical apertures (NAs). The large electric field gradient in the light spot will polarize small dielectric particles such as polystyrene beads and attract them to the location of highest field intensity, the center of the optical trap. For a small displacement of the bead from the trap center (typically <300 nm for a 2-μm-diameter bead), the restoring force exerted on the bead is proportional to the displacement (Greenleaf, Woodside, Abbondanzieri, & Block, 2005). Thus, the optical trap serves as a harmonic potential for the bead. The force constant of the trap is proportional to the total intensity of the trapping light and can be experimentally measured. Furthermore, optical tweezers contain optoelectronic modules to measure the bead displacement with high resolution based upon an interference method. When attaching the macromolecule of interest to two trapped beads, the beads can serve as excellent force and displacement sensors to report the structure and dynamics of the macromolecule. Therefore, optical tweezers extend our hands and allow us to manipulate single molecules and detect their movements and responses to external forces in real time. As a result, optical tweezers have been widely applied to study molecular motors and structures and dynamics of macromolecules (Bustamante, Cheng, & Mejia, 2011).

In this chapter, we detail single-molecule experiments of remodeler translocation using high-resolution optical tweezers. After an introduction to the optical tweezers, we focus on the translocation of remodelers on bare DNA and nucleosomal DNA.

2. INSTRUMENT

2.1. Instrumentation of high-resolution dual-trap optical tweezers

The instrument was assembled on an optical table using off-the-shelf and custom-made parts, mainly as previously described (Fig. 1.1A) (Moffitt, Chemla, Izhaky, & Bustamante, 2006). Briefly, a single laser beam of 1064 nm is expanded, collimated (by the telescope T1), and split into two orthogonally polarized beams (by PBS2) that are reflected by a rotary mirror (NM) and a stationary mirror, respectively. The two beams are then combined by a polarizing beam splitter (PBS3), further expanded (by the telescope T2), and finally focused deep into water to form two optical traps in a flow cell by a water-immersion microscope objective (FO, Nikon 60× NA=1.2). The separation between the two traps can be precisely controlled to adjust the force applied to the molecule of interest, by turning the piezo-electrically controlled rotary mirror (Mad City Labs). To detect bead

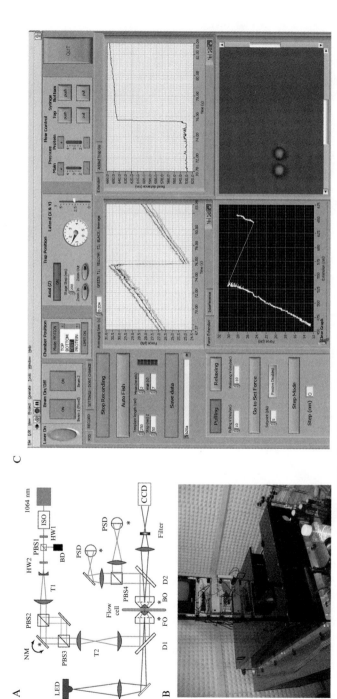

Figure 1.1 High-resolution optical tweezers. (A) Optical diagram of the dual-trap optical tweezers. ISO, optical isolator; HW, half-wave plate; BD, beam dumper; T, telescope; PBS, polarizing beam splitter; NM, piezo mirror; D, dichroic mirror; FO, front objective; BO, back objective. The positions marked by stars represent optically conjugated planes. (B) The enclosed optical paths and components of the tweezers on an optical table in an environmentally controlled room. (C) Labview interface used to remotely operate the optical tweezers.

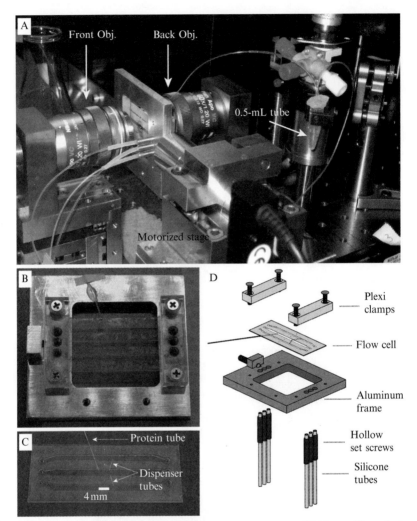

Figure 1.2 Microfluidic system used in the tweezer instrument. (A) Flow cell installed on a motorized stage and positioned between two water-immersion objectives. Tubing connected to the flow cell, including the 0.5-mL microcentrifuge tube containing the remodeler solution, can be seen. (B) Flow cell mounted on a 2.5″ × 2.75″ aluminum frame and held in place by two plexi clamps. (C) Picture of an assembled flow cell. It is constructed by sandwiching two nescofilm (Alfresa Pharma Co.) gaskets and three glass tubes between two microscope cover slides (60 × 24 × 0.17 mm). The nescofilm gasket is cut to create three channels by a laser engraver (Epilog Zing 16 laser system with 30W CO_2 laser, Epilog Laser, CO). The bottom coverslip containing six holes for fluid delivery is cut by the same engraver. The flow cell is then placed onto a hot plate and heated to 75 °C. At this temperature, the nescofilm will not melt, but air bubbles can be removed with small amounts of pressure. The temperature is then increased to 110 °C,

displacement, the outgoing laser light is collimated again by a second identical objective (BO, Fig. 1.2A) and projected to two position-sensitive detectors (PSD, Fig. 1.1A). The detectors have voltage outputs that are proportional to the small lateral bead displacements and can be converted to displacement and force measurements after certain calibrations (Gittes & Schmidt, 1998).

The design of dual-trap optical tweezers has been proven to be robust and immune to many noises. Because the biomolecule of interest is attached between two optically levitated beads and isolated from the microscope stage, experimental measurements, such as the end-to-end extension and tension of the biomolecule, are not affected by stage drift that is sensitive to environmental changes (Nugent-Glandorf & Perkins, 2004). Furthermore, the extension and tension can be measured by the difference between outputs of two PSDs. This differential detection scheme greatly reduces noises common to both traps, such as laser pointing errors, in-phase bead fluctuations, and even residue flows in the fluidic channel. Moreover, the environment for the optical tweezers is carefully controlled (Fig. 1.1B). The optical components of the tweezers are located in an acoustically isolated room with stringent controls in temperature and air flow (Abbondanzieri, Greenleaf, Shaevitz, Landick, & Block, 2005; Moffitt et al., 2006). Noisy equipment is put outside of the tweezer room, including the computer that controls the tweezers and the pumping light source for the diode-pumped solid state laser for optical trapping. As a result, optical tweezers are remotely operated through a Labview interface (Fig. 1.1C) after the sample is loaded. Finally, we found that the quality of the trapping laser is crucial for optical tweezers. Commercially available lasers at 1064 nm should be tested in the lab for good Gaussian beam profiles and stability, using a beam profiler and a power meter, respectively. The beam quality can also be checked by shining part of the beam to a PSD and assessed by taking the power spectrum density of the PSD signal, as is demonstrated in

---

covered with a hot block, and left for 30 min. The assembled flow cell should be clear, particularly on the edges of the channels and near the holes. The glass tubes are custom made by King Precision Glass, CA, and have diameters of 80 μm (OD) and 40 μm (ID) for the protein injection tube and of 100 μm (OD) and 25 μm (ID) for the dispenser tube. The channels in the flow cell have a thickness of around 180 μm. (D) Diagram showing the assembly of the microfluidic system. Silicone tubes are used to snuggly fit to the holes into the three microfluidic channels when screwed into the aluminum frame.

Fig. 1.3. Using the design described above and taking precautions for the environment and quality of the trapping laser, we and others have built high-resolution optical tweezers reaching base-pair resolution with high baseline stability (Moffitt et al., 2006; Sirinakis et al., 2011). High-resolution and stable optical tweezers are essential for single-molecule studies of remodeler translocation. Because of remodeler's relatively small processivities ($\lesssim 100$ bp) and velocities (<30 bp/s) (Sirinakis et al., 2011; Zhang et al., 2006), the resultant small changes in DNA extension and tension due to remodeler translocation can only be well distinguished from background noises or baseline drifts by high-resolution optical tweezers.

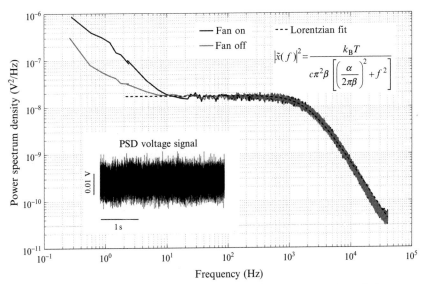

Figure 1.3 Power spectrum density distributions of Brownian motion of the same trapped bead measured when the fan is on (black curve) and off (gray curve). Part of the corresponding PSD output is shown as an inset. The distributions can be well fit with a Lorentzian function (dashed line shown for fan off) in the range of 10–10^4 Hz. The Lorentzian function is expressed as an inset, where k_B is the Boltzmann constant, T the temperature, c the conversion constant from PSD voltage to bead displacement, α the force constant of the trap, and f the frequency. Here $\beta = 6\pi r\eta$ is the drag coefficient of the bead, which can be calculated based on a known bead diameter (r) and buffer viscosity (η). The power spectrum density distribution at a low frequency range (<10 Hz) is sensitive to the air flow through the optical path, as well as other low frequency noises, which can be decreased by turning off the ventilation fan for the tweezer room and enclosing the optical path and components (Fig. 1.1B).

2.2. Flow cell assembly

The dual-trap optical tweezers have a horizontal optical layout (Fig. 1.1A and B) that contributes to the excellent mechanical stability and high spatio-temporal resolution of the instrument. However, this design necessitates a thin and vertical flow cell mounted on a motorized stage (Fig. 1.2A). We use a flow cell with one central and two auxiliary channels (Fig. 1.2B and C). The channels are laser engraved in two layers of nescofilm, sandwiched between two glass coverslips and sealed by heating (Fig. 1.2D). The central channel is used for optical trapping, while the auxiliary channels supply the two different kinds of beads required for the single-molecule experiments, typically streptavidin (SA)- and anti-digoxigenin (αDIG)-coated polystyrene beads. The beads are delivered from the auxiliary channels to the central channel through glass tubing. A different glass tube is also used to directly inject the remodeler solution into a small reaction area in the central channel. This method of protein injection conserves the precious remodeler sample and allows fast protein addition and removal, compared to an alternative approach in which the protein solution is flowed through the entire central channel. It also offers the advantage to add remodeler solution after a single DNA molecule is attached to two trapped beads.

Flow in the central channel and the protein injection tube is controlled by the pressure in the corresponding solution vials through a combination of solenoid valves. These valves regulate the influx or efflux of pressurized nitrogen in each vial. Flow of bead solutions in the auxiliary channels is achieved with the help of home-built, computer-controlled syringe pumps. However, during the single-molecule experiment, the flow in the central channel should be minimal (but not zero) to avoid perturbation of the experiment due to differential flow dragging forces applied to both trapped beads. The background flow in the central channel can be judged by the drifting velocity of a bead after released from the trap by turning off the trap. In general, a downstream background flow should be set before the tweezer experiment such that free beads only slowly drift in the direction from the tip of the protein injection tube to the tips of dispenser tubes. This default background flow (typically < 1 μm/s) prevents the beads exiting from the dispenser tubes from drifting to the test area between tips of the two kinds of tubes and interfering with the single-molecule experiment.

Liquid in the flow cell and its connecting tubes should be free of air bubbles and contaminant particles. Air bubbles trapped in the flow cell can disturb the flow and introduce noise in the measured extension and force signals. Therefore, air bubbles should be removed from the flow cell and connecting tubing before any tweezers experiments. Solution injected into the central channel should be clear and free of particles and other contaminants because they can be caught by the optical traps and disturb the measurements. Thus, buffers should be filtered and degassed. Protein samples often contain large contaminants or gel debris left from purification processes, which can be removed by spinning the protein solution after dilution into final reaction buffers. Filtering protein solution may cause protein depletion due to its absorption to the filter and is not recommended for cleaning remodeler solutions. Finally, at the end of tweezer experiment, the flow cell and all tubing should be thoroughly washed and filled with a solution of 0.01% sodium azide to prevent bacterial growth.

2.3. Tweezer calibration

The optical tweezers need to be calibrated to determine coefficients for the linear conversions from the mirror rotation angle to the trap separation and from the measured PSD voltages to bead displacements and forces (Moffitt et al., 2006). To calibrate trap separation, two polystyrene beads are held by the optical traps and separated in a step-wise manner. During this process, images of the two beads are taken by the CCD camera at each mirror position (Fig. 1.1A and C). Then the trap separation at each mirror position is measured as the distance between the centroids of the two bead images. Finally, the resultant plot of trap separation versus mirror rotation angle is linearly fit to get the slope and offset required to convert the mirror rotation angle to the trap separation. Such calibration is performed regularly (typically weekly) to correct any spontaneous drift of optical components.

In contrast, trap stiffness and the voltage-to-displacement conversion constants are generally determined for each trapped bead at the beginning of each experiment because these parameters slightly change among the beads with variable diameters. One first records the PSD response to the Brownian motion of the bead in each trap with a high bandwidth (typically 80 kHz) and then calculates the power spectrum density distribution of the acquired voltage signal after its baseline is subtracted (Fig. 1.3). For a particle confined in a harmonic potential, its displacement, in principle, should have a power spectrum density of Lorentzian distribution that is a function of the trap stiffness and the voltage-to-displacement conversion constant. A nonlinear fitting of the measured power spectrum density with the

Lorentzian function yields the two constants required for the calibration. In our experiments, typical trap stiffness and the conversion constant are ~0.2 pN/nm and 0.6 nm/mV, respectively, for a ~2 μm diameter polystyrene bead and ~300 mW laser power per trap. Note that it is beneficial to keep the stiffness of and bead diameters in both traps approximately equal to maximize the spatial resolution of the dual-trap optical tweezers.

3. REMODELER TRANSLOCATION ON BARE DNA

3.1. Experimental setup

To investigate motor translocation on bare DNA using optical tweezers, we have developed a tethered motor assay in collaboration with Cairns' group (Fig. 1.4A) (Sirinakis et al., 2011). Crucial to this assay is a DNA translocase fused with a tetracycline repressor TetR that can specifically bind to the tetO site in the middle of a DNA molecule stretched by the optical tweezers. Wild-type TetR is a homodimer and binds tetO with an association constant of $\sim 10^{11}$ M^{-1} (Orth, Schnappinger, Hillen, Saenger, & Hinrichs, 2000), which provides a strong anchor for motor translocation and force generation. This strong association can be reversed by adding tetracycline, which can serve as a control experiment to test the role of motor anchoring in the observed translocation signal. Once anchored on the DNA, motor translocation is accompanied by the formation of a DNA loop between the remodeler motor domain and TetR, which shortens the DNA end-to-end distance and increases the DNA tension detected by optical tweezers (Fig. 1.4B). Thus, motor translocation can be detected in real time (Fig. 1.4C).

3.2. Preparation of the DNA substrate

The 5063 bp DNA molecule contains a tetO site (tctatcattg atagg) incorporated into a pUC19 plasmid containing nine nucleosome position sequences (NPSs) "601" (pUC-N9). Construction of pUC-N9 will be described in Section 4.3. A polynucleotide containing a tetO site and its complementary strand are chemically synthesized, hybridized, and inserted into the modified plasmid between *Eco*RI and *Ban*II restriction sites. The resultant plasmid containing tetO site (tet-N9) is transformed into DH5α cells, amplified, and purified. The plasmid DNA is then digested with a *Sty*I restriction enzyme and labeled as detailed in the later section.

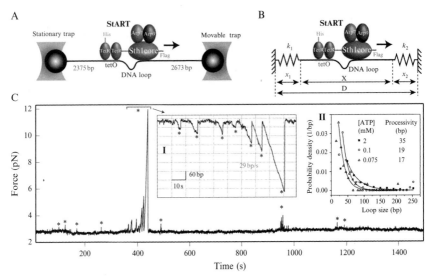

Figure 1.4 Tethered motor assay for remodeler translocation on bare DNA based on dual-trap high-resolution optical tweezers (Sirinakis et al., 2011). (A) Illustration of the experimental setup and the composition of the tethered minimal RSC complex (StART). Not drawn to scale. (B) Simplified diagram showing the principle of force and extension detection by optical tweezers. The two optical traps are equivalent to two microscopic springs with force constants k_1 and k_2. As the tethered motor (Sth1 core here) translocates DNA, it shortens the DNA end-to-end extension (X) and pulls the two beads away from their corresponding trap centers (increases in x_1 and x_2) while the trap separation (D) is kept fixed. (C) Time-dependent tension or length (inset I) of a single DNA molecule showing remodeler translocation and loop formation. The StART activities (marked by stars) contain continuous increases in force corresponding to remodeler translocation, followed by sudden drops to the baseline indicating disengagement of the remodeler translocation domain from the DNA. Signals tend to appear in a row, suggesting that the translocation domain can undergo multiple rounds of translocation and disengagement before the complex completely dissociates from the DNA molecule. Inset I shows the plot of DNA length (Section 5) versus time corresponding to the indicated force–time region. Translocation velocity and distance can be calculated based on this plot. Inset II shows probability density distributions of the translocation distance of StART at different ATP concentrations. The ATP-dependent processivity of the motor is listed in the inset.

3.3. Preparation of the tethered remodeler system

We first demonstrated the tethered motor assay using the minimal RSC complex (Sirinakis et al., 2011). The protein complex contains four subunits: Sth1 core, Arp7, Arp9, and TetR, designated as StART (Figs. 1.4A and 1.5A). Here, the two monomers in TetR are expressed from genes in the same plasmid under control of the same promoter, one fused with Sth1$_{(301-1097)}$-FLAG and the other labeled with an N-terminal His tag.

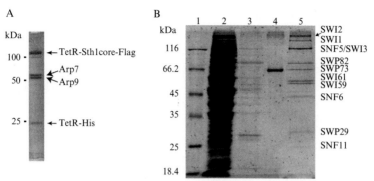

Figure 1.5 SDS gels of remodelers in different purification steps. Protein subunits in each purified complex are labeled on the right. (A) SDS gel of the purified tethered minimal RSC complex. (B) SDS gel of protein samples in different steps of SWI/SNF purification. The samples in different lanes are the protein marker with their molecular weights indicated on the left (lane 1), flow through of the IgG-Sepharose (lane 2), supernatant after the SWI/SNF-bound IgG-Sepharose is treated with TEV protease (lane 3), 100 ng BSA (lane 4), and final purified SWI/SNF complex (lane 5), respectively. The proteins are stained with SYPRO red (Invitrogen).

The two actin-related proteins (Arp7 and Arp9) are expressed from a second plasmid. Both plasmids are cotransformed into *Escherichia coli* BL21(DE3) RIL. The StART complexes are assembled in the cell and purified successively using Ni–NTA agarose resin (Qiagen) and anti-Flag M2 affinity gel (Sigma). The complexes are eluted with 3 × FLAG peptide (Sigma) and further refined by gel filtration on S200GL 10/300 (Amersham, GE) column. The pure complex is confirmed by SDS gel electrophoresis (Fig. 1.5A).

3.4. Procedure of single-molecule experiments on optical tweezers

3.4.1 Buffers

PBS buffer for bead dilution: 137 mM NaCl, 2.7 mM KCl, 8.1 mM Na$_2$HPO$_4$, 1.8 mM KH$_2$PO$_4$, pH 7.4.

The DNA translocation buffer for StART: 20 mM Tris–acetate, 10 mM magnesium acetate, 50 mM potassium acetate, 1 mM DTT, 6% glycerol, 0.1 mg/mL BSA, supplemented with typically 2 mM ATP.

3.4.2 Remodeler solution preparation

An aliquot of remodeler stock solution is diluted in the DNA translocation buffer just before the tweezer experiment and then spun at 4 °C at the highest speed of a bench-top centrifuge for 20 min. The supernatant is collected for the tweezer experiment.

3.4.3 Bead preparation

We typically use 1.87-μm-diameter streptavidin-coated and 2.17-μm-diameter anti-digoxigenin-coated polystyrene beads (Spherotech SVP-15-5 and DIGP-20-2, respectively). For applications requiring higher spatiotemporal resolution, beads with ∼1 μm diameter or smaller should be used. However, these small beads tend to be more difficult to be visualized and trapped and lead to shorter tether lifetime than the big beads. The limited tether lifetime results from the photo damage induced by the high-intensity (>10 MW/cm^2) trapping light that produces free radicals to cleave the DNA-bead linkages (Landry, McCall, Qi, & Chemla, 2009). The photo damaging can be alleviated by adding an oxygen-scavenging system in the reaction buffer. But the oxygen-scavenging system tends to lower the pH of the protein solution, thus affecting the enzymatic activity. Fortunately, using the ∼2-μm-diameter beads, we find that the DNA tether is generally stable enough to hold a low tension (<5 pN) for more than 20 min, without the presence of the oxygen-scavenging system. Therefore, these beads are mainly used in our remodeler translocation assays.

To prepare the beads for optical trapping, aliquots of the bead stock solutions are dispersed by either vortexing for 1 min or sonicating for 20 min. The DNA molecules can be bound to either SA beads or αDIG beads, with the latter as our choice here. The amount of DNA added to the bead solution should be optimized to allow pulling single DNA molecules, as detailed below.

3.4.4 Pull a single DNA

1. Wash and fill the central channel of the flow cell with the translocation buffer. To minimize protein absorption, the inner surface of the central channel and the protein injection tubing may be passivated by washing the channel and tubing with 0.5 mg/mL bovine serum albumin (5 × BSA) or 0.05% powdered milk and then thoroughly rinsing with the translocation buffer.
2. Dilute 2 μL of dispersed SA beads in 1 mL PBS buffer.
3. Bind the DNA molecules (tet-N9) to αDIG beads. About 5 ng of the DNA molecule is mixed with 20 μL dispersed αDIG beads and incubated for 20 min at room temperature to allow DNA binding. Then the DNA-bead solution is diluted in 1 mL of PBS buffer.
4. Inject the diluted DNA-αDIG bead solution and the SA bead solution to the auxiliary channels of the flow cell using 1 mL syringes.
5. Set the trap separation to the maximum.

6. Turn the stationary trap on and the movable trap off.
7. Move the flow cell through the motorized stage to position the tip of the dispenser tube for the DNA-bound beads in the vicinity of the stationary trap and capture a single DNA-αDIG bead.
8. Turn on the movable trap and similarly grab a single streptavidin-coated bead in this trap near the other dispenser tube.
9. Move the flow cell to position the trapped beads approximately in the middle of the central channel and between the dispenser tubes and the protein injection tube, an area designated as the test area.
10. Record Brownian motions for both trapped beads for about 20 s and calibrate the traps as described in Section 2.3.
11. Fish the DNA attached to the αDIG bead by moving the SA bead first close to and then away from the αDIG bead. In this process, the interacting force between the two beads is recorded. If one or multiple DNA molecules are captured by the SA bead, the force should increase as two beads are moving apart. Otherwise, the force remains zero during the separation. In this case, the fishing process will be repeated by moving the two beads successively closer, until formation of a DNA tether between the two beads is confirmed. Note that the fishing process can be facilitated by adding flow to stretch the DNA molecule on the αDIG bead toward the SA bead. However, if no DNA tether is found within a reasonable time period or about eight rounds of fishing, the αDIG bead may have no DNA molecule attached in the region accessible to the SA bead. Then both beads are released and Steps 5–11 will be repeated until a DNA tether is formed. We have automated this fishing process to facilitate the tweezer experiment (Fig. 1.1C).
12. Pull the DNA tether by separating the two traps at a uniform speed, typically 10–500 nm/s, until the tether breaks. If the DNA tether breaks in one step with a force drop to around zero or show a B-to-S transition around 65 pN (Smith, Cui, & Bustamante, 1996), the DNA tether is a single DNA molecule. Otherwise, the DNA tether may have multiple DNA molecules.
13. Test a total of around 10 different pairs of beads for DNA tether formation and perform the pulling experiment by repeating Steps 5–12.
14. Optimize the DNA amount bound to the αDIG beads if necessary by repeating Steps 3–13. Under the conditions for single-molecule experiments, the number of DNA molecules bound on a single bead (or more specifically the surface region accessible by the SA bead) follows a Poisson distribution, which has a maximum probability of 0.37 to have

a single DNA molecule for all bead pairs tested. Under this optimal condition, the probabilities to have zero and more than one DNA molecules are 0.37 and 0.26, respectively. If the probability to tether a single DNA molecule between a pair of beads is significantly lower than 0.37, one should increase or decrease the amount of DNA bound to the αDIG beads to make the single-molecule experiments more efficient.

3.4.5 Single remodeler translocation on bare DNA

15. Add the clean remodeler solution prepared in Section 3.4.2 into the 0.5-mL microcentrifuge tube connected to the protein injection tube (Fig. 1.2A).
16. Stretch a DNA tether to a tension of interest, typically around 3 pN. Then inject the protein solution into the test area in the central channel. The protein solution flow can exert a drag force on two beads (typically a few pN) and can be used to judge the flow rate ($\sim$50 μm/s/pN). Stop the flow after about 30 s when the test area is filled with the remodeler solution.
17. Measure the activity of the enzyme in real time while keeping the trap separation fixed for typically 20 min. The translocation signals are seen as spikes in which the force linearly increases followed by a sudden drop to the baseline (Fig. 1.4C). Record data at 5 kHz to a hard disk.
18. Confirm a single DNA molecule by pulling the tether to high forces as described in Step 12.
19. Collect more translocation signals by repeating Steps 16–18 using different pairs of beads.

4. NUCLEOSOME-DEPENDENT REMODELER TRANSLOCATION

4.1. Experimental design

In contrast with the tethered minimal RSC complex, full complexes of SWI/SNF and RSC can target histones and pump DNA around the histone octamer (Zhang et al., 2006). Both remodelers are comprised of more than 10 different subunits and contain histone binding domains that serve as DNA anchors to constrain the DNA loop while their translocases pump DNA toward the histone octamer (Fig. 1.6A). Thus, DNA translocation properties of the full remodeler complexes can be similarly measured on nucleosomal DNA templates and compared with those obtained on bare DNA templates. These comparisons can reveal the effects of the nucleosome substrate and

DNA Translocation by Single Remodelers 19

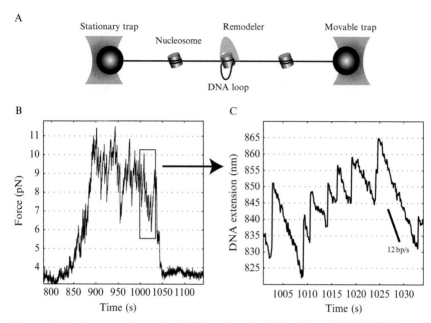

Figure 1.6 Nucleosome-dependent RSC translocation (Sirinakis et al., 2011). (A) Diagram of the experimental setup. (B) Time-dependent tension of the nucleosomal DNA template showing RSC translocation. The repetitive RSC-induced DNA loop formation and dissipation is caused by DNA translocation of a single RSC complex after bound on the nucleosome. (C) Time-dependent DNA length corresponding to the boxed area in (B).

other protein subunits of remodelers on DNA translocation (Fig. 1.6B and C). In the following, we will describe the single-molecule experiment on the nucleosomal DNA template.

It may be advantageous to put a single nucleosome on a long DNA molecule to test the nucleosome-dependent remodeler translocation. This single-nucleosome DNA template can avoid remodeler collisions into neighboring nucleosomes during translocation. Moreover, the experimental noise is generally lower on the single-nucleosome template than on the nucleosomal array template because fast nucleosomal DNA unwrapping–rewrapping fluctuations on histone surfaces add to the measured noises in a nucleosome number-dependent manner. However, the single-nucleosome template is difficult to make (Zhang et al., 2006), partly because a single nucleosome on a long DNA molecule is not so stable under the single-molecule conditions. Furthermore, the chance to observe translocation signals is relatively low on a single nucleosome, probably because the nucleosome-dependent loop formation is only one of many pathways in a remodeler-catalyzed remodeling reaction. To overcome these difficulties, we use a special nucleosomal array in our single-molecule experiment. This array consists of nine tandem repeats of 258 bp DNA containing the "601" nucleosome-positioning sequence (Lowary & Widom, 1998). The large spacing between nucleosomes deposited on this DNA molecule allows enough bare DNA on both sides of a nucleosome for remodeler translocation.

4.2. Purification of SWI/SNF and RSC

SWI/SNF and RSC are purified endogenously using the tandem affinity purification (TAP) method as previously reported (Smith, Horowitz-Scherer, Flanagan, Woodcock, & Peterson, 2003; Wittmeyer, Saha, & Cairns, 2004). In this method, one of the protein subunits is added with a TAP-tag at its carboxyl-terminal. The TAP-tag contains a calmodulin-binding peptide, a TEV protease cleavage site, and protein A. The TAP-tag is fused to the SWI2 subunit of SWI/SNF and the Rsc2 subunit of RSC, respectively. Starting from yeast cell extracts, both complexes are first bound to IgG-Sepharose, cut from the resin by TEV protease, then bound to calmodulin beads with calcium, and finally eluted from the beads in the presence of EGTA, yielding pure remodeler complexes (Fig. 1.5B).

4.3. Preparation of the DNA containing NPSs (Zhang et al., 2006)

1. Make the DNA molecule with the following multicloning sequence:

   EcoRI   BanII   AvaI   PstI   StyI   HindIII
   CAGT**GAATTC**ACCGTCTAAGATCTGATTC**GAGCCC**GGTAC**TCGGG**GATCCTCTAGAGTAGACC**TGCAG**GCTCAG**CCTTGG**ATGATGC**AAGCTT**GGCG
   GTCACTTAAGTGGCAGATTCTAGACTAAGCTCGGGCCATGAGCCCCTAGGAGATCTCATCTGGACGTCCGAGTCGGAACCTACTACGTTCGAACCGC

2. Clone the sequence into the pUC19 plasmid between the *Eco*RI and *Hin*dIII sites, generating a pUC19 plasmid variant with a modified multicloning site.
3. Synthesize the DNA molecules containing the 258 bp sequence repeat with nonpalindromic *Ava*I restriction sites (CTCGGG) on both ends. Clone the DNA fragment into the pUC19 variant through the *Ava*I site, generating a sufficient amount (>50 μg) of the new plasmid pUC-N1 containing one NPS repeat.
4. Digest the pUC-N1 plasmid with *Ava*I and purify the NPS repeat sequence using agarose gel electrophoresis.

5. Ligate the NPS repeat sequence and separate the resultant DNA ladder by agarose gel electrophoresis (Fig. 1.7A). Purify the tandem repeats containing the required number of NPSs from the gel.
6. Clone the purified tandem NPS repeats back into the pUC19 plasmid variant through the *Ava*I site, generating the required plasmid DNA sequences containing defined number of NPS repeats. Purify the plasmid DNA using the standard plasmid purification method (such as the Qiagen plasmid purification kits) and further concentrate the DNA by ethanol precipitation.

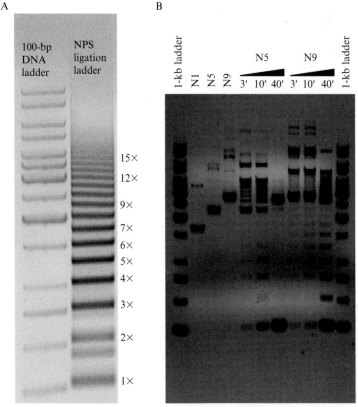

Figure 1.7 Construction of the tandem repeats of nucleosome-positioning sequences (NPSs). (A) Agarose gel of the NPS ladder formed by ligation of NPS monomers (258 bp). The number of NPS repeats is indicated on the right. (B) Agarose gel of the plasmids containing 1 (N1), 5 (N5), and 9 (N9) NPS repeats purified from *E. coli* and their limited digested products by *Ava*I restriction enzyme. *Ava*I digestion for a time period of 3, 10, and 40 min gradually releases multiple NPS repeats or monomers. The maximum number of bands in the NPS ladder indicates the number of NPS repeats in the plasmid.

7. Confirm the number of NPS repeats in the plasmids by limited AvaI digestion (Fig. 1.7B).

4.4. DNA labeling with biotin and digoxigenin by DNA polymerase extension

1. Digest 40 μg of pUC-N9 DNA plasmid with 20 units of StyI restriction enzyme (New England Biolabs, MA) in 50 μL of 1 × NEB Buffer 3 (50 mM Tris–HCl, 100 mM NaCl, 10 mM MgCl$_2$, and 1 mM DTT, pH 7.9) containing 0.1 mg/mL BSA. Incubate overnight at 37 °C.
2. Add to the above reaction mixture 4 μL of 1 mM digoxigenin-dUTP (Roche), 8 μL of 0.4 mM biotin-dATP (Roche), 0.7 μL of 10 mM dCTP, 0.7 μL of 10 mM dGTP, 3 μL of 10 × EcoPol Buffer (NEB), 2.6 μL ddH$_2$O, and 1 μL Klenow (Exo$^-$) (NEB). Incubate the 70 μL mixture at 37 °C for 1 h. The reaction adds two biotin moieties at one end of the DNA and two digoxigenin moieties at the other end.
3. Purify the labeled DNA by phenol:chloroform extraction and ethanol precipitation. Store the DNA in 20 μL TE buffer (10 mM Tris–HCl, 1 mM EDTA, pH 8.0).

4.5. Nucleosome reconstitution by salt-dialysis method

1. Mix 10 μg of the labeled pUC-N9 DNA with 2.5, 2.8, and 3.2 μg of histone octamer in three 2 M NaCl solutions with a final 100 μL volume. The histone octamer is purified from chicken erythrocytes.
2. Transfer the mixtures to "Slide-a-Lyzer" mini dialysis buckets (Pierce) and dialyze at 4 °C against 1500 mL Tris–EDTA buffer (10 mM Tris–HCl, pH 8.0, 0.25 mM EDTA) containing 1.5 M NaCl for 3 h.
3. Dialyze the reconstitution mixtures against the Tris–EDTA buffer containing successively lower concentrations of NaCl, that is, 1.0 M, 0.8 M, 0.6 M, and finally 2.5 mM, each for 3 h at 4 °C.
4. Collect the sample and store at 4 °C.

The quality of the reconstituted nucleosomal arrays is crucial for chromatin structure and remodeling studies. In the following, we will use atomic force microscopy (AFM) imaging and single-molecule manipulation method to assess the nucleosome number distribution on the DNA molecules. Whereas undersaturated nucleosome arrays can be used in some experiments, oversaturated arrays should be avoided because noncanonical nucleosomal structures may be formed on these templates. We normally make at least three reconstituted nucleosomal arrays at a time with slightly different histone-to-DNA ratios, which allow selection of the arrays with proper nucleosome occupancy for single-molecule experiments.

4.6. AFM imaging of nucleosomal arrays

The nucleosomal array is fixed by glutaraldehyde before AFM imaging. To fix the array, the reconstituted sample is further dialyzed. First, the array is dialyzed against 1 mM EDTA, pH 8.0; then against 1 mM EDTA, 0.1% fresh glutaraldehyde, pH 7.7; finally against 1 mM EDTA, pH 7.6. Each step of dialysis should be at 4 °C for 6 h. The nucleosomal array is imaged by a MultiMode NanoScope V AFM (Veeco Instruments) with a type E scanner. First, the fresh mica surface is treated with spermidine by adding 10 μL of 100 mM spermidine, rinsed with ddH$_2$O, and dried with nitrogen flow. Then about 5 ng of fixed nucleosomal DNA in 20 μL EDTA buffer, pH 7.6 is added to the treated mica surface and imaged using the AFM in a tapping mode. The silicon cantilever (Nanosensors) used for imaging has a resonance frequency of 260–410 kHz and a force constant of 21–78 N/m. Representative images are shown in Fig. 1.8, from which the average number and positioning of nucleosomes on the DNA sequence can be scored.

4.7. Pulling nucleosomal arrays

The nucleosomal DNA is attached to the two beads and pulled similar to the bare DNA described in Section 3.4.4. A representative force–extension curve (FEC) is shown in Fig. 1.9A, in which each rip corresponds to the mechanical disruption of a nucleosome core particle on the DNA molecule. Thus, the number of nucleosomes on each nucleosomal array can be counted as the number of rips in the measured FECs. Furthermore, canonical nucleosomes show extension changes centered at 70 bp (Fig. 1.9B) and disruption forces around 23 pN (Fig. 1.9C), consistent with the previous report (Brower-Toland et al., 2002). Significant deviation from these distributions may indicate improper nucleosome reconstitution.

4.8. Nucleosome-dependent SWI/SNF and RSC translocation

The translocation is performed in different buffers for both remodelers: 20 mM Tris–HCl (pH 8.0), 125 mM NaCl, 5 mM MgCl$_2$, 1% glycerol, 0.2 mM DTT, 0.1 mg/mL BSA, and 0.05% NP-40 for SWI/SNF, and 20 mM HEPES (pH 7.6), 100 mM potassium acetate, 5 mM magnesium acetate, 1% glycerol, 0.2 mM DTT, 0.1 mg/mL BSA, and 0.05% NP-40 for RSC. The typical nominal remodeler concentration (the remodeler

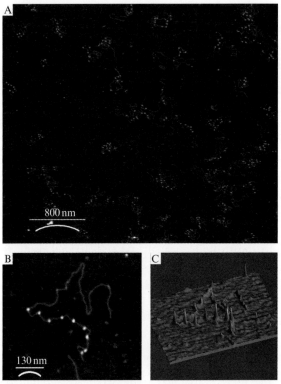

Figure 1.8 AFM images of the reconstituted nucleosomal arrays. (A) AFM image of the nucleosome arrays on the plasmid DNA containing nine tandem repeats of NPSs (pUC-N9). Nucleosomes (bright spots) are mainly formed on NPSs located at half of the DNA molecule, with the other half being plasmid DNA free of nucleosomes. (B) Close-up view of a single nucleosomal array containing nine uniformly spaced nucleosomes. (C) The 3D image corresponding to the image in (B). The average height of nucleosomes is 3.1 nm. (See Color Insert.)

concentration before adding to the flow cell) is 2–20 nM. DNA translocation activities of SWI/SNF and RSC are tested on the nucleosomal DNA in a way similar to that of StART on bare DNA (Figs. 1.4A and 1.6A). Because histones can dissociate from DNA over time in a force-dependent manner, the experiment is normally carried out at a low tension (2–5 pN) to minimize spontaneous nucleosome disassembly and to facilitate loop formation. We found that SWI/SNF and RSC have similar nucleosome-dependent translocation properties (Zhang et al., 2006), with a trace shown for RSC in Fig. 1.6B.

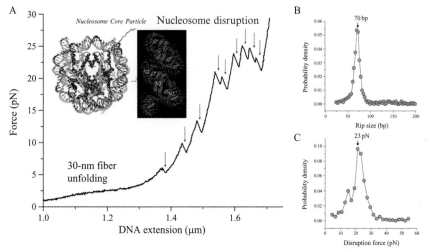

Figure 1.9 Mechanical disruption of single nucleosomal arrays. (A) Force–extension curve (FEC) of a single nucleosomal array with nine nucleosomes obtained by pulling the array at a rate of 100 nm/s. Individual events of nucleosome disruption are indicated by arrows. Here, each rip corresponds to discontinuous nucleosomal DNA unwrapping from the histone surface in a high-force regime (>5 pN). In the low-force regime (<5 pN), the FEC shows a force plateau around 2.5 pN. This plateau is reversible in the low-force regime and distinct only in the presence of approximately saturated nucleosomal array and >2 mM magnesium in the buffer. Compared to previous results (Kruithof et al., 2009), we conclude that this plateau is a signature of reversible folding and unfolding of the 30-nm fiber. The inset shows the crystal structure of the nucleosome core particle (Luger, Mader, Richmond, Sargent, & Richmond, 1997) and a model of the 30-nm fiber (Schalch, Duda, Sargent, & Richmond, 2005). (B) Histogram distribution of the DNA length released when nucleosomes are disrupted. The average length is 70 bp as indicated. (C) Histogram distribution of the force to disrupt nucleosomes measured from the rips in the FECs. The average force is 23 pN as indicated.

5. DATA ANALYSIS

Data corresponding to individual tethers are saved in a binary format in separate files, including calibration parameters, PSD signals, rotary mirror angles, and other experimental information such as pulling velocities. These data are read by a MATLAB program to calculate the tether extension and tension and displayed in plots of the FEC and the time-dependent extension, force, and trap separation. Only the data obtained on single DNA molecules are further analyzed. To determine the actual translocation distance of a remodeler along the DNA contour from the measured DNA extension and tension, we calculate the contour length of the portion of the DNA directly stretched by optical traps, based on the worm-like-chain model of a DNA molecule. The time-dependent contour length trace is typically mean filtered to 20 Hz and presented (Figs. 1.4C and 1.6C). To unambiguously identify remodeler translocation signals, the instantaneous velocity of remodeler translocation is calculated by a linear regression of the contour length trace with a moving Gaussian function as weight. The standard deviation of the Gaussian function varies with the noise level of the baseline in the range of 0.5–2 s. Contour length changes are considered as signals only when their corresponding absolute instantaneous velocities exceed a threshold value (1–4 bp/s). This approach typically identifies looping signals > 10 bp and smoothes out possible smaller signals. Once a signal is identified, the corresponding translocation velocities are calculated by linear regression of the pause-free regions, and the translocation distance is scored as the length difference between the starting and ending points of the translocation signal.

ACKNOWLEDGMENTS

We thank Drs. Brad Cairns, Cedric Clapier, and Craig Peterson for contributing to this research. The research in the Zhang lab has been funded by the Albert Einstein College of Medicine, the Alexandrine and Alexander L Sinsherimer Fund, the Kingsley Fund, and Yale University.

REFERENCES

Abbondanzieri, E. A., Greenleaf, W. J., Shaevitz, J. W., Landick, R., & Block, S. M. (2005). Direct observation of base-pair stepping by RNA polymerase. *Nature, 438*, 460–465.

Amitani, I., Baskin, R. J., & Kowalczykowski, S. C. (2006). Visualization of Rad54, a chromatin remodeling protein, translocating on single DNA molecules. *Molecular Cell, 23*, 143–148.

Boeger, H., Griesenbeck, J., Strattan, J. S., & Kornberg, R. D. (2004). Removal of promoter nucleosomes by disassembly rather than sliding in vivo. *Molecular Cell, 14*, 667–673.

Bowman, G. D. (2010). Mechanisms of ATP-dependent nucleosome sliding. *Current Opinion in Structural Biology, 20*, 73–81.

Brower-Toland, B. D., Smith, C. L., Yeh, R. C., Lis, J. T., Peterson, C. L., & Wang, M. D. (2002). Mechanical disruption of individual nucleosomes reveals a reversible multistage release of DNA. *Proceedings of the National Academy of Sciences of the United States of America, 99*, 1960–1965.

Bustamante, C., Cheng, W., & Mejia, Y. X. (2011). Revisiting the central dogma one molecule at a time. *Cell, 144*, 480–497.

Cairns, B. R. (2005). Chromatin remodeling complexes: Strength in diversity, precision through specialization. *Current Opinion in Genetics & Development, 15*, 185–190.

Clapier, C. R., & Cairns, B. R. (2009). The biology of chromatin remodeling complexes. *Annual Review of Biochemistry, 78*, 273–304.

Flaus, A., Martin, D. M., Barton, G. J., & Owen-Hughes, T. (2006). Identification of multiple distinct Snf2 subfamilies with conserved structural motifs. *Nucleic Acids Research, 34*, 2887–2905.

Gittes, F., & Schmidt, C. F. (1998). Interference model for back-focal-plane displacement detection in optical tweezers. *Optics Letters*, *23*, 7–9.

Gkikopoulos, T., Schofield, P., Singh, V., Pinskaya, M., Mellor, J., Smolle, M., et al. (2011). A role for Snf2-related nucleosome-spacing enzymes in genome-wide nucleosome organization. *Science*, *333*, 1758–1760.

Greenleaf, W. J., Woodside, M. T., Abbondanzieri, E. A., & Block, S. M. (2005). Passive all-optical force clamp for high-resolution laser trapping. *Physical Review Letters*, *95*, 2081021–2081024.

Hamiche, A., Sandaltzopoulos, R., Gdula, D. A., & Wu, C. (1999). ATP-dependent histone octamer sliding mediated by the chromatin remodeling complex NURF. *Cell*, *97*, 833–842.

Hargreaves, D. C., & Crabtree, G. R. (2011). ATP-dependent chromatin remodeling: Genetics, genomics and mechanisms. *Cell Research*, *21*, 396–420.

Konev, A. Y., Tribus, M., Park, S. Y., Podhraski, V., Lim, C. Y., Emelyanov, A. V., et al. (2007). CHD1 motor protein is required for deposition of histone variant h3.3 into chromatin in vivo. *Science*, *317*, 1087–1090.

Kruithof, M., Chien, F. T., Routh, A., Logie, C., Rhodes, D., & van Noort, J. (2009). Single-molecule force spectroscopy reveals a highly compliant helical folding for the 30-nm chromatin fiber. *Nature Structural & Molecular Biology*, *16*, 534–540.

Landry, M. P., McCall, P. M., Qi, Z., & Chemla, Y. R. (2009). Characterization of photoactivated singlet oxygen damage in single-molecule optical trap experiments. *Biophysical Journal*, *97*, 2128–2136.

Langst, G., Bonte, E. J., Corona, D. F. V., & Becker, P. B. (1999). Nucleosome movement by CHRAC and ISWI without disruption or trans-displacement of the histone octamer. *Cell*, *97*, 843–852.

Lowary, P. T., & Widom, J. (1998). New DNA sequence rules for high affinity binding to histone octamer and sequence-directed nucleosome positioning. *Journal of Molecular Biology*, *276*, 19–42.

Luger, K., Mader, A. W., Richmond, R. K., Sargent, D. F., & Richmond, T. J. (1997). Crystal structure of the nucleosome core particle at 2.8 angstrom resolution. *Nature*, *389*, 251–260.

Mizuguchi, G., Shen, X. T., Landry, J., Wu, W. H., Sen, S., & Wu, C. (2004). ATP-driven exchange of histone H2AZ variant catalyzed by SWR1 chromatin remodeling complex. *Science*, *303*, 343–348.

Moffitt, J. R., Chemla, Y. R., Izhaky, D., & Bustamante, C. (2006). Differential detection of dual traps improves the spatial resolution of optical tweezers. *Proceedings of the National Academy of Sciences of the United States of America*, *103*, 9006–9011.

Moffitt, J. R., Chemla, Y. R., Smith, S. B., & Bustamante, C. (2008). Recent advances in optical tweezers. *Annual Review of Biochemistry*, *77*, 205–228.

Nugent-Glandorf, L., & Perkins, T. T. (2004). Measuring 0.1-nm motion in 1 ms in an optical microscope with differential back-focal-plane detection. *Optics Letters*, *29*, 2611–2613.

Orth, P., Schnappinger, D., Hillen, W., Saenger, W., & Hinrichs, W. (2000). Structural basis of gene regulation by the tetracycline inducible Tet repressor-operator system. *Nature Structural Biology*, *7*, 215–219.

Papamichos-Chronakis, M., & Peterson, C. L. (2008). The Ino80 chromatin-remodeling enzyme regulates replisome function and stability. *Nature Structural & Molecular Biology*, *15*, 338–345.

Pyle, A. M. (2008). Translocation and unwinding mechanisms of RNA and DNA helicases. *Annual Review of Biophysics*, *37*, 317–336.

Saha, A., Wittmeyer, J., & Cairns, B. R. (2002). Chromatin remodeling by RSC involves ATP-dependent DNA translocation. *Genes & Development*, *16*, 2120–2134.

Schalch, T., Duda, S., Sargent, D. F., & Richmond, T. J. (2005). X-ray structure of a tetranucleosome and its implications for the chromatin fibre. *Nature*, *436*, 138–141.
Sinha, M., Watanabe, S., Johnson, A., Moazed, D., & Peterson, C. L. (2009). Recombinational repair within heterochromatin requires ATP-dependent chromatin remodeling. *Cell*, *138*, 1109–1121.
Sirinakis, G., Clapier, C. R., Gao, Y., Viswanathanc, R., Cairns, B. R., & Zhang, Y. L. (2011). The RSC chromatin remodeling ATPase translocates DNA with high force and small step size. *The EMBO Journal*, *30*, 2364–2372.
Smith, C. L., Cui, Y., & Bustamante, C. (1996). Overstretching B-DNA: The elastic response of individual double-stranded and single-stranded DNA molecules. *Science*, *271*, 795–799.
Smith, C. L., Horowitz-Scherer, R., Flanagan, J. F., Woodcock, C. L., & Peterson, C. L. (2003). Structural analysis of the yeast SWI/SNF chromatin remodeling complex. *Nature Structural Biology*, *10*, 141–145.
Smith, C. L., & Peterson, C. L. (2005). ATP-dependent chromatin remodeling. *Current Topics in Developmental Biology*, *65*, 115–148.
Whitehouse, I., Stockdale, C., Flaus, A., Szczelkun, M. D., & Owen-Hughes, T. (2003). Evidence for DNA translocation by the ISWI chromatin-remodeling enzyme. *Molecular and Cellular Biology*, *23*, 1935–1945.
Wittmeyer, J., Saha, A., & Cairns, B. (2004). DNA translocation and nucleosome remodeling assays by the RSC chromatin remodeling complex. *Methods in Enzymology*, *377*, 322–343.
Zhang, Y. L., Smith, C. L., Saha, A., Grill, S. W., Mihardja, S., Smith, S. B., et al. (2006). DNA translocation and loop formation mechanism of chromatin remodeling by SWI/SNF and RSC. *Molecular Cell*, *24*, 559–568.

CHAPTER TWO

Unzipping Single DNA Molecules to Study Nucleosome Structure and Dynamics

Ming Li[*], Michelle D. Wang[†,‡,1]

[*]Department of Chemistry and Chemical Biology, Cornell University, Ithaca, New York, USA
[†]Department of Physics, Laboratory of Atomic and Solid State Physics, Cornell University, Ithaca, New York, USA
[‡]Howard Hughes Medical Institute, Cornell University, Ithaca, New York, USA
[1]Corresponding author: e-mail address: mwang@physics.cornell.edu

Contents

1. Introduction 30
2. Sample Preparation 33
   2.1 DNA unzipping template design 33
   2.2 Nucleosome reconstitution 36
   2.3 Formation of the final unzipping template 37
   2.4 Preparation of experimental sample chambers 38
3. Instrumentation and Data Collection 39
   3.1 Layout of single-beam optical trapping apparatus 39
   3.2 Calibration of the optical trapping system 41
   3.3 Experimental control: loading rate clamp and force clamp 41
   3.4 Data acquisition 42
4. Data Processing 43
   4.1 DNA elastic parameter determination 43
   4.2 Trap height determination for individual unzipping curves 43
   4.3 Data conversion to force and extension based on geometry 44
   4.4 Conversion to number of base pairs unzipped (j) 44
   4.5 Unzipping curve alignment 45
5. Determination of Unzipping Accuracy and Precision 46
6. Unzipping in Nucleosome Studies 48
   6.1 Nucleosome unzipping signature 48
   6.2 High-resolution mapping of histone–DNA interactions in a nucleosome 50
   6.3 Nucleosome remodeling 54
7. Conclusions 56
Acknowledgments 56
References 56

Abstract

DNA unzipping is a powerful tool to study protein–DNA interactions at the single-molecule level. In this chapter, we provide a detailed and practical guide to performing this technique with an optical trap, using nucleosome studies as an example. We detail protocols for preparing an unzipping template, constructing and calibrating the instrument, and acquiring, processing, and analyzing unzipping data. We also summarize major results from utilization of this technique for the studies of nucleosome structure, dynamics, positioning, and remodeling.

1. INTRODUCTION

As the fundamental units of eukaryotic chromatin, nucleosomes are responsible for packaging the genome into the nucleus and regulating access to genetic information during various cellular processes. The nucleosome core particle consists of 147 bp of DNA wrapped ~ 1.7 times around a histone octamer, containing two copies of H2A, H2B, H3, and H4 (Luger, Mader, Richmond, Sargent, & Richmond, 1997). The nonuniform distribution of histone–DNA interactions within a nucleosome governs its dynamic role in regulating access to nucleosomal DNA during transcription, replication, and DNA repair (Andrews & Luger, 2011; Korber & Becker, 2010). The position of nucleosomes along the genome is not only influenced by the properties of the underlying DNA sequence (Kaplan et al., 2009) but is also regulated by various histone chaperones (Das, Tyler, & Churchill, 2010; Park & Luger, 2008; Ransom, Dennehey, & Tyler, 2010), ATP-dependent chromatin remodeling complexes (Bowman, 2010; Clapier & Cairns, 2009), and other DNA-binding proteins, such as transcription factors (Bell, Tiwari, Thoma, & Schubeler, 2011; Zhang et al., 2009). In addition, several types of epigenetic marks, including covalent histone modifications, affect both the structure and stability of nucleosomes, as well as higher-order chromatin structure (Bannister & Kouzarides, 2011; Campos & Reinberg, 2009; Ray-Gallet & Almouzni, 2010). Thus, a detailed understanding of nucleosome structure and dynamics, as well as the relationship between nucleosomes and relevant regulatory factors, is of great interest to multiple fields and can enhance our knowledge of the basic tenets of biology.

Single-molecule techniques offer the unique ability to both detect the inherent heterogeneities of biomolecules and directly monitor dynamic processes in real time and are thus important complementary to ensemble studies for understanding various biological systems (Joo, Balci, Ishitsuka, Buranachai, & Ha, 2008; Killian, Li, Sheinin, & Wang, 2012; Moffitt, Chemla, Smith, & Bustamante, 2008). In particular, DNA stretching experiments utilizing single-molecule manipulation techniques, such as magnetic tweezers or optical tweezers, allow for the direct investigation of the mechanical properties of both single nucleosomes and nucleosome arrays (Brower-Toland et al., 2002; Gemmen et al., 2005; Mihardja, Spakowitz, Zhang, & Bustamante, 2006; Simon et al., 2011). However, stretching experiments are unable to directly determine the location of a nucleosome on a long DNA template or directly probe the absolute locations of specific histone–DNA interactions in a nucleosome. To overcome these limitations, we developed an optical trapping-based single-molecule unzipping technique as a versatile tool to probe a variety of protein–DNA interactions (Dechassa et al., 2011; Hall et al., 2009; Jiang et al., 2005; Jin et al., 2010; Koch, Shundrovsky, Jantzen, & Wang, 2002; Koch & Wang, 2003; Shundrovsky, Smith, Lis, Peterson, & Wang, 2006). The unzipping technique is a straightforward concept and may be incorporated into different optical trapping configurations (an example is shown in Fig. 2.1). Briefly, a single double-stranded DNA (dsDNA) is unzipped in the presence of DNA-binding proteins. Mechanical force is applied to separate dsDNA into two single strands (Fig. 2.1A). DNA-bound proteins or protein complexes act as barriers to the unzipping fork so that resistance to unzipping provides a measure of the strengths of protein–DNA interactions, while the amount of DNA unzipped reveals the locations of these interactions along the DNA. These locations may be mapped to near base-pair precision and accuracy, making unzipping a powerful high-resolution technique for mapping these interactions.

Unzipping is a unique and extremely powerful single-molecule technique with many advantages: (1) The interaction map of a protein–DNA complex, as characterized by the strengths and locations of interactions, provides important structural information about the complex (Dechassa et al., 2011; Hall et al., 2009; Jin et al., 2010; Shundrovsky et al., 2006); (2) The footprint of a bound complex can be directly measured by unzipping DNA molecules from both directions (Dechassa et al., 2011; Hall et al., 2009; Jiang et al., 2005; Jin et al., 2010); (3) Unzipping directly reveals the presence or absence of a bound protein, making it an ideal method for measuring its equilibrium dissociation constant, even for tight binding in the pM range (Jiang et al., 2005; Koch et al., 2002); (4) Using dynamic force

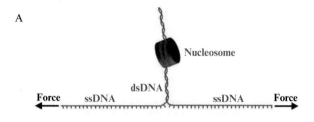

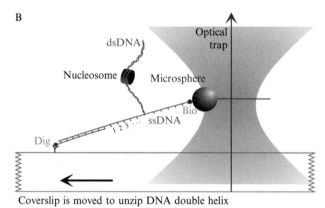

Coverslip is moved to unzip DNA double helix

Figure 2.1 Experimental unzipping configuration. (Adapted from Hall et al., 2009 and Shundrovsky et al., 2006, with permissions from the publishers.) (A) A simplified cartoon of the unzipping configuration. A DNA double helix is mechanically unzipped in the presence of DNA-binding proteins, such as a nucleosome, by the application of opposing forces on the two strands. (B) A typical experimental configuration for unzipping. An optical trap is used to apply a force necessary to unzip through the DNA as the coverslip is moved away from the trapped microsphere. (See Color Insert.)

measurements, it is possible to differentiate different bound species which may bind to the same DNA sequence (Koch & Wang, 2003); (5) Unzipping is capable of determining the location of a protein on a very long DNA molecule with near base-pair accuracy, making it ideal for studying the positioning and/or repositioning of proteins and protein complexes along DNA (Shundrovsky et al., 2006).

Unzipping has been successfully utilized to study the binding affinity of restriction enzymes (Koch et al., 2002; Koch & Wang, 2003), mismatch detection by DNA repair enzymes (Jiang et al., 2005), the dynamics of nucleosome structure and positioning (Dechassa et al., 2011; Hall et al., 2009; Shundrovsky et al., 2006), and how RNA polymerase overcomes a nucleosome barrier (Jin et al., 2010). For clarity and brevity, we focus below on the experimental procedures utilizing our particular single-beam optical trapping system to study nucleosome structure and dynamics (Fig. 2.1B).

2. SAMPLE PREPARATION
2.1. DNA unzipping template design

Here, we detail the construction of unzipping templates that can be used with the optical trapping system shown in Fig. 2.1B. Although different experimental configurations may require somewhat different templates, the general template designs share common features (Bockelmann, Essevaz-Roulet, & Heslot, 1998; Koch et al., 2002). The template generally consists of two segments: an anchoring segment and an unzipping segment, separated by a nick (Fig. 2.2). At one end of the anchoring segment is a tag that will bind to the coverslip, and near the nick is a different tag that will bind to a microsphere. By moving the coverslip away from the trapped microsphere, the unzipping segment can be unzipped.

The anchoring segment is generally 1–2 kb long and consists of a dsDNA linker arm with an end-labeled digoxigenin tag. This length will provide sufficient distance between the anchor point and the unzipping segment to ensure that the trapped microsphere does not contact the coverslip surface. This will facilitate data analysis. However, an anchoring segment that is too long will lead to increased Brownian noise of the trapped microsphere and compromise the accuracy of position measurements.

The unzipping segment consists of an experiment-specific target sequence with an internal biotin tag near the nick. This segment may also contain bound proteins or protein complexes to be studied. The length of the unzipping segment can vary from hundreds to a few thousands of base pairs. Note that a segment that is too short will not allow for optimal data alignment (detailed in a later section). Bound proteins or protein complexes can be located in any region of the unzipping segment, although we typically leave at least 200 bp of flanking DNA on either side of a bound protein to ensure accuracy during data alignment. In some experiments, unzipping should be conducted from both directions (forward and reverse) along the same unzipping segment to investigate possible asymmetric binding. Therefore, both forward and reverse unzipping segments should be prepared.

As an example, we provide below a detailed protocol for constructing a typical forward unzipping template for studying histone–DNA interactions

A   Construction of nucleosome unzipping template

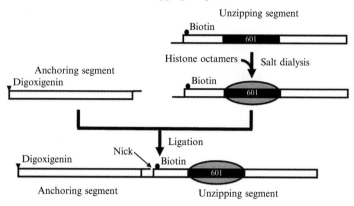

B   Construction of hairpin unzipping template

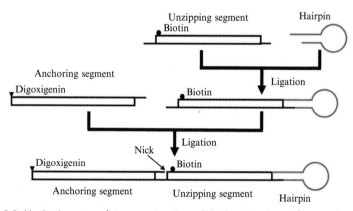

Figure 2.2 Unzipping template construction. (A) Construction of a nucleosome unzipping template. A DNA template for nucleosome unzipping experiments consists of a digoxigenin-labeled anchoring segment and a biotin-labeled nucleosome unzipping segment. As an example, a nucleosome unzipping segment is shown containing a 601 nucleosome positioning sequence. (B) Construction of a hairpin-capped unzipping template. This template consists of a digoxigenin-labeled anchoring segment and a biotin-labeled unzipping segment with a hairpin capped at the distal end.

in a single nucleosome (Dechassa et al., 2011; Hall et al., 2009; Shundrovsky et al., 2006). In this template, the anchoring segment is 1.1 kb and the unzipping segment is 774 bp. The nucleosome is located near the center of the unzipping template and is well positioned on a 147-bp "Widom 601" nucleosome positioning element (601), which has an extremely high affinity for a nucleosome (Fig. 2.2) (Lowary & Widom, 1998). Labeling and producing the DNA templates are accomplished by standard enzymatic reactions and purification methods with biotin and digoxigenin-labeled nucleic acids. These two labels are especially convenient due to the ease of covalent attachment of streptavidin (Sigma) to carboxylated polystyrene microspheres (Polysciences, Inc.) and the availability of high-affinity antidigoxigenin antibodies (Roche Applied Sciences, Indianapolis, IN). The reverse nucleosomal unzipping segment is prepared using methods nearly identical to those of the forward unzipping segment, except that the entire segment is flipped by the use of different primers, such that the ligatable overhang is located on the opposite end.

Anchoring segment preparation

1. PCR amplify the anchoring segment from plasmid pRL574 (Schafer, Gelles, Sheetz, & Landick, 1991). The forward primer contains a 5′-digoxigenin label, designed to be ∼1.1 kb away from the single BstXI cutting site located on the plasmid.
2. BstXI (New England Biolabs) digest the PCR product to generate a 3′-overhang for ligation of the unzipping segment.

Unzipping segment preparation

1. PCR amplify the unzipping segment from plasmid p601 (Lowary & Widom, 1998). The forward primer was designed to be ∼200 bp upstream of the 601 element and to contain a BstXI cutting site. This site will be utilized to generate a 3′-overhang complementary to the one produced on the anchoring segment. The forward primer contains an internal biotin label near the 3′-overhang. The reverse primer is located ∼400 bp downstream of the 601 element.
2. BstXI (New England Biolabs) digest the PCR product. Follow the digestion with the addition of a stoichiometric amount of calf intestinal alkaline phosphatase (CIP, New England Biolabs) in the same buffer, to remove the phosphate from the 3′-overhang. This allows for the generation of a nick after subsequent ligation (discussed below in Section 2.3).

To characterize the precision and accuracy of the unzipping method in locating a bound protein along the DNA (discussed later), we have also designed multiple unzipping templates of varying lengths, capped with hairpins at distal ends (Fig. 2.2B). These hairpins act as strong binding sites by preventing further unzipping at well-defined locations along the DNA and allow for a direction comparison with measured locations. In addition, these unzipping templates are also used to determine the elastic parameters of single-stranded DNA (ssDNA) under experimental conditions discussed below. Here, we have outlined a protocol for constructing hairpin-capped unzipping segments of various lengths.

1. PCR amplify the unzipping segment from p601 using the same forward primer specified above in "unzipping segment preparation." The reverse primers are located at various distances downstream from the forward primer to generate templates of different lengths. The reverse primer is also designed to contain an EarI cutting site that will generate a 5′-overhang.
2. EarI (New England Biolabs) digest the PCR products to generate a 5′-overhang.
3. The oligonucleotide (Integrated DNA Technologies) utilized to generate the hairpin is designed to form a three-base hairpin loop and a short dsDNA stem (~12 bp) with a 5′-overhang which is complementary to the overhang in the unzipping segment.
4. Ligate the unzipping segment with a hairpin oligonucleotide (1:10 molar ratio) using T4 Ligase (New England Biolabs). Overnight ligation is normally necessary to maximize the ligation yield. Purify the ligated products using agarose gel purification.
5. BstXI (New England Biolabs) digest the gel purified product to generate an overhang near the biotin label. Following the digestion, add a stoichiometric amount of CIP (New England Biolabs) in the same digestion buffer to remove the phosphate from the 3′-overhang. This allows for the generation of a nick after ligation with the anchoring segment (discussed below in Section 2.3).

2.2. Nucleosome reconstitution

It has been well established that *in vitro* assembly of nucleosomes and chromatin arrays from highly purified DNA and histone components can be achieved by either salt-gradient dialysis (Luger, Rechsteiner, & Richmond, 1999) or a chaperone-mediated approach (Fyodorov & Kadonaga, 2003). We employ a salt-gradient dialysis method for nucleosome reconstitution. Different types of individual histones can be prepared as previously described (Dyer et al., 2004). Purified histone octamers from several species are also commercially available in forms suitable for reconstitution (Protein Expression/Purification Facility, Colorado State University). In our previous publications (Dechassa et al., 2011; Hall et al., 2009; Jin et al., 2010;

Shundrovsky et al., 2006), a well-established salt dialysis method (Dyer et al., 2004; Thastrom, Lowary, & Widom, 2004) was modified to reconstitute a single nucleosome on a long piece of DNA containing one 601 positioning element. The modified protocol uses a small total volume and requires a low concentration of DNA template and histones (100 nM or even lower) that are suitable for single-molecule studies. The modifications are listed below.

1. The dialysis button is constructed following the procedure from Thastrom et al. (2004). This allows us to work with assembly volumes of ∼30 µL, which is much smaller than commercially available dialysis chambers.
2. We include 0.2 g/L sodium azide (Sigma-Aldrich) in both the high and low salt buffers to remove bacteria which may contaminate the buffers and decrease assembly efficiency.
3. We include 0.1 mg/mL acetylated BSA (acBSA) (Ambion) to each dialysis button as a crowding agent to assist with assembly.
4. The dialysis is performed at 4 °C and the dialysis pump is set to 1.2 mL/min for ∼18 h. The flow is then changed to 2.5 mL/min for an additional ∼4 h.
5. After dialysis, the samples are transferred to zero salt buffer and incubated for 2–3 h at 4 °C.

Fine-tuning the molar ratio between the histone octamer and DNA template is critical to achieve a high reconstitution yield and avoid over assembly. In addition, it is also important to remove bubbles from the dialysis button, because bubbles can prevent buffer exchange between the high salt solution and the sample. After reconstitution, the nucleosome samples may be stored for a few weeks at 4 °C.

2.3. Formation of the final unzipping template

The unzipping segment (containing either a nucleosome or a hairpin) is directly ligated to the anchoring segment (in a 1:1 molar ratio) immediately prior to use (Fig. 2.2). The CIP dephosphorylation of the unzipping segment ensures that only one strand of the DNA is ligated and a nick is generated on the complementary strand during the ligation step. This complete unzipping template is labeled with a single digoxigenin tag at the 5′-end of the anchor segment and a biotin tag located near the nick on the unzipping segment. A complete template lacking a nucleosome (naked DNA) is stable for a few days without DNA nicking at 4 °C; a template containing a nucleosome should be used within a few hours of ligation.

2.4. Preparation of experimental sample chambers

For single-molecule studies in general, individual DNA tethers need to be immobilized in a single-molecule sample chamber or a flow cell. These allow the user to sequentially flow in different solutions for use with the optical trapping system. In our nucleosome unzipping studies, sample chambers with an ~ 15 μL volume are prepared at room temperature and then mounted onto an optical trapping setup. By performing incubations in a humid chamber prior to mounting onto the optical setup, buffer evaporation can be minimized.

Buffer solutions
1. Sample buffer (SB): 10 mM Tris HCl (pH 7.5), 1 mM Na$_2$EDTA, 150 mM NaCl, 3% (v/v) glycerol, 1 mM DTT, 0.1 mg/mL acBSA.
2. Blocking buffer (BB): SB + 5 mg/mL casein sodium salt from bovine milk.
3. Nucleosome unzipping buffer (NUB): 10 mM Tris HCl (pH 8.0), 1 mM Na$_2$EDTA, 100 mM NaCl, 1.5 mM MgCl$_2$, 3% (v/v) glycercol, 1 mM DTT, 0.02% (v/v) Tween20, 2 mg/mL acBSA.

Blocking agents are used to coat the surface of the sample chamber to prevent unwanted protein attachment to the surface. The blocking agent that has been the most successful for us is casein sodium salt from bovine milk (Sigma-Aldrich Co.). AcBSA is thought to mimic conditions of a higher protein concentration in the sample buffer, creating a more "crowded" environment, and thus prevents protein dissociation from the template (Gansen, Hauger, Toth, & Langowski, 2007). We utilize a polyclonal sheep antidigoxigenin (Roche Applied Sciences) for attachment of digoxigenin-labeled DNA samples to sample chamber surfaces. The complete unzipping template is diluted to a desired concentration in SB immediately before introduction to sample chamber. The detailed procedure of creating a tethered DNA sample chamber is given below:
1. Apply two thin pieces of double-stick tape (~ 0.1 mm thick) to a coverslip (24 mm × 40 mm × 0.15 mm). Orient the pieces parallel to one another and separate them by ~ 5 mm.
2. Place a glass slide on top of the coverslip and perpendicular to it, to create an ~ 15-μL channel down the center.
3. Flow in one volume (~ 15 μL) of antidigoxigenin solution (20 ng/μL in H$_2$O).
4. Incubate for 5 min. Wash with five volumes of BB. Incubate with residual blocker for 5 min.
5. Wash with five volumes of SB. Immediately flow in one volume of diluted unzipping template in SB. Incubate for 10 min.

6. Wash with five volumes of SB. Flow in one volume of streptavidin-coated polystyrene microspheres (5 pM in BB). Incubate for 10 min.
7. Wash with 10 volumes of NUB.

The concentration to which the unzipping sample is diluted prior to being added to the sample chamber is critical for achieving an optimal tether density under single-molecule conditions. A concentration that is too high will lead to multiple tethers (one microsphere attached to multiple DNA molecules) and a concentration that is too low will make it difficult to locate a suitable unzipping tether. In theory, 10 pM of DNA template is needed to achieve an acceptable tether density. However, as the ligated templates are directly diluted without purification, the appropriate "flow-in" concentration depends heavily on the ligation efficiency. Therefore, for each new template, several chambers with different template concentrations are often made and evaluated to establish an appropriate flow-in concentration.

Consistency in all aspects of sample preparation and utilization is critical to achieve reproducible results among different sample chambers. In this regard, unzipping experiments are conducted in a temperature- and humidity-controlled soundproof room. Once prepared, the sample chambers should be utilized in a timely fashion, typically within 1 h, to avoid unnecessary complications such as protein dissociation or sample sticking to the surface of the chamber.

3. INSTRUMENTATION AND DATA COLLECTION

3.1. Layout of single-beam optical trapping apparatus

Since the pioneering work by Arthur Ashkin over 20 years ago (Ashkin, Dziedzic, Bjorkholm, & Chu, 1986), the optical trapping field has grown tremendously due to the unique ability of optical tweezers to monitor and manipulate biological targets with high temporal and spatial resolution. In addition, refinements of established methods and the integration of this tool with other forms of single-molecule manipulation or detection have made this technique of great interest in both physics and biology. The single-beam optical trapping instrument that we use has a very straightforward design (Brower-Toland & Wang, 2004; Koch et al., 2002), containing the minimal set of optical components required for the operation of a high-precision instrument of its kind (Fig. 2.3). A 1064-nm laser (Spectra-Physics Lasers, Inc. Mountain View, CA) is transmitted through

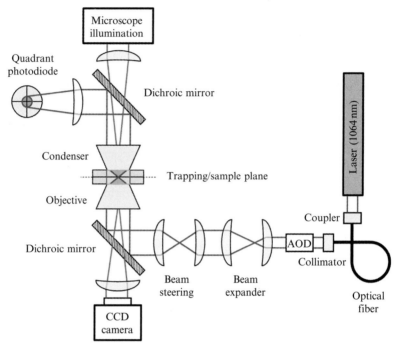

Figure 2.3 Layout of the optical trapping apparatus. See text for a detailed description of the setup. (See Color Insert.)

a single-mode optical fiber (Oz Optics, Carp, ON), expanded by a telescope lens pair, and focused onto the back focal plane of a 100×, 1.4 NA oil-immersion microscope objective that is mounted in a modified Eclipse TE 200 DIC inverted microscope (Nikon USA, Melville, NY). The focused beam serves as a trap for a 500-nm polystyrene microsphere (Polysciences, Inc.). Forward-scattered light is collected by a condenser lens and imaged onto a quadrant photodiode (Hamamatsu, Bridgewater, NJ). A displacement of a trapped microsphere imparts a deflection of the forward-scattered light and is captured as a differential voltage signal at the quadrant photodiode. The laser intensity is adjusted by modulating the voltage amplitude applied to an acoustic optical deflector (AOD) (NEOS Technologies, Inc., Melbourne, FL) placed between the laser aperture and the beam expander. Samples are manipulated manually via a microstage or via a high-precision 3D piezoelectric stage (Mad City Labs, Madison, WI). Analog voltage signals from the position detector and stage position sensor are anti-alias filtered at 5 kHz (Krohn-Hite, Avon, MA) and digitized at 7–13 kHz for each channel using a multiplexed analog to digital conversion PCIe board (National Instruments Corporation, Austin, TX).

3.2. Calibration of the optical trapping system

The instrument calibration methods for our optical trapping setup were detailed in a previous publication (Wang, Yin, Landick, Gelles, & Block, 1997). In brief, they include (1) the determination of the position detector sensitivity and the trap stiffness, (2) the determination of the position of the trap center relative to the beam waist and the height of the trap center relative to the coverslip, and (3) the location of the anchor position of the unzipping tether on the coverslip, which is determined prior to each measurement by stretching the anchoring segment laterally at a low load (< 5 pN). These calibrations are subsequently used to convert raw data into force and extension values.

3.3. Experimental control: loading rate clamp and force clamp

In nucleosome unzipping experiments, we often use two approaches to disrupt a nucleosome: loading rate clamp unzipping and force clamp unzipping. The advantages and disadvantages of these two methods are discussed below.

A loading rate clamp allows the force to increase linearly at a specific rate until the disruption of an interaction. This approach generates distinct force unzipping signatures (unzipping force as a function of the number of base pair unzipped) which can be used to distinguish a nucleosome from other DNA-binding proteins. Because the disruption is a thermally activated process (Evans, 2001), the force needed to disrupt a specific interaction in the nucleosome is dependent on the loading rate as well as the starting force. After a disruption, the force naturally drops but is not allowed to recover to the naked DNA unzipping baseline. Consequently, the starting force for a subsequent disruption is higher than for the initial disruption. Thus, weak interactions in a nucleosome may be detected if they are first encountered by the unzipping fork. This method may be used to highlight weak histone–DNA interactions near the entrance and exit sites. It will of course detect all strong interactions.

A force clamp allows for the disruption of all interactions in a nucleosome under the same force. This is well suited to a quantitative analysis of the strength of the detected interactions (Forties et al., 2011). However, more experimentation is normally required to determine an appropriate range of desired unzipping forces. A force that is too small will make the time to disrupt the nucleosome too long to be experimentally accessible, and a force that is too large may overlook specific interactions in the nucleosome. A force clamp is usually implemented with loading rate clamps before and after it to simplify data alignment (discussed further below). A loading rate clamp is most suitable to study the interactions around the periphery of a nucleosome, while the force-clamp mode is optimal for determining the interactions around the dyad. More importantly, a loading rate clamp provides a clear force unzipping signature, while a force clamp is more convenient in quantitative analysis of the energy landscape of histone–DNA interactions in a nucleosome.

3.4. Data acquisition

Here, we detail the process of data collection using the two aforementioned approaches.

When utilizing the loading rate clamp mode (Dechassa et al., 2011; Hall et al., 2009; Koch & Wang, 2003; Shundrovsky et al., 2006), the microscope coverslip velocity is adjusted to produce a constant force-loading rate by controlling the position of the piezo stage, while the position of the microsphere in the trap is kept constant by modulating the light intensity (trap stiffness) of the trapping laser. Unzipping through a nucleosome is visualized as a group of force peaks up to 30–40 pN ramping up linearly above the naked DNA unzipping baseline (13–16 pN). In the force-clamp mode (Hall et al., 2009), the unzipping begins with a loading rate clamp until the desired force (threshold force) is reached within a nucleosome. The unzipping force is then held constant via feedback control of the coverslip position. The threshold force is carefully selected so that it is much higher than the baseline unzipping force of the naked DNA, but is still low enough to allow sufficient dwell time at most histone–DNA interactions for detection. Upon passing through the nucleosome region, the unzipping reverts to the original loading rate clamp. The distinctive naked DNA unzipping signatures, detected by the loading rate clamp before and after the nucleosome, are important for data alignment (see below).

Apart from the two unzipping modes presented above, we can also modulate the unzipping process to allow the DNA to be unzipped and rezipped multiple times by controlling the unzipping forces and the corresponding position of the piezo stage. This modulation allows unzipping experiments to potentially mimic important biological process, such as a motor protein progressing into a nucleosome (Hall et al., 2009).

4. DATA PROCESSING

Data acquired by the optical trapping setup need to be processed according to these steps: (1) determine the elastic parameters of the dsDNA and ssDNA, (2) determine trap height, (3) perform data conversion to force and extension based on geometry, (4) convert DNA extension to number of base pairs unzipped, (5) perform data alignment against a theoretical curve. These steps are detailed below.

4.1. DNA elastic parameter determination

As unzipping experiments involve the extension of dsDNA (the anchoring segment) in series with ssDNA (unzipped DNA) (Fig. 2.1B), elastic parameters of both dsDNA and ssDNA are necessary for data analysis. These parameters are strongly dependent on the buffer conditions used in unzipping experiments. We obtain elastic parameters of dsDNA by stretching dsDNA and fitting the resulting force versus extension curve to a modified worm-like chain model under the same buffer conditions as the actual unzipping experiment following the detailed procedures published previously (Wang et al., 1997). To obtain the elastic parameters of ssDNA, we unzip a template capped with a hairpin at the distal end under the same buffer conditions as the actual unzipping experiment (Koch et al., 2002; Hall et al., 2009). Once completely unzipped, the unzipped DNA is then stretched to a high force (up to 50 pN) to obtain the force-extension curve, which reflects elastic contributions from both the dsDNA and ssDNA. Given the elastic parameters of dsDNA under these conditions, this curve allows for the determination of the elastic properties of ssDNA using an extensible freely jointed chain (FJC) model (Smith et al., 1996).

4.2. Trap height determination for individual unzipping curves

Prior to nucleosome unzipping experiments, we calibrate the height of the trap center relative to the surface of the coverslip when the objective is focused on the coverslip surface (Wang et al., 1997); it is typically found to be ~ 600 nm. However, the actual trap height during a nucleosome unzipping experiment may differ from the calibrated height by as much as 100 nm due to limited focusing precision. We have therefore implemented a technique to obtain trace-specific trap height of the unzipping data. In this method, we analyze the initial segment of the data prior to strand separation (0–10 pN). Because DNA is not yet unzipped, the expected force-extension curve has been fully characterized as described in Section 4.1 and is simply that of the dsDNA anchoring segment of known contour length. The trap height is determined when the difference between the converted force-extension curve and the expected curve is minimized.

4.3. Data conversion to force and extension based on geometry

Once the trap height is determined, force and extension of DNA as a function of time may be obtained following a method that has been previously described in detail (Wang et al., 1997). At a given time point, a number of parameters must be detected and/or calibrated: DNA anchor point on the coverslip, the position of the trapped microsphere relative to the trap center, and the stiffness of the trap. The results of the conversion are the force (F) and extension (x) along the direction of the stretched DNA molecule.

4.4. Conversion to number of base pairs unzipped (j)

Once the unzipping data are converted to force (F) and extension (x) for a given time point (t), the number of base pairs unzipped (j) at each time point may be obtained. The extension of the DNA (x) contains contributions from both the dsDNA (x_{ds}) and ssDNA (x_{ss}) under the same force:

$$x(F) = x_{ds}(F) + x_{ss}(F) \qquad (2.1)$$

$x_{ds}(F)$ is determined because the force-extension curve of the anchoring segment is fully characterized. $x_{ss}(F)$ is thus obtained from Eq. (2.1) and is proportional to the number of ssDNA nucleotides. Using the extensible FJC model, $x_{ss}(F)$ is converted to the number of base pairs unzipped (j).

As the DNA is extended but prior to strand separation, $j = 0$ bp, resulting in a vertical rise in the F versus j plot (Fig. 2.4A). Once strand separation starts, a characteristic force signature, determined by the underlying DNA sequence as discussed below, is detected with an increase in j. In the presence of bound proteins, this gently varying baseline is interrupted by sharp force rises. When the unzipping fork encounters a strong protein–DNA interaction, force increases linearly, while the number of base pair unzipped remains unchanged until the sudden dissociation of the bound protein, leading to a sudden reduction in the force. Therefore, the F versus j plot

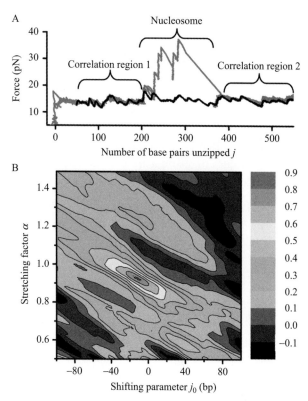

Figure 2.4 Unzipping curve alignment. (A) An example of force versus number of base pairs unzipped plot for a nucleosome unzipping curve (red) after alignment with the corresponding theoretical curve for naked DNA of the same sequence (black). Unzipping was carried out at a loading rate of 8 pN/s. Regions 1 and 2 flanking the nucleosome were used for correlation. Note also that the initial rise of force is located at $j = 0$ bp, corresponding to stretching of the anchoring segment before strand separation. (B) A two-dimensional intensity graph of the generalized correlation function $R(a, j_0)$ for the trace shown in A. The peak $R = 0.80$ is located at stretching factor $a = 0.93$ and shifting parameter $j = -14$ bp. (See Color Insert.)

provides a direct measurement of (1) the location of the bound protein on the DNA (j at which the force rise starts) and (2) the strength of the interaction (force magnitude).

4.5. Unzipping curve alignment

Although the raw F versus j plot already contains critical information about a bound protein, the precision and accuracy of locating a bound protein are limited to ~ 10 bp, due to small but significant uncertainties in a number of parameters (trap height, microshpere size, and trapped microshpere position and force). In order to improve on this, we take advantage of the characteristic unzipping force signatures that depend strongly on the DNA sequence and align an experimental unzipping force curve $F_{\exp}(j)$ against a theoretical curve $F_{\text{theo}}(j)$. $F_{\text{theo}}(j)$ is computed based on an equilibrium statistical mechanics model that considers sequence-dependent base pairing energy and DNA elasticity(Bockelmann et al., 1998). During the correlation, the argument of $F_{\exp}(j)$ is both shifted by j_0 number of base pairs and stretched by a factor of a. The best values of j_0 and a are obtained by maximizing the following generalized cross-correlation function:

$$R(a, j_0) = \frac{\int dj [F_{\text{theo}}(j) - \bar{F}_{\text{theo}}][F_{\exp}(aj + j_0) - \bar{F}_{\exp}]}{\sqrt{\int dj [F_{\text{theo}}(j) - \bar{F}_{\text{theo}}]^2 \int dj [F_{\exp}(aj + j_0) - \bar{F}_{\exp}]^2}} \quad (2.2)$$

where $\bar{F}_{\text{theo}}$ and $\bar{F}_{\exp}$ are the mean values of the $\bar{F}_{\text{theo}}(j)$ and $\bar{F}_{\exp}(aj + j_0)$, respectively. The search for optimal j_0 and a may be also facilitated by the use of a SIMPLEX search algorithm.

When using this method to align a trace taken from DNA containing a nucleosome against the known DNA sequence, regions of naked DNA, ~ 100–200 bp, adjacent to the nucleosome should be used for correlation. We found that once unzipping passes a nucleosome, the unzipping curve immediately following the nucleosome did not always show the expected naked DNA pattern (Shundrovsky et al., 2006). Instead, in some traces we observed random high-force peaks that were not present when unzipping naked DNA. We attribute this effect to nonspecific interactions between the end of the DNA and the histone proteins removed from the disrupted nucleosome. For those traces, only the naked DNA preceding a nucleosome can be used for correlation, which may result in somewhat lower precision.

Figure 2.4 is an example of the application of this method to a nucleosome unzipping trace. The correlation was performed using two regions of naked DNA flanking the nucleosome (Fig. 2.4A). The generalized correlation function (Fig. 2.4B) shows a maximum of 0.80 at $j_0 = -14$ bp and $a = 0.93$.

5. DETERMINATION OF UNZIPPING ACCURACY AND PRECISION

To characterize the ability of the unzipping technique to locate the absolute position of an interaction, we unzip naked DNA templates capped with hairpins at distal ends (Fig. 2.5A) and analyze the measured locations of

Unzipping Single DNA Molecules to Study Nucleosome Structure and Dynamics 47

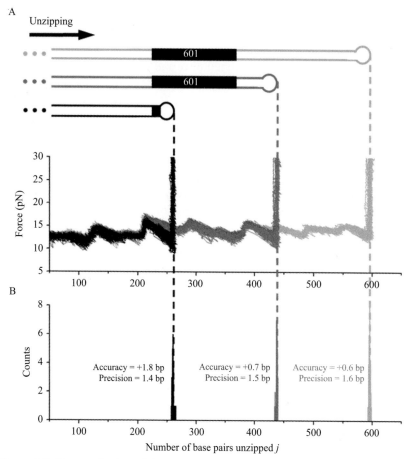

Figure 2.5 Characterization of the accuracy, and precision, of the unzipping method. (Adapted from Hall et al., 2009, with permission from the publisher.) (A) Three hairpin-capped unzipping templates were unzipped using a loading rate clamp (8 pN/s): 258 bp (black, 21 traces), 437 bp (red, 27 traces), and 595 bp (green, 33 traces). (B) For each template, a histogram was generated from the data points in the vertically rising section only. The measured hairpin location of each template was taken as the mean of the histogram. The accuracy was determined by the difference between the mean of the histogram and the expected value (dashed vertical line). The precision was determined by the standard deviation of the histogram. (See Color Insert.)

hairpins (Hall et al., 2009). These hairpins mimic strong binding sites at well-defined locations on DNA. As shown in Fig. 2.5A, three unzipping templates of varying length, each with a hairpin located near where a nucleosome could be assembled, are unzipped. Each unzipping curve follows that of the naked DNA until it reaches the hairpin where the force rises sharply, providing a clear indication of the hairpin location. These unzipping curves are aligned as described above. Figure 2.5B shows histograms of the detected binding locations for each hairpin template and a comparison with expected locations. Note that accuracy is a measure of the closeness of the measured value with the true value, whereas precision is a measure of the repeatability of measurements. For each template, accuracy is given by the difference between the mean measured location and the expected location, while precision is given by the standard deviation of the histogram. For all three templates, the accuracy is within 1 bp and the precision is within 2 bp. Therefore, we conclude that the unzipping technique has the capability to determine the absolute sequence position of an interaction with near single base pair accuracy and precision.

6. UNZIPPING IN NUCLEOSOME STUDIES

Unzipping is ideally suited for the manipulation of protein–DNA interactions and the detection of their dynamics at the single-molecule level. Below, we provide a very brief summary of various studies on nucleosome structure, positioning, and remodeling which we have explored using our unzipping technique. For more specific details regarding these experiments or data analysis, we refer the reader to the original publications (Dechassa et al., 2011; Hall et al., 2009; Jin et al., 2010; Shundrovsky et al., 2006).

6.1. Nucleosome unzipping signature

A nucleosome has the most distinctive signature when unzipped with a loading rate clamp. As an example, we have unzipped a DNA containing a positioned nucleosome in both forward and reverse directions (Dechassa et al., 2011; Hall et al., 2009). In either direction, two regions of strong interactions are detected: one preceding the dyad and one near the dyad (Fig. 2.6C). When results from both directions are combined, the unzipping force signatures reveal three distinct regions of interactions: one located around the dyad axis and the other two ~ 40 bp on either side of the dyad axis. Within each region, interactions are discretely spaced with ~ 5 bp periodicity (Fig. 2.6B). By comparison with the crystal structure of the nucleosome, the dyad region should correspond to contacts from the $(H3/H4)_2$ tetramer at superhelical location (SHL) -2.5 to $+2.5$, and the two off-dyad regions should correspond to contacts from the two H2A/H2B dimers between SHL -3.5 to -6.5 and $+3.5$

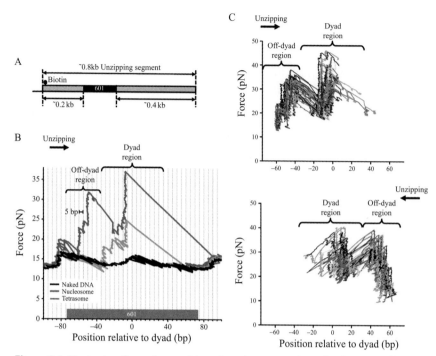

Figure 2.6 Unzipping through a positioned nucleosome using a loading rate clamp at 8 pN/s. (Adapted from Dechassa et al., 2011, with permission from the publisher.) (A) A sketch of the forward nucleosome unzipping segment. (B) Representative force unzipping signatures of naked DNA (black), DNA containing a nucleosome (red), and DNA containing a tetrasome (green). Both the nucleosome and the tetrasome were assembled onto an unzipping segment containing the 601 positioning element. The arrow indicates the unzipping direction. Two distinct regions of interactions, as well as a 5-bp periodicity within each region, were observed for the nucleosome. The tetrasome signature exhibits only a single region of interactions, which substantially overlaps the dyad region identified in the nucleosome. (C) Multiple traces of unzipping through a nucleosome from both forward (upper panel, 31 traces) and reverse (lower panel, 28 traces) directions. Each color represents data obtained from a single nucleosomal DNA molecule. Distinct regions of interactions and a 5-bp periodicity within each region are highly reproducible. (See Color Insert.)

to $+6.5$, respectively. The absence of the last region for each direction of unzipping also indicates that after the first and second regions are disrupted, the nucleosome structure likely becomes unstable and histone dissociation occurs before the last region can be probed. These features are further discussed below.

We have also verified that the unzipping method could clearly distinguish a nucleosome from a tetrasome consisting only of a $(H3/H4)_2$ tetramer (Fig. 2.6B). Unzipping through a tetrasome exhibits only a single region of strong interactions near the dyad and this region substantially overlaps with the dyad region of interactions for canonical nucleosomes (Dechassa et al., 2011).

The nucleosome unzipping signature characteristic of a positioned nucleosome is also shared by nucleosomes on arbitrary sequences (Hall et al., 2009). This was demonstrated by assembling nucleosomes onto a DNA segment that does not contain any known positioning elements (Fig. 2.7A). The assembly condition was controlled to achieve a relatively low saturation level so that each DNA molecule had at most one nucleosome. When such nucleosomal DNA molecules were unzipped with a loading rate clamp using the same conditions as those of Fig. 2.6, nucleosomes were found at various locations on the template (Fig. 2.7B), likely due to a lack of known nucleosome positioning elements on this DNA sequence. Each unzipping trace contains two major regions of strong interaction, with the second region presumably located near the dyad. These nucleosome unzipping signatures possessed essentially identical characteristics to those of the 601 sequence, except that their peak forces within each region were typically smaller by a few pN, reflecting weaker interactions of histone with nonpositioning DNA sequences. The key features remained essentially identical: the three regions of strong interactions with the strongest at the dyad, the 5 bp periodicity, and the loss of nucleosome stability upon dyad disruption (Fig. 2.7B and C).

6.2. High-resolution mapping of histone–DNA interactions in a nucleosome

To quantitatively assay the strengths of the histone–DNA interactions, we unzipped through individual nucleosomal DNA molecules with a constant unzipping force (Hall et al., 2009). Under a force clamp (Johnson, Bai, Smith, Patel, & Wang, 2007), the dwell times at different sequence positions measure the strengths of interactions at those positions. Thus this method allows direct mapping of the strengths of interactions. Figure 2.8A shows example traces for unzipping DNA through a nucleosome under a constant force. DNA molecules were unzipped from both directions along the DNA. In both cases, the unzipping fork did not move through the nucleosomal DNA at a constant rate but instead dwelled at specific locations within the nucleosome, indicating the presence of strong interactions. In particular,

Unzipping Single DNA Molecules to Study Nucleosome Structure and Dynamics 51

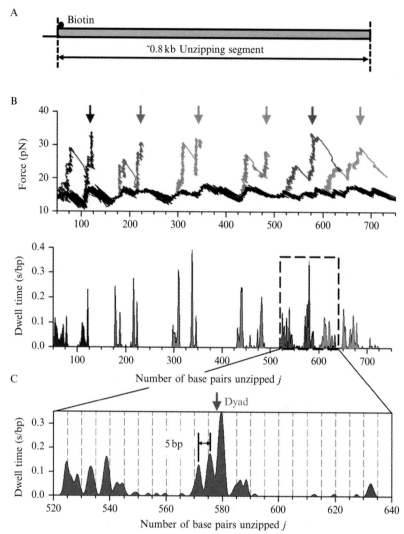

Figure 2.7 Unzipping through a nucleosomes on an arbitrary sequence using a loading rate clamp at 8 pN/s. (Adapted from Hall et al., 2009, with permission from the publisher.) (A) A sketch of the unzipping segment. (B) Force unzipping signature of a nucleosome at different locations on a DNA template lacking known strong positioning elements. Each color was obtained from a single nucleosome unzipping trace, with the unzipping force shown in the top panel and the corresponding dwell time histogram shown in the bottom panel. The unzipping signature of a naked DNA molecule of the same sequence is also shown (black), as a reference. Vertical arrows indicate the observed dyad locations of these nucleosomes. (C) Close-up of the dwell time histogram for a specific unzipping trace (red) to emphasize the 5-bp periodicity observed in each interaction region of the unzipping signature. (See Color Insert.)

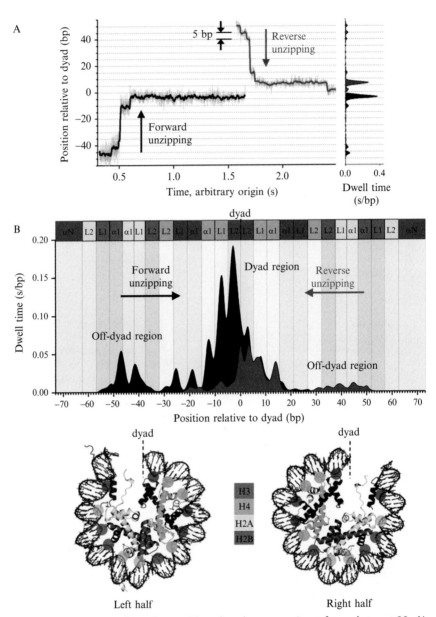

Figure 2.8 Unzipping through a positioned nucleosome using a force clamp at 28 pN. (Adapted from Hall et al., 2009, with permission from the publisher.) (A) Representative traces of forward (black) and reverse (red) unzipping through a nucleosome under a constant applied force (~ 28 pN). The unzipping fork paused at specific locations when passing through a nucleosome, which are evident from both the traces (left) and their corresponding dwell time histograms (right). (B) A histone–DNA interaction map is

these traces revealed that the fork dwelled with discrete steps spaced by ~5 bp and the longest dwell times tended to occur near the dyad.

An interaction map was generated by averaging dwell time histogram measurements from many traces from both forward and reverse unzipping, as shown in Fig. 2.8B. Several features, consistent with findings using the loading rate clamp, are evident from these plots:

(1) Histone–DNA interactions are highly nonuniform within a nucleosome. There are three broad regions of strong interactions: one located at the dyad and two $\sim\pm40$ bp from the dyad. The locations of all three regions are strongly correlated with those estimated from the crystal structure of the nucleosome (Davey, Sargent, Luger, Maeder, & Richmond, 2002; Luger et al., 1997). The locations of these interactions are also consistent with estimates from our nucleosome stretching experiments (Brower-Toland et al., 2002, 2005; Brower-Toland & Wang, 2004), although results from those studies are less direct in identifying the absolute locations of strong interactions and are more difficult to interpret.

(2) An ~5 bp periodicity occurs within each region of interaction. According to the crystal structure of the nucleosome, histone core domains are expected to make strong contacts with the DNA minor groove every 10 bp (Davey et al., 2002; Luger et al., 1997). The observed 5 bp periodicity demonstrated that two distinct interactions at each minor groove contact, one from each strand, could be disrupted sequentially rather than simultaneously.

(3) The interactions near the entry and exit DNA are particularly weak. The unzipping fork did not dwell at a 20-bp region of both entry and exit DNA, indicating that the histones are only loosely bound to the DNA. Note that these weaker interactions are detected by the loading rate clamp described above.

---

constructed by using a total of 27 traces from the forward direction and 30 traces from the reverse direction. Each peak corresponds to an individual histone–DNA interaction and the heights are indicative of their relative strengths. Three regions of strong interactions are indicated: one located at the dyad and two located off-dyad. The bottom panel is the crystal structure of the nucleosome core particle (Luger et al., 1997), where dots indicate individual histone binding motifs that are expected to interact with DNA. The two halves of the nucleosome are shown separately for clarity. On the top panel, these predicted interactions are shown as colored boxes. (See Color Insert.)

(4) For unzipping in both the forward and reverse directions, the first two regions of interactions encountered were always detected, but not the last region. This indicates that once the dyad region of interactions was disrupted, the nucleosome became unstable and histones dissociated from the 601 sequence.

(5) The total dwell time in the nucleosome was longer in the forward direction compared with that in the reverse direction, indicating that nucleosomes were more difficult to disrupt when unzipped in the forward direction, likely reflecting the nonpalindromic nature of the 601 sequence.

These mechanical unzipping experiments resemble the action of RNA polymerase which opens up a transcription bubble and unzips the downstream DNA while advancing into a nucleosome. The histone–DNA interaction map has significant implications for how RNA polymerases or other motor proteins may gain access to DNA associated with a nucleosome (Hall et al., 2009; Jin et al., 2010).

6.3. Nucleosome remodeling

We have also applied the nucleosome unzipping method to investigate nucleosome remodeling dynamics (Shundrovsky et al., 2006). A major advantage of this method is its ability to accurately locate a nucleosome on a long DNA template such that undesired effects of close proximity to DNA ends may be minimized. We have used the unzipping technique to probe the structure of individual nucleosomes after SWI/SNF remodeling. A mononucleosome was initially positioned by 601 in the middle of an ~ 800-bp unzipping segment (Fig. 2.9A) and was remodeled by yeast SWI/SNF. We used a loading rate clamp to detect the nucleosome after remodeling. For convenience, we defined the nucleosome position simply as the mean location of the off-dyad region. The precision of this method for nucleosome position determination was 2.6 bp, calculated by fitting a Gaussian function to the position histogram of nucleosomes assembled onto the 601 sequence (Fig. 2.9B and C). This precision is somewhat lower than that for the detection of a hairpin because determining a nucleosome position requires the consideration of multiple histone–DNA interaction peaks that vary in amplitude from trace-to-trace due to the nature of a thermally activated disruption process.

We observed that under our experimental conditions SWI/SNF remodeling does not alter the overall nucleosome structure: the histone octamer

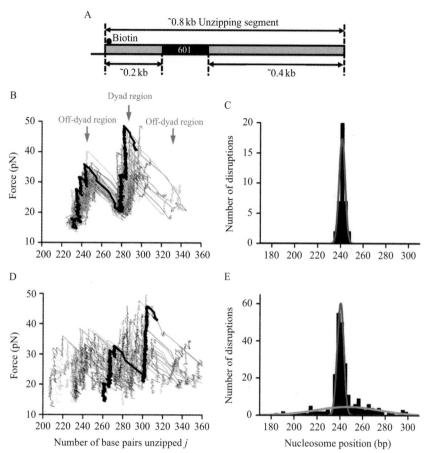

Figure 2.9 Unzipping through nucleosomes before and after remodeling. (Adapted from Shundrovsky et al., 2006, with permission from the publisher.) (A) A sketch of the nucleosome unzipping segment. (B) Unzipping signatures for 30 unremodeled data curves. A single representative curve is highlighted in black. (C) Histogram of unremodeled nucleosome positions on the DNA. A nucleosome position is defined as the mean position of interaction of the first off-dyad region. Data (black) and their Gaussian fit (red) are shown. The distribution is centered at 241 bp, on this particular template, with a standard deviation (SD) of 2.6 bp. (D) Force unzipping signature for 30 data curves obtained after SWI/SNF remodeling. A single representative curve is highlighted in black. (E) Histogram of remodeled nucleosome positions on the DNA after remodeling reaction times <1 min. Data (black) and their fit to two Gaussians (red and green) are shown. The red fit curve represents the nucleosome population that remained at the original 601 position (unremodeled; center=240 bp, SD=2.8 bp), while the green curve corresponds to those moved by the action of yeast SWI/SNF (remodeled; center=247 bp, SD=28 bp). (See Color Insert.)

remains intact and the overall strength and position of histone–DNA interactions within the nucleosome are essentially unchanged. However, nucleosomes were moved bidirectionally along the DNA with a characteristic spreading of 28 bp per remodeling event (Fig. 2.9D and E). Taken together, these results on SWI/SNF mediated nucleosome remodeling generated by unzipping provide direct measurements of the structure and location of remodeled nucleosomes.

7. CONCLUSIONS

An increasing interest in the nucleosome as a key regulator of chromatin structure and many cellular processes has inspired the development of a variety of novel techniques, particularly at the single-molecule level (for a recent review, see Killian et al., 2012). The unzipping method detailed here offers high accuracy and precision in locating a nucleosome as well as the ability to elucidate both structural and dynamic features of protein–DNA interactions, complementing more traditional techniques. We anticipate that the unzipping method will continue to play an important role in the study of nucleosome structure, regulation, and remodeling and is readily extendable to studies of a wide variety of DNA-based activities.

ACKNOWLEDGMENTS

We thank Dr. Shanna M. Fellman, Dr. Robert A. Forties, Dr. Robert M. Fulbright, James T. Inman, Jessie L. Killian, and Maxim Y. Sheinin for critical comments on the manuscript. We wish to acknowledge support from National Institutes of Health grants (GM059849 to M. D. W.) and National Science Foundation grant (MCB-0820293 to M. D. W.).

REFERENCES

Andrews, A. J., & Luger, K. (2011). Nucleosome structure(s) and stability: Variations on a theme. *Annual Review of Biophysics, 40*, 99–117.

Ashkin, A., Dziedzic, J. M., Bjorkholm, J. E., & Chu, S. (1986). Observation of a single-beam gradient force optical trap for dielectric particles. *Optics Letters, 11*, 288.

Bannister, A. J., & Kouzarides, T. (2011). Regulation of chromatin by histone modifications. *Cell Research, 21*, 381–395.

Bell, O., Tiwari, V. K., Thoma, N. H., & Schubeler, D. (2011). Determinants and dynamics of genome accessibility. *Nature Reviews Genetics, 12*, 554–564.

Bockelmann, U., Essevaz-Roulet, B., & Heslot, F. (1998). DNA strand separation studied by single molecule force measurements. *Physical Review E, 58*, 2386–2394.

Bowman, G. D. (2010). Mechanisms of ATP-dependent nucleosome sliding. *Current Opinion in Structural Biology, 20*, 73–81.

Brower-Toland, B. D., Smith, C. L., Yeh, R. C., Lis, J. T., Peterson, C. L., & Wang, M. D. (2002). Mechanical disruption of individual nucleosomes reveals a reversible multistage

release of DNA. *Proceedings of the National Academy of Sciences of the United States of America, 99,* 1960–1965.
Brower-Toland, B., Wacker, D. A., Fulbright, R. M., Lis, J. T., Kraus, W. L., & Wang, M. D. (2005). Specific contributions of histone tails and their acetylation to the mechanical stability of nucleosomes. *Journal of Molecular Biology, 346,* 135–146.
Brower-Toland, B., & Wang, M. D. (2004). Use of optical trapping techniques to study single-nucleosome dynamics. *Methods in Enzymology, 376,* 62–72.
Campos, E. I., & Reinberg, D. (2009). Histones: Annotating chromatin. *Annual Review of Genetics, 43,* 559–599.
Clapier, C. R., & Cairns, B. R. (2009). The biology of chromatin remodeling complexes. *Annual Review of Biochemistry, 78,* 273–304.
Das, C., Tyler, J. K., & Churchill, M. E. (2010). The histone shuffle: Histone chaperones in an energetic dance. *Trends in Biochemical Sciences, 35,* 476–489.
Davey, C. A., Sargent, D. F., Luger, K., Maeder, A. W., & Richmond, T. J. (2002). Solvent mediated interactions in the structure of the nucleosome core particle at 1.9 a resolution. *Journal of Molecular Biology, 319,* 1097–1113.
Dechassa, M. L., Wyns, K., Li, M., Hall, M. A., Wang, M. D., & Luger, K. (2011). Structure and Scm3-mediated assembly of budding yeast centromeric nucleosomes. *Nature Communications, 2,* 313.
Dyer, P. N., Edayathumangalam, R. S., White, C. L., Bao, Y., Chakravarthy, S., & Muthurajan, U. M. (2004). Reconstitution of nucleosome core particles from recombinant histones and DNA. *Methods in Enzymology, 375,* 23–44.
Evans, E. (2001). Probing the relation between force—lifetime—and chemistry in single molecular bonds. *Annual Review of Biophysics and Biomolecular Structure, 30,* 105–128.
Forties, R. A., North, J. A., Javaid, S., Tabbaa, O. P., Fishel, R., Poirier, M. G., et al. (2011). A quantitative model of nucleosome dynamics. *Nucleic Acids Research, 39,* 8306–8313.
Fyodorov, D. V., & Kadonaga, J. T. (2003). Chromatin assembly in vitro with purified recombinant ACF and NAP-1. *Methods in Enzymology, 371,* 499–515.
Gansen, A., Hauger, F., Toth, K., & Langowski, J. (2007). Single-pair fluorescence resonance energy transfer of nucleosomes in free diffusion: Optimizing stability and resolution of subpopulations. *Analytical Biochemistry, 368,* 193–204.
Gemmen, G. J., Sim, R., Haushalter, K. A., Ke, P. C., Kadonaga, J. T., & Smith, D. E. (2005). Forced unraveling of nucleosomes assembled on heterogeneous DNA using core histones, NAP-1, and ACF. *Journal of Molecular Biology, 351,* 89–99.
Hall, M. A., Shundrovsky, A., Bai, L., Fulbright, R. M., Lis, J. T., & Wang, M. D. (2009). High-resolution dynamic mapping of histone-DNA interactions in a nucleosome. *Nature Structural & Molecular Biology, 16,* 124–129.
Jiang, J., Bai, L., Surtees, J. A., Gemici, Z., Wang, M. D., & Alani, E. (2005). Detection of high-affinity and sliding clamp modes for MSH2-MSH6 by single-molecule unzipping force analysis. *Molecular Cell, 20,* 771–781.
Jin, J., Bai, L., Johnson, D. S., Fulbright, R. M., Kireeva, M. L., Kashlev, M., et al. (2010). Synergistic action of RNA polymerases in overcoming the nucleosomal barrier. *Nature Structural & Molecular Biology, 17,* 745–752.
Johnson, D. S., Bai, L., Smith, B. Y., Patel, S. S., & Wang, M. D. (2007). Single-molecule studies reveal dynamics of DNA unwinding by the ring-shaped T7 helicase. *Cell, 129,* 1299–1309.
Joo, C., Balci, H., Ishitsuka, Y., Buranachai, C., & Ha, T. (2008). Advances in single-molecule fluorescence methods for molecular biology. *Annual Review of Biochemistry, 77,* 51–76.
Kaplan, N., Moore, I. K., Fondufe-Mittendorf, Y., Gossett, A. J., Tillo, D., Field, Y., et al. (2009). The DNA-encoded nucleosome organization of a eukaryotic genome. *Nature, 458,* 362–366.

Killian, J. L., Li, M., Sheinin, M. Y., & Wang, M. D. (2012). Recent advances in single molecule studies of nucleosomes. *Current Opinion in Structural Biology, 22,* 80–87.

Koch, S. J., Shundrovsky, A., Jantzen, B. C., & Wang, M. D. (2002). Probing protein-DNA interactions by unzipping a single DNA double helix. *Biophysical Journal, 83,* 1098–1105.

Koch, S. J., & Wang, M. D. (2003). Dynamic force spectroscopy of protein-DNA interactions by unzipping DNA. *Physical Review Letters, 91,* 028103.

Korber, P., & Becker, P. B. (2010). Nucleosome dynamics and epigenetic stability. *Essays in Biochemistry, 48,* 63–74.

Lowary, P. T., & Widom, J. (1998). New DNA sequence rules for high affinity binding to histone octamer and sequence-directed nucleosome positioning. *Journal of Molecular Biology, 276,* 19–42.

Luger, K., Mader, A. W., Richmond, R. K., Sargent, D. F., & Richmond, T. J. (1997). Crystal structure of the nucleosome core particle at 2.8 A resolution. *Nature, 389,* 251–260.

Luger, K., Rechsteiner, T. J., & Richmond, T. J. (1999). Preparation of nucleosome core particle from recombinant histones. *Methods in Enzymology, 304,* 3–19.

Mihardja, S., Spakowitz, A. J., Zhang, Y., & Bustamante, C. (2006). Effect of force on mononucleosomal dynamics. *Proceedings of the National Academy of Sciences of the United States of America, 103,* 15871–15876.

Moffitt, J. R., Chemla, Y. R., Smith, S. B., & Bustamante, C. (2008). Recent advances in optical tweezers. *Annual Review of Biochemistry, 77,* 205–228.

Park, Y. J., & Luger, K. (2008). Histone chaperones in nucleosome eviction and histone exchange. *Current Opinion in Structural Biology, 18,* 282–289.

Ransom, M., Dennehey, B. K., & Tyler, J. K. (2010). Chaperoning histones during DNA replication and repair. *Cell, 140,* 183–195.

Ray-Gallet, D., & Almouzni, G. (2010). Nucleosome dynamics and histone variants. *Essays in Biochemistry, 48,* 75–87.

Schafer, D. A., Gelles, J., Sheetz, M. P., & Landick, R. (1991). Transcription by single molecules of RNA polymerase observed by light microscopy. *Nature, 352,* 444–448.

Shundrovsky, A., Smith, C. L., Lis, J. T., Peterson, C. L., & Wang, M. D. (2006). Probing SWI/SNF remodeling of the nucleosome by unzipping single DNA molecules. *Nature Structural & Molecular Biology, 13,* 549–554.

Simon, M., North, J. A., Shimko, J. C., Forties, R. A., Ferdinand, M. B., Manohar, M., et al. (2011). Histone fold modifications control nucleosome unwrapping and disassembly. *Proceedings of the National Academy of Sciences of the United States of America, 108,* 12711–12716.

Smith, S. B., Cui, Y., & Bustamante, C. (1996). Overstretching B-DNA: The elastic response of individual double-stranded and single-stranded DNA molecules. *Science, 271,* 795–799.

Thastrom, A., Lowary, P. T., & Widom, J. (2004). Measurement of histone-DNA interaction free energy in nucleosomes. *Methods, 33,* 33–44.

Wang, M. D., Yin, H., Landick, R., Gelles, J., & Block, S. M. (1997). Stretching DNA with optical tweezers. *Biophysical Journal, 72,* 1335–1346.

Zhang, Y., Moqtaderi, Z., Rattner, B. P., Euskirchen, G., Snyder, M., Kadonaga, J. T., et al. (2009). Intrinsic histone-DNA interactions are not the major determinant of nucleosome positions in vivo. *Nature Structural & Molecular Biology, 16,* 847–852.

CHAPTER THREE

Monitoring Conformational Dynamics with Single-Molecule Fluorescence Energy Transfer: Applications in Nucleosome Remodeling

Sebastian Deindl*,[†], Xiaowei Zhuang*,[†,‡,1]
*Howard Hughes Medical Institute, Harvard University, Cambridge, Massachusetts, USA
[†]Department of Chemistry and Chemical Biology, Harvard University, Cambridge, Massachusetts, USA
[‡]Department of Physics, Harvard University, Cambridge, Massachusetts, USA
[1]Corresponding author: e-mail address: zhuang@chemistry.harvard.edu

Contents

1. Introduction 60
2. Preparation of Fluorescently Labeled Sample for Single-Molecule FRET Imaging 62
   2.1 Preparation and purification of fluorescently labeled DNA 63
   2.2 Site-specific labeling of proteins with fluorescent dyes 64
   2.3 Preparation of histone octamer from purified core histone proteins 66
   2.4 Nucleosome reconstitution and purification 67
3. Preparing PEG-Coated Slides with Sample Chamber 69
   3.1 Preparing and cleaning quartz slides 69
   3.2 Surface coating with PEG 69
   3.3 Sample chamber preparation 71
4. Optical Setup and Single-Molecule FRET Data Acquisition 71
   4.1 Optical setup 71
   4.2 Sample immobilization and data acquisition 73
5. Data Analysis 77
   5.1 Calculating the FRET efficiency 78
   5.2 Constructing and analyzing FRET histograms 78
   5.3 Analysis of single-molecule FRET time traces 79
6. Summary 83
Acknowledgments 83
References 83

Abstract

Due to its ability to track distance changes within individual molecules or molecular complexes on the nanometer scale and in real time, single-molecule fluorescence resonance energy transfer (single-molecule FRET) is a powerful tool to tackle a wide range of important biological questions. Using our recently developed single-molecule FRET assay to monitor nucleosome translocation as an illustrative example, we describe here in detail how to set up, carry out, and analyze single-molecule FRET experiments that provide time-dependent information on biomolecular processes.

1. INTRODUCTION

Essential cellular processes such as interactions between biomolecules or conformational changes often involve distance changes at a nanometer length scale (1–10 nm). Fluorescence resonance energy transfer (FRET) (Forster, 1949; Stryer & Haugland, 1967) is a spectroscopic technique that enables the observation of distance changes at this length scale with high sensitivity and in real time. In this technique, a donor and an acceptor fluorophore are introduced at sites whose distance is to be monitored. Upon excitation of the donor fluorophore, a fraction of its energy can be transferred to the acceptor fluorophore in a nonradiative process. The efficiency of energy transfer, E, is highly sensitive to the distance R between the two fluorescent molecules: $E = 1/1 + (R/R_0)^6$, where R_0 is the Förster radius at which $E=0.5$ (Fig. 3.1A). FRET measurements at the single-molecule level allow the observation of dynamics on a molecular scale that would be inaccessible in ensemble measurements due to random averaging (Ha, 2001; Ha et al., 1996; Zhuang et al., 2000). The ability to monitor conformational dynamics of individual molecules in real time makes single-molecule techniques highly effective in the study of a wide range of mechanistic questions.

Recently, several groups have reported the application of single-molecule techniques to directly observe DNA or nucleosome translocation by ATP-dependent chromatin remodeling enzymes (Amitani, Baskin, & Kowalczykowski, 2006; Blosser, Yang, Stone, Narlikar, & Zhuang, 2009; Lia et al., 2006; Nimonkar, Amitani, Baskin, & Kowalczykowski, 2007; Prasad et al., 2007; Shundrovsky, Smith, Lis, Peterson, & Wang, 2006; Sirinakis et al., 2011; Zhang et al., 2006). These enzymes help regulate DNA accessibility by altering the structure of chromatin, which is based on the nucleosome as a fundamental repeating unit (Kornberg, 1974; Kornberg & Lorch, 1999). There are several subfamilies of chromatin remodeling enzymes, SWI/SNF, ISWI, CHD/Mi2, and INO80, that differ in composition and function (Becker & Horz, 2002; Clapier & Cairns,

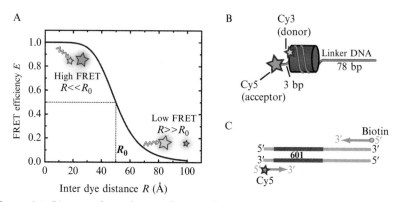

Figure 3.1 Distance dependence of FRET efficiency and example construct design for single-molecule FRET imaging of nucleosome remodeling. (A) The FRET efficiency as a function of the distance between donor and acceptor dye molecules. The donor and acceptor fluorophores are represented as green and red stars, respectively. At the Förster distance, R_0, the donor fluorophore transfers 50% of its energy to the acceptor fluorophore. (B) Example nucleosome with a single donor dye on the proximal H2A subunit. The histone octamer and the nucleosomal DNA are shown in blue and brown, respectively. The Cy5 (acceptor) and Cy3 (donor) fluorophores are depicted as red and green stars, respectively. (C) The Cy5 dye (red star) and the biotin moiety (orange dot) are included in the PCR primers used to generate the nucleosomal DNA. (See Color Insert.)

2009; Flaus & Owen-Hughes, 2004; Narlikar, Fan, & Kingston, 2002; Smith & Peterson, 2005; Tsukiyama & Wu, 1997). Members of the ISWI subfamily reposition nucleosomes along the DNA while preserving the canonical nucleosome structure (Hamiche, Sandaltzopoulos, Gdula, & Wu, 1999; Kassabov, Henry, Zofall, Tsukiyama, & Bartholomew, 2002; Langst, Bonte, Corona, & Becker, 1999; Tsukiyama, Palmer, Landel, Shiloach, & Wu, 1999). Recently, we have used single-molecule FRET measurements to monitor in real time the translocation of individual nucleosomes by an ISWI family enzyme (Blosser et al., 2009), the ATP-dependent chromatin assembly and remodeling factor (ACF) (Ito, Bulger, Pazin, Kobayashi, & Kadonaga, 1997). These experiments allowed us to reveal previously unknown kinetic intermediates and mechanistic aspects of this remodeling enzyme (Blosser et al., 2009). The dynamic nature of chromatin structure and its regulation makes single-molecule FRET ideally suited for this type of mechanistic study. Here, we provide a detailed step-by-step guide for how to set up and carry out single-molecule FRET measurements, including experiments that require buffer exchange during data acquisition. As an illustrative example, we describe single-molecule FRET measurements on fluorescently labeled mononucleosomes during

ACF-induced remodeling. With the exception of sections regarding the preparation of mononucleosomes, all protocols described here are generally applicable, with minimal adaptations, to the single-molecule FRET study of other biological systems.

2. PREPARATION OF FLUORESCENTLY LABELED SAMPLE FOR SINGLE-MOLECULE FRET IMAGING

The attachment sites for donor and acceptor dyes are chosen such that their distance changes are relatively large during the process under investigation. For processes that are associated with relatively small distance changes, the FRET dyes are ideally positioned at a distance close to the R_0 value of the FRET pair, because at this interdye distance, the FRET efficiency is most sensitive to distance changes. Common labeling approaches include NHS ester or maleimide chemistry, and several different FRET dye pairs can be used (Roy, Hohng, & Ha, 2008). The labeling of nucleic acids with fluorescent dyes is generally straightforward, which greatly facilitates the study of nucleic acid–protein interactions. Because single-molecule FRET measurements of a single donor–acceptor pair yield a one-dimensional signal, finding the best labeling choice in terms of labeling sites and fluorophores is crucial in understanding conformational dynamics that typically occur in three dimensions. As an example, we describe here how to generate nucleosomes that are labeled with a Cy3–Cy5 FRET pair, but the procedures can readily be adapted to other nucleic acid and protein substrates under study.

For single-molecule measurements of chromatin remodeling, it is typically desired to carry out experiments with a homogenous population of nucleosomes where the histone octamer is initially placed at a specific location on the DNA substrate. This can be achieved by incorporating a strong nucleosome positioning sequence, such as the 601 positioning sequence (Li, Levitus, Bustamante, & Widom, 2005), into the DNA substrate used for nucleosome assembly. One should, however, be cautious about the possibility of a sequence-specific bias, and control experiments with distinct and nonrelated positioning sequences should be performed to test whether the observed phenomenon is sequence-specific. Depending on the exact mechanistic question under study, a variety of nucleosomal substrates can be used that differ in the flanking DNA length at entry and exit sides of the nucleosome, in the location of the FRET dyes on the DNA and/or the histone octamer, and in the specific composition of nucleosomal histone proteins.

In our example, initially end-positioned mononucleosomes such as shown in Fig. 3.1B are prepared to observe the real-time dynamics of nucleosome remodeling. The double-stranded DNA molecule used to reconstitute these nucleosomes contains the FRET acceptor dye (Cy5) and a biotin moiety for immobilization at opposing ends. Histone octamers are labeled with the FRET donor dye (Cy3) on histone H2A and positioned at a defined number of base pairs (bp) away from the acceptor dye (Cy5)-labeled exit end of the DNA. As an illustrative example, we describe a construct with DNA linkers of 3 bp and 78 bp on the exit and entry sides of the nucleosome, respectively.

2.1. Preparation and purification of fluorescently labeled DNA

Nucleic acid molecules used for single-molecule FRET imaging are typically assembled by PCR or annealing reactions. This approach has the advantage that molecules such as fluorophores can be conveniently introduced by using commercially available custom-synthesized oligonucleotides that contain the desired moieties. In the example described here, the nucleosomal DNA is prepared by PCR, with the Cy5 dye and the biotin label included in custom-synthesized PCR primers (Fig. 3.1C).

2.1.1 Preparation of fluorophore-labeled DNA by PCR

1. At 4 °C, prepare an amplification mix with a total volume of 10 ml and final concentrations of 0.02 U/μl Phusion Hot Start High-Fidelity Polymerase (New England Biolabs), 200 μM of each deoxynucleotide triphosphate (dNTP), 1× Phusion HF Buffer, 0.5 μM forward and reverse primers, 0.2 ng/μl template plasmid (containing positioning and desired flanking DNA sequences). Using a multichannel pipette, transfer 100 μl of the amplification mix into each well of a 96-well standard thin-walled PCR plate, seal the plate, and amplify the DNA in a total of 35 cycles with a final extension of 10 min.
2. With a multichannel pipette, combine all reactions, add 1/10 volume of 3 M sodium acetate pH 5.2 and 2.5 volumes (calculated after addition of sodium acetate) of >95% (v/v) ethanol and precipitate the DNA overnight at −20 °C. Pellet the DNA by centrifugation at >14,000 × g and 4 °C for 15 min and discard the supernatant. Rinse the DNA pellet with 70% (v/v) ethanol, repeat the centrifugation, discard the supernatant, and air-dry the pellet.
3. Dissolve the DNA in 60 μl T50 buffer (10 mM Tris pH 7.5 and 50 mM NaCl). Incubation at 37 °C for 10 min facilitates this process. Pellet any denatured, insoluble protein by centrifugation at >14,000 × g and room

temperature for 15 min and keep the supernatant that contains the DNA. Add glycerol to a final concentration of 5% (v/v) for subsequent preparative polyacrylamide gel electrophoresis (PAGE).

2.1.2 Native PAGE purification of dye-labeled DNA

1. For the purification of dye-labeled DNA, pour a nondenaturing polyacrylamide, 0.5 × TBE (final concentrations 45 mM Tris–borate and 1 mM EDTA pH 8.0) gel (dimensions 20 cm × 16 cm × 3 mm). Choose the polyacrylamide percentage according to the size of the DNA sample. For example, a 5% polyacrylamide gel works well for the purification of a ∼150–250 bp sample.
2. Electrophorese the gel in 0.5 × TBE at 10 V/cm and 4 °C for 1 h, rinse wells, and load sample. Resume electrophoresis for ∼1.5 h at 4 °C. The dye label allows direct observation of DNA migration through the gel. Excise the region of the gel containing the desired DNA with a razor blade.
3. Transfer into 5 ml of 10 mM Tris pH 7.5 and shake at 37 °C for at least 12 h to allow the DNA to diffuse out of the gel slice and into the buffer. Take off the DNA-containing supernatant, replace with another 5 ml of 10 mM Tris pH 7.5, and repeat the incubation at 37 °C for another 12 h. Remove the supernatant and combine with the supernatant from the first incubation step.
4. Concentrate the DNA by adding an equal volume of 2-butanol to the combined supernatant. Mix, then separate organic and aqueous phases by brief centrifugation. Aspirate off and discard the upper organic layer. Repeat the extraction with 2-butanol until the volume of the aqueous phase is less than ∼500 µl. Add 1/10 volume of 3 M sodium acetate pH 5.2 and 2.5 volumes (calculated after addition of sodium acetate) of >95% (v/v) ethanol. After overnight precipitation at −20 °C, pellet the DNA by centrifugation at >14,000 × g and 4 °C for 15 min and discard the supernatant. Rinse the DNA pellet with 70% (v/v) ethanol, repeat the centrifugation, discard the supernatant, and air-dry the pellet. Dissolve the DNA in TE buffer (10 mM Tris, 1 mM EDTA, pH 8.0).

2.2. Site-specific labeling of proteins with fluorescent dyes

A variety of different approaches for site-specific labeling of proteins have been reported (Fernandez-Suarez & Ting, 2008; Heyduk, 2002). For example, in an *in vitro* transcription/translation reaction, a tRNA loaded with a fluorophore-derivatized amino acid can be used to incorporate it at a specific site defined by a nonsense codon (Mendel, Cornish, &

Schultz, 1995). Alternatively, peptide sequences that are specifically recognized by a fluorophore-labeled ligand can be introduced into the protein (Griffin, Adams, & Tsien, 1998; Kapanidis, Ebright, & Ebright, 2001; Miller, Cai, Sheetz, & Cornish, 2005). Recently, various enzymes have been used to catalyze the ligation of the fluorophore to its target site (Baruah, Puthenveetil, Choi, Shah, & Ting, 2008; Fernandez-Suarez et al., 2007; Popp, Antos, Grotenbreg, Spooner, & Ploegh, 2007; Tanaka, Yamamoto, Tsukiji, & Nagamune, 2008). Further, chemoselective ligation strategies can be applied (Dawson, Muir, Clarklewis, & Kent, 1994; Mekler et al., 2002; Muir, Sondhi, & Cole, 1998).

One of the most commonly used approaches for site-specific labeling of proteins involves using a protein without any reactive cysteine side chains other than at the site of interest, referred to as a "Cys-lite" (Rice et al., 1999) protein. A Cys-lite protein can be obtained by site-directed mutagenesis: Endogenous cysteine residues other than the target residue are mutated to nonreactive amino acid residues, and the amino acid residue at the target site is mutated to a cysteine residue. Once established that the Cys-lite mutant retains full physiological function, its cysteine thiol moiety can be derivatized by nucleophilic substitution, for example, with a monofunctional maleimide ester of a fluorescent dye. As an example for this approach, we provide here a protocol for the site-specific labeling of a core histone protein H2A with Cy3 at a single cysteine side chain (residue 120, H2A-120C) (Yang, Madrid, Sevastopoulos, & Narlikar, 2006).

1. At room temperature, dissolve 1 mg of lyophilized H2A-120C in 800 μl of labeling buffer (20 mM Tris pH 7.0, 7 M guanidinium HCl, 5 mM EDTA) for a protein concentration of $\sim$100 μM.
2. In order to stabilize the free sulfhydryls, add 2 μl of 0.5 M TCEP (final concentration: 1.25 mM) and incubate for 2 h at room temperature in the dark.
3. Dissolve dried Cy3-maleimide ester (GE Healthcare Amersham) in anhydrous DMSO at a concentration of 100 mM, taking into account the fraction of reactive dye indicated by the manufacturer.
4. Add 25 μl of 100 mM Cy3-maleimide ester ($\sim$3 mM final concentration) to the histone protein and incubate for 3 h in the dark at room temperature.
5. Quench the labeling reaction by adding 4.6 μl of β-mercaptoethanol (final concentration: 80 mM). Set a small aliquot ($\sim$ 2 μl) aside for analysis via diagnostic gel.

6. Excess dye can be removed by dialysis, gel filtration, or the use of dye removal resin. For example, to remove excess Cy3 dye by dialysis, transfer the quenched labeling reaction into a 3-ml dialysis cassette (Slide-A-Lyzer, 7000 MWCO; Thermo Scientific) and dialyze at room temperature in the dark three times against 1000 ml of fresh dialysis buffer (20 mM Tris pH 7.0, 7 M guanidinium HCl, 1 mM DTT) for at least 3 h per iteration.
7. Use 1 μl of the undialyzed labeled histone sample (set aside in step 5) and dilute by a factor of 25 with water. Dilute 1 μl of the dialyzed labeled histone sample analogously. Add SDS loading dye to both dilutions and incubate at 95 °C for 2 min. In separate lanes, analyze 2–10 μl of each dilution on a 15% SDS analytical polyacrylamide gel. Quantify Cy3 intensities from the histone band in both lanes using an imaging system such as a Molecular Dynamics Typhoon. Stain the gel with SYPRO Red (Sigma-Aldrich) and quantify the protein content of both bands based on the SYPRO stain. The comparison of the SYPRO band intensities yields the concentration of the dialyzed Cy3-labeled sample, since the concentration of the undialyzed sample is known. The labeling efficiency can be roughly estimated by comparing the concentrations obtained from the SYPRO and Cy3 fluorescence signals.
8. Concentrate the dialyzed Cy3-labeled H2A as needed using a centrifugal filter device and use to prepare Cy3-labeled histone octamer.

2.3. Preparation of histone octamer from purified core histone proteins

Histone octamer composed of unlabeled H3, H4, H2B, and Cy3-labeled H2A subunits can be assembled by dialysis and subsequent size exclusion purification (Dyer et al., 2004; Luger, Rechsteiner, & Richmond, 1999; Shahian & Narlikar, 2012):

1. From each of the three lyophilized and unlabeled core histone proteins (H3, H4, H2B), make a 2 mg/ml solution in unfolding buffer (20 mM Tris pH 7.5, 7 M guanidinium hydrochloride, 10 mM DTT). Dissolve histone proteins by gently pipetting up and down. Incubate the solutions for at least 30 min to ensure complete unfolding but do not exceed an incubation time of 3 h.
2. Using Cy3-labeled H2A (from Section 2.2, in dialysis buffer), and the H2B, H3, and H4 solutions from step 1, prepare a mixture of the four core histone proteins with a molar ratio of 6:6:5:5 (H2A:H2B:H3:H4). The slight molar excess of H2A and H2B is used to shift the

equilibrium toward more complete octamer formation. Any excess H2A/H2B dimer can be removed by subsequent size exclusion chromatography. Adjust the total protein concentration to 1 mg/ml with unfolding buffer.

3. Transfer the core histone mix into a dialysis bag with a 6–8 kDa molecular weight cutoff and dialyze three times against 1 l of fresh refolding buffer (2 M NaCl, 10 mM Tris pH 7.5, 1 mM EDTA, 5 mM β-mercaptoethanol). One of the dialysis steps should be carried out overnight, while the other two steps should proceed for at least 3 h each.

4. Recover the dialyzed sample and pellet any precipitate by ultracentrifugation at maximum speed for 20 min. Concentrate the supernatant to a volume of ∼200 μl using a centrifugal filter device (Millipore YM-10, 10,000 NMWL). Set the filtrate and ∼5 μl of the concentrated sample (retentate) aside for analytical purposes.

5. Subject the concentrated sample to size exclusion chromatography using a Superdex 200 HR 10/30 column (Amersham Pharmacia Biotech). Wash the column (typically stored in 20% (v/v) ethanol) with two column volumes of water (∼50 ml) and then equilibrate with at least two column volumes of refolding buffer. Inject the sample and collect 0.5-ml fractions at a flow rate of 0.5 ml/min. The histone octamer elutes first, with a smaller elution volume than that of any excess H2A/H2B dimer. Monitor absorption at 280 nm (protein) and 550 nm (Cy3) to identify octamer peak fractions.

6. Analyze ∼10 μl of the filtrate (flow-through) and 5 μl of the concentrated sample (load) from step 4 as well as 5 μl of each octamer peak fraction on a 15% SDS polyacrylamide gel. Stain the gel with SYPRO Red gel stain and combine those octamer peak fractions that display approximately equimolar quantities of all four histone core proteins.

7. Concentrate the pooled fractions to a concentration of at least 1 mg/ml using a centrifugal filter device (Millipore YM-10, 10,000 NMWL). To determine the final concentration more accurately, analyze several dilutions of the concentrated sample together with a series of bovine serum albumin (BSA) standards on a 15% SDS polyacrylamide gel and quantify band intensities of standards and sample upon SYPRO staining.

2.4. Nucleosome reconstitution and purification

Nucleosomes for single-molecule FRET studies can be reconstituted by combining Cy3-labeled histone octamer and Cy5-labeled DNA at a molar ratio of 6:5 using a salt gradient procedure (Dyer et al., 2004; Lee & Narlikar, 2001; Luger et al., 1999).

1. For a typical small-scale reconstitution mixture of 150 µl, combine 200 pmol Cy3-histone octamer and ~167 pmol dye-labeled DNA in 20 mM Tris pH 7.5, 1 mM EDTA pH 8.0, 10 mM DTT, and 2 M KCl at 4 °C. Add the histone octamer last to the reconstitution mixture in order to avoid the formation of aggregates at salt concentrations <2 M.

2. Prepare 250 ml of high salt buffer (10 mM Tris pH 7.5, 2 M KCl, 1 mM EDTA pH 8.0, 1 mM DTT, 0.5 mM Benzamidine) and 1000 ml of low salt buffer (10 mM Tris pH 7.5, 250 mM KCl, 1 mM EDTA pH 8.0, 1 mM DTT, 0.5 mM Benzamidine) and chill to 4 °C. At 4 °C, transfer the reconstitution mixture into a 0.5-ml dialysis cassette (Slide-A-Lyzer, 7000 MWCO; Thermo Scientific) and place the cassette into a beaker containing the high salt buffer. Under constant stirring, use a peristaltic pump setup to continually remove buffer from the dialysis beaker and to replace it with low salt buffer. Both inlet and outlet should be calibrated and adjusted to yield closely matching flow rates of ~215 µl/min. Perform the salt gradient dialysis at 4 °C and in the dark (to prevent photobleaching of fluorophores) for 60 h, then transfer the dialysis cassette into a beaker containing 1 l of TCS buffer (20 mM Tris pH 7.5, 1 mM EDTA pH 8.0, 1 mM DTT), prechilled to 4 °C, and dialyze for at least 2 h.

3. To purify nucleosomes from free DNA, prepare a 10% glycerol buffer (20 mM Tris pH 7.5, 1 mM EDTA, 0.1% NP-40, 10% glycerol) and a 30% glycerol buffer (20 mM Tris pH 7.5, 1 mM EDTA, 0.1% NP-40, 30% glycerol). Using a gradient forming instrument such as a Gradient Master (BioComp), form a linear 10–30% glycerol gradient in a 5-ml polycarbonate tube (Beckman Polyallomer No. 326819, 0.5 × 2.5″) and layer the contents of the dialysis cassette on top of the gradient. Centrifuge at 4 °C at 35,000 rpm in a SW55Ti rotor for 16.5 h.

4. At 4 °C, puncture the centrifuge tube with a butterfly needle ~1 cm from the bottom and collect 30–40 fractions of ~5 drops each. Analyze fractions on a 5% polyacrylamide, 0.5 × TBE gel and combine fractions that contain correctly assembled nucleosomes but no detectable free DNA. Concentrate to a volume of ~20–50 µl using a centrifugal filter device (Millipore YM-100, 100,000 NMWL). Upon recovery of the retentate (concentrated nucleosome preparation), store both nucleosomes and the filtrate at 4 °C in the dark. The filtrate is useful for preparing nucleosome dilutions as needed for single-molecule FRET imaging.

3. PREPARING PEG-COATED SLIDES WITH SAMPLE CHAMBER

The observation of conformational dynamics at the single-molecule level often requires the ability to rapidly exchange the buffer during data acquisition. For example, an enzymatic reaction such as nucleosome remodeling can be initiated at a well-defined time point by perfusing a sample chamber (Fig. 3.2A) with remodeling enzyme and ATP. The quartz slide on which the sample chamber is assembled is typically passivated with a polyethylene glycol (PEG) polymer brush to minimize nonspecific protein adsorption to the quartz surface (Ha et al., 2002). A PEG mixture can be used in which a small fraction of polymer molecules is terminally coupled to a biotin moiety in order to tightly bind streptavidin, which in turn serves to immobilize the biotinylated sample.

3.1. Preparing and cleaning quartz slides

For the following steps, use only MilliQ water (18.5 MΩ).
1. In order to assemble sample chambers, drill two holes (0.75 mm diameter) into each quartz slide as shown in Fig. 3.2A. The holes will serve as connection points for the inlet (connected to a syringe pump; KD Scientific) and outlet of the sample chamber.
2. Thoroughly scrub the slides with an aqueous slurry of Alconox detergent powder (Alconox Inc.) and rinse generously with water. Repeat the scrubbing procedure twice and then rinse with copious amounts of water. Sonicate the slides in a glass staining dish for 20 min in 1 M KOH. Next, rinse slides with water and transfer to another clean glass staining dish. Add >95% (v/v) ethanol and sonicate for 20 min. Rinse slides with water, sonicate for 20 min in acetone, and rinse again with water.
3. Flame the surface to be PEG coated and imaged with a propane torch in order to minimize fluorescent background caused by residual organic molecules.
4. Transfer the slides into a plasma cleaner (Harrick Plasma) and treat the surface to be imaged with argon plasma.

3.2. Surface coating with PEG

1. Rinse a clean and dry glass staining dish with >99.9% (w/v) acetone to remove any residual water. Transfer the slides into the glass staining dish and amino-modify them by treatment with 1% (w/v) Vectabond (Vector

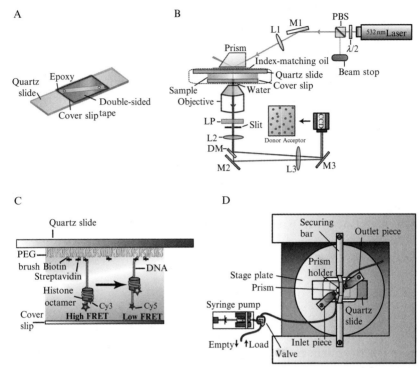

Figure 3.2 Sample chamber and optical setup for single-molecule FRET imaging with buffer exchange during data acquisition. Individual components are not drawn to scale. (A) Components used to assemble a flow chamber on a quartz slide. (B) Schematic of the optical setup for single-molecule FRET imaging using the prism-type TIR geometry. The excitation laser beam is focused into a Pellin-Broca fused silica prism with a shallow incident angle (<23°) such that an evanescent field is created at the quartz–water interface on the quartz slide. Fluorescence emission from donor and acceptor dyes is collected by the objective, and scattered excitation laser light is removed by a long pass filter. Donor and acceptor emission is separated by a dichroic mirror. The dichroic mirror and an additional mirror are used to slightly offset the paths of donor and acceptor fluorescence such that they can be imaged onto two halves of the CCD camera. The image is reduced by a vertical slit in the imaging plane to fit onto half of the CCD chip area. λ/2, half waveplate; PBS, polarizing beam splitter; M1–M3, mirrors; L1–L3, lenses; LP, long pass filter; DM, dichroic mirror. (C) Larger view of the boxed region (dotted line) in (B). The slide is covered with a PEG brush to minimize nonspecific sticking of proteins to the quartz glass, and the sample is immobilized via a biotin–streptavidin linkage. The flow channel is formed between the quartz slide and the cover slip. (D) Top view of the microscope stage. A stage plate holds the sample cell. The prism is attached to the prism holder and placed on top of the quartz slide using a securing bar. Flow in the sample chamber is introduced with a motorized syringe pump that is connected to the inlet piece of the flow channel. (See Color Insert.)

Laboratories) in >99.9% (w/v) acetone for 5 min. Rinse slides by repeated dipping into a beaker with water and dry under nitrogen stream.
2. Apply a mixture of 20% (w/v) methoxy-PEG (M_r 5000; Nektar Therapeutics) and 0.2% biotin-PEG (M_r 5000; Nektar Therapeutics) in 0.1 M sodium bicarbonate pH 8.4 to the plasma-cleaned surface. Cover with a glass coverslip to spread PEG mixture as an even film while avoiding bubble formation. Place slides in a wet box and allow PEG coating to proceed for at least 4 h. Rinse the PEG-coated surface thoroughly with copious amounts of water and then dry the slides under nitrogen stream.

3.3. Sample chamber preparation

1. Attach two pieces of double-sided tape on the PEG-coated side of the quartz slide parallel to the line connecting inlet and outlet holes to form a flow channel of ~0.5 cm width (Fig. 3.2A).
2. Place a plasma-cleaned coverslip on top to form a sample chamber with a volume of ~20 μl.
3. Using a 5-min two-component epoxy resin, seal the boundaries between the cover slip and the quartz slide on all four sides.

PEG-coated sample chambers should be used within a few days of their assembly. Once sample has been immobilized on the slide for an experiment, PEG-coated sample chambers cannot be reused. However, the relatively expensive quartz slides can be recycled. To prepare new sample chambers, remove the coverslip, epoxy, and adhesive tape and repeat all of the above steps starting with the Alconox scrubbing. Storage of used quartz slides in water facilitates the removal of tape and epoxy residue.

4. OPTICAL SETUP AND SINGLE-MOLECULE FRET DATA ACQUISITION

4.1. Optical setup

In total internal reflection (TIR) microscopy (Axelrod, 2003), an evanescent field of light is generated that decays exponentially with distance from the surface onto which the sample is immobilized. This approach greatly reduces background fluorescence, as only fluorophores within ~100–200 nm of the surface are excited. There are two commonly used types of TIR microscopy: objective-type TIR and prism-type TIR.

In prism-type TIR microscopy, a focused laser beam is introduced into the sample by means of a prism attached to a quartz slide surface, onto which the imaged specimen is immobilized. The fluorescence signal is collected by an objective focused on the quartz slide surface. In objective-type TIR microscopy, the excitation laser light is introduced through the microscope objective, and the specimen is immobilized on the surface of a cover glass at the focus of the objective. The fluorescence signal is collected by the same objective. Both prism-type and objective-type geometries can be used for single-molecule FRET experiments.

A prism-type TIR setup can be assembled around a commercial inverted light microscope (e.g., Olympus IX70) as shown in Fig. 3.2B. In this setup, the FRET donor dye molecules (Cy3) are excited by a 532-nm Nd:YAG (neodymium-doped yttrium aluminum garnet) laser (CrystaLaser). The excitation laser intensity is attenuated and polarization-cleaned by a half-wave plate and a polarizing beamsplitter. A TIR geometry is achieved by focusing the excitation beam into a Pellin-Broca fused silica prism with a shallow incident angle ($<23°$). The prism holder is rigidly attached to the body of the microscope through a custom-built securing bar (Fig. 3.2D), which allows lateral translation of the sample on the stage while keeping the prism immobile. In order to accurately measure kinetic parameters such as transition times and rates, the temperature of the sample chamber often needs to be carefully controlled during data acquisition. This can be achieved by using a water-circulating bath that flows water at a defined temperature through tubing connected to a custom-built brass collar on the objective, a brass plate that holds the sample cell, and the metal pieces embracing the prism. The fluorescence emission from donor and acceptor molecules is collected by a water immersion objective lens (1.2 NA, 60×, Olympus) and filtered with a 550-nm long pass filter (Chroma Technology) to block scattered excitation light. Cy3 and Cy5 fluorescence signals are spectrally split by a 630-nm dichroic mirror (Chroma Technology) and imaged onto two halves of a CCD camera (Andor iXonEM$^+$ 888). Donor and acceptor signals of the same molecule are identified by aligning these two channels using fluorescent beads whose fluorescence emission can be detected in both channels. The CCD chip is cooled to $-75\ °C$ for acquisition of single-molecule FRET data.

In objective-type TIR microscopy, identical emission optics are used but the evanescent field is created with an oil immersion objective of high numerical aperture (for example, 1.4 NA). The excitation laser beam is focused at the rear focal plane of the objective and emerges from the objective front lens in a parallel beam. TIR at the glass–water interface on the coverslip is then achieved by steering the beam to the periphery of the objective. The fluorescence emission from sample molecules immobilized on the coverslip is collected using the same objective.

4.2. Sample immobilization and data acquisition

4.2.1 Imaging buffer and reagents

The buffer system used for single-molecule FRET imaging must not only ensure the integrity of sample and enzymes but also promote high photostability of the fluorescent dyes. Photoblinking of many fluorescent dyes is reduced in the presence of a triplet-state quenching agent such as Trolox (6-hydroxy-2,5,7,8-tetramethylchroman-2-carboxylic acid), a water-soluble derivative of vitamin E (Rasnik, McKinney, & Ha, 2006). Often, BSA is added to the imaging buffer in order to reduce nonspecific binding to the surfaces of tubes and pipette tips. In order to increase the photostability of Cy5 and Cy3 dye molecules, the sample chamber can be infused with an oxygen scavenger system to reduce photobleaching upon sample immobilization and prior to single-molecule FRET data acquisition. The oxygen scavenger system consumes molecular oxygen by the coupled enzymatic action of glucose oxidase and catalase (Benesch & Benesch, 1953). First, glucose oxidase catalyzes the reaction of molecular oxygen with β-D-glucose to produce D-glucono-1,5-lactone (which hydrolyzes to gluconic acid) and hydrogen peroxide. The second enzyme, catalase, then catalyzes the disproportionation of hydrogen peroxide into water and oxygen. The two coupled enzymatic reactions result in a net decrease of molecular oxygen concentrations. Glucose, a substrate for the glucose oxidase reaction, is provided in the imaging buffer. The enzymatic generation of gluconic acid leads to an acidification of the imaging buffer over time. As many biological samples are sensitive to changes in pH, the buffer strength of the imaging buffer must be sufficiently high to maintain appropriate pH, and the oxygen scavenger system should be added at the beginning of single-molecule FRET data acquisition. Additionally, proper sealing of the slide will minimize exposure of the sample chamber contents to air during the course of the experiment.

1. Prepare imaging buffer consisting of 12 mM HEPES, 40 mM Tris pH 7.5, 60 mM KCl, 0.32 mM EDTA, 3 mM MgCl$_2$, 10% (v/v) glycerol, 10% (w/v) glucose, 0.1 mg/ml BSA (acetylated; Promega), 2 mM Trolox, and 0.02% Igepal (Sigma-Aldrich). Filter the imaging buffer with a 0.2-μm bottle filter (polyethersulfone; Nalgene) and chill to 4 °C. When stored at 4 °C, the imaging buffer can typically be used for several weeks.

2. To prepare a 100× stock solution of the oxygen scavenger system (Gloxy), dissolve 10 mg of glucose oxidase in 100 μl T50 (10 mM Tris,

50 mM NaCl, pH 7.5) by gently pipetting up and down. Resuspend 20 mg/ml catalase stock (Sigma-Aldrich C100) and add 30 µl to the glucose oxidase solution. Gently mix by pipetting up and down and centrifuge at > 14,000 × g at 4 °C for 5 min. Use the supernatant as a 100 × Gloxy stock. Store the stock at 4 °C and use within a few weeks. Alternatively, flash freeze 100× Gloxy stock and store them at −80 °C.

4.2.2 Immobilization of the sample

The sample can be conveniently immobilized on the quartz slide via a biotin–streptavidin linkage (Fig. 3.2C). Solutions/reagents are flowed into the sample chamber (Fig. 3.2A) using a pipette with the pipette tip snuggly plugged into one of the two holes. If the sample chamber is already filled and its contents are to be replaced with a new solution, gently apply a Kimwipe (Kimberly Clark) to the outlet hole at the same time. This will generate capillary suction to slowly remove the sample chamber contents. Simultaneously, deliver new solution from the inlet hole with a pipette. Care must be taken to deliver new solution at the right rate so as to avoid sucking the sample chamber dry. In general, the formation of air bubbles in the sample chamber should be avoided at all times.

1. Apply ∼35 µl of 0.2 mg/ml streptavidin (Molecular Probes) in T50 to the slide and incubate for ∼2 min at room temperature to allow for quantitative binding of streptavidin to the surface-immobilized biotin moieties.
2. Wash out excess streptavidin by flowing ∼100 µl of imaging buffer through the sample cell.
3. To immobilize the sample, apply ∼70 µl of an appropriate dilution of biotinylated and dye-labeled sample to the sample cell. In the case of a typical nucleosome preparation as described above, dilute by a factor of ∼5000–10,000. Make a first 1:10 dilution with the filtrate from the concentration step and further dilutions with the imaging buffer. If the right sample dilution is unknown, start out with a dilution that likely generates too low rather than too high a sample density. Check the density on the microscope, and if necessary, increase the sample density by flowing a more concentrated dilution onto the slide. Repeat this procedure until the desired density (∼1 molecule/5 µm^2 imaging area) is achieved. In general, using higher sample densities is beneficial for two reasons. First, a high sample density will reduce the relative concentration of fluorescent debris molecules that may be present on the slide. Second, a higher sample density allows more single-molecule FRET

traces to be analyzed per field of view so that adequate statistics can be generated more readily. Sample density should not be too high, however, since on the CCD, the images of individual fluorescent molecules need to be well separated to prevent misinterpretation of single-molecule FRET traces.

4.2.3 Collecting data for FRET histograms
1. Before imaging the actual sample, record images of fluorescent beads for the alignment of donor and acceptor channels as described above and in Section 5. Prepare a slide with fluorescent beads (0.1 μm TetraSpeck microspheres; Invitrogen Molecular Probes) as follows: Dilute the fluorescent beads by a factor of 100 with 4 M KCl. Assemble a flow channel on a quartz slide prepared essentially as described above, with the exception that no inlet/outlet holes need to be drilled and the channel formed by the double-sided tape and the cover slip is not sealed with epoxy glue at this point. From one edge of the channel, apply ∼50 μl of the diluted microspheres to the sample chamber and incubate at room temperature for ∼10 min to allow the beads to adhere to the quartz surface. Rinse channel with ∼100 μl of 4 M KCl to remove excess beads. Should the density of adhered beads on the slide not be sufficient, repeat the initial steps with a larger concentration of microspheres. Seal all four sides of the bead slide with epoxy glue for future use.
2. For channel alignment, image the bead slide with 532 nm excitation in a TIR geometry as described above. Adjust the position of the lens (L1, Fig. 3.2B) closest to the prism to generate a beamspot that evenly illuminates the imaged area. Adjust the focal plane and record a few short movies of different imaging areas of the bead slide.
3. Immobilize the sample in a fresh sample chamber as described above. Prepare 100 μl of a 1:100 dilution of the 100 × Gloxy stock with imaging buffer. Flow the 1 × Gloxy dilution onto the slide and proceed immediately with the data acquisition.
4. Record movies from several different fields of view for construction of the FRET histogram.

4.2.4 Acquiring single-molecule FRET traces in flow experiments
In order to obtain single-molecule FRET traces where an enzymatic reaction is initiated at a well-defined starting time, flow experiments can be carried out. The buffer in the sample cell is exchanged during data acquisition using a syringe pump setup (Fig. 3.2D), for example, to deliver enzyme and

ATP. These types of experiments are particularly well suited for the determination of kinetic properties such as transition rates and intermediate states.

1. Immobilize the sample on a quartz slide with sample chamber as described above. Connect the syringe to a two-way valve that will, depending on its setting, allow for flow through the inlet piece into the sample chamber or bypass the sample chamber for direct loading/emptying of the syringe (Fig. 3.2D). Connect one port of the valve with tubing to the inlet piece that consists of a custom-made brass plate with a rubber O-ring. Fill the syringe with $\sim$300–500 µl of 1 $\times$ Gloxy dilution in imaging buffer and push a small volume through the inlet piece to remove air bubbles inside the tubing. Connect the inlet piece to the sample chamber by screwing it onto the stage such that the O-ring creates a tight connection between syringe pump tubing and the inlet of the sample chamber. Analogously, attach a custom-made brass plate with O-ring to the outlet hole of the sample cell. Direct a short piece of tubing connected to the outlet brass piece to a small disposal vessel on the microscope stage.
2. Using the syringe pump, infuse the sample chamber with $\sim$75–100 µl of 1 $\times$ Gloxy dilution. If the seal on both inlet and outlet of the sample chamber is tight and air was properly removed from the syringe and tubing, no air bubbles should be introduced into the sample chamber during the flow.
3. Record a few movies from different imaging areas to construct a FRET histogram of the sample before the buffer exchange.
4. At 4 °C, set up the reaction mix containing all components required for enzymatic action on the sample. For example, to observe nucleosome remodeling dynamics, prepare a 400 µl remodeling mix in imaging buffer that contains 1 µM to 2 mM ATP, an equal concentration of $MgCl_2$, 1 $\times$ Gloxy, and remodeling enzyme. Load the reaction mix into the syringe using the syringe pump motor and the valve setting that bypasses the sample chamber (Fig. 3.2D). Switch the syringe valve to the other setting and start the data acquisition. After 10 frames, flow the remodeling reaction mix through the sample chamber. Refocus to compensate for any slight defocusing that the flow might introduce and continue data acquisition until a large portion of the fluorescent dye molecules have bleached.
5. Stop data acquisition and collect movies of several different and unbleached imaging areas to construct a FRET histogram after the reaction end-point.

5. DATA ANALYSIS

We use a home-written LabView code to control the CCD camera. Fluorescence intensities are read out from the CCD chip for each pixel in each frame and stored on the hard drive in a single binary movie file.

The calculation of FRET efficiencies requires each fluorescence intensity spot in one channel to be unambiguously paired with its corresponding spot in the other channel. However, due to imperfections in the alignment of optical elements in the setup, the two halves of the CCD camera cannot be overlaid by a mere translation. Instead, an unambiguous alignment is achieved by calculating a polynomial warping map between the two channels that accounts for translocation, shear distortion, rotation, and resizing. The map coefficients are determined by analyzing a short movie (20–30 frames) of a slide with immobilized fluorescent beads that display detectable fluorescence emission in both channels. Using an IDL script, an image is calculated by averaging the first 20 frames of the movie. In this average image, the corresponding positions in donor and acceptor channel are interactively determined for three beads and used to generate an initial mapping between the two channels. The map coefficients for a more rigorous polynomial warping between the two channels are then determined by taking into account all beads on the slide.

Once a map for alignment of the two channels is determined, IDL scripts are used to analyze movie files from the single-molecule FRET experiments in a multistep procedure. First, an overlay movie is generated by mapping, frame by frame, the image of one channel onto the image of the other channel. In this overlay movie, corresponding donor and acceptor signals from the same molecule are found at the same coordinates in donor and acceptor channels. Next, to identify peaks from individual molecules, an average sum image is generated by averaging together the first 20 frames of the overlay movie. After background subtraction, the average sum image is used to identify peaks from individual molecules in either channel. Peaks are initially identified as local intensity maxima. For each peak, the average intensity of all pixels on a ring with a given radius around the maximum is calculated and compared to the maximum intensity at the peak center. Peaks are only accepted if the ratio of average ring intensity to maximum intensity is below a threshold ratio, such that accepted peaks are relatively bright inside and relatively dim outside the ring. The coordinates of all accepted peaks are stored for further analysis. The threshold ratio and ring radius for the peak finding algorithm are determined empirically. An ideal choice of parameters allows for the identification of a large number of individual molecules while overlapping peaks or dim peaks caused by fluorescent debris on the slide are excluded. Once the peak coordinates are identified based on the average sum image, the following analysis uses the overlay movie described above. In each frame of the overlay movie and at each peak position, the integrated fluorescence intensity signal of all pixels within a ring of a given, empirically determined radius is determined for each channel. This procedure yields the donor and acceptor signals for every molecule in each frame and stores them in a single traces file. The intensity information in the traces file can then be used to calculate a FRET time trace for each imaged molecule.

5.1. Calculating the FRET efficiency

The FRET efficiency is calculated from the fluorescence emission intensities measured in donor and acceptor channels. A fraction of the donor signal is expected to leak into the acceptor channel, which causes an apparent increase in the acceptor emission. Therefore, the measured raw acceptor signal $I_{A,raw}$ needs to be corrected by an empirically determined leakage fraction α to yield the acceptor intensity I_A: $I_A = I_{A, raw} - \alpha I_D$, where I_D is the raw donor signal. The leakage fraction α is typically determined by measuring donor and acceptor signals of a molecule that is labeled with a donor dye only, but does not possess an acceptor fluorophore. The FRET efficiency can then be approximated as the ratio of acceptor intensity to the sum of acceptor and donor intensities, $I_A/(I_A + I_D)$. This ratio provides a good estimate of the absolute FRET value if the quantum yields and detection efficiencies of the two fluorescent dyes are very similar, which is the case for the Cy3–Cy5 FRET pair. FRET values cannot typically be used directly to extract absolute distance information, but often an experimental calibration is possible, as outlined below.

5.2. Constructing and analyzing FRET histograms

Before single-molecule FRET traces from individual molecules are analyzed, data from several imaging areas on the slide (fields of view) are combined to yield a distribution of FRET values among many molecules, usually displayed as a histogram. Using a Matlab script, the histogram is constructed by calculating the mean FRET efficiency of the first ~ 10 frames for each individual molecule in each of the imaged areas. FRET histograms are convenient for an initial analysis of a new sample because they aid in determining the different FRET populations present in the sample. In the case of only one FRET state, the FRET histogram will exhibit a single peak. The peak position is then determined in good approximation by fitting the FRET distribution to a single Gaussian. If the sample contains multiple species with different FRET values, the resulting histogram will be more complex. Such a situation could arise if multiple conformations of the same molecule or a mixture of distinct molecules with different FRET efficiencies are present in the sample. For example, the FRET histogram of nucleosomes assembled as described above and shown in Fig. 3.1B displays three distinct peaks (Fig. 3.3A). Since the FRET donor Cy3 is attached to the H2A subunit of the histone octamer, the presence of two H2A subunits in each octamer leads to three distinct populations of nucleosomes with different labeling configurations: (1) Nucleosomes with a single donor on the H2A subunit proximal to the acceptor, giving rise to a peak at high FRET, (2) nucleosomes with a single donor on the distal H2A, yielding a lower FRET level, and (3) nucleosomes with a donor dye on each of the two H2A subunits, yielding an intermediate FRET value. For this reason, the FRET histogram is best fit with three Gaussian peaks. These FRET peaks are well separated in the histogram, allowing us to clearly identify these populations and select a single population for further analysis. FRET distributions often represent a powerful tool to examine overall changes in the sample, for example, induced by enzymatic activity or by alterations in buffer composition or temperature. For example, the FRET histogram obtained from aforementioned nucleosomes displays a dramatic shift toward very low FRET values upon incubation of the slide with the chromatin remodeling enzyme ACF and ATP (Fig. 3.3A). Such a shift is to be expected from the remodeling activity of ACF, which centers mononucleosomes on the DNA substrate and thus reduces the FRET values by increasing the distance between donor and acceptor dyes.

5.3. Analysis of single-molecule FRET time traces

For visual inspection of single-molecule FRET time traces, we use a Matlab script that displays the time evolution of acceptor and donor fluorescence signals as well as the FRET efficiency. In our nucleosome example, a flow experiment where the sample chamber is infused with ATP and ACF yields time traces in which the FRET efficiency decreases over time as a consequence of remodeling. In this case, the analysis can be significantly simplified by selecting time traces from nucleosomes containing a single donor on the proximal H2A (Fig. 3.3B). These nucleosomes display an initially high FRET value and therefore provide a large dynamic range to observe FRET

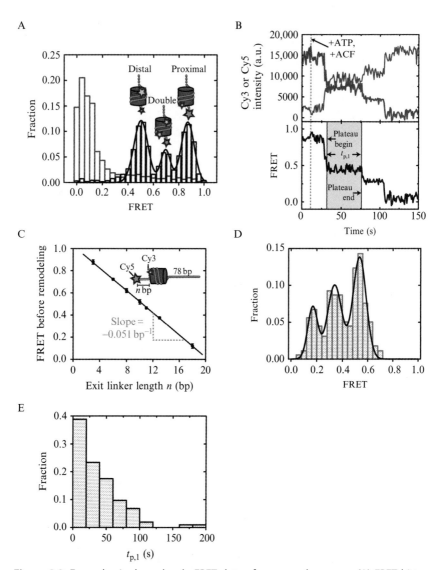

Figure 3.3 Example single-molecule FRET data of mononucleosomes. (A) FRET histogram constructed from many nucleosomes before (blue bars) and after (red bars) ACF-catalyzed remodeling. The three initial peaks (blue bars) arise from the three distinct donor dye labeling configurations (single proximal donor, single distal donor, proximal and distal donor present as described in the text). The black line represents a fit with three Gaussians. Upon remodeling with ACF, the FRET values are shifted toward very low values (red bars). (B) Donor fluorescence (green), acceptor fluorescence (red), and FRET (blue) traces depicting ACF-induced remodeling of an individual nucleosome with a single proximal donor fluorophore. The first intermediate FRET plateau is shaded in yellow. The duration of the first translocation pause is labeled as $t_{p,1}$, and

decrease due to remodeling. The selection process involves the selection of traces that exhibit a mean FRET >0.75 in the first 10 frames prior to infusion of the sample chamber with enzyme and ATP. Among these traces, those displaying a single donor bleaching step are identified. The traces selected this way comprise the entire population of nucleosomes with a single donor on the proximal H2A. Conversely, traces from nucleosomes with donor dyes on both H2A subunits, which exhibit two donor bleaching steps, are excluded.

The observed FRET efficiencies are not absolute but instead depend on the photophysical properties of the fluorophores and on the specific set of filter elements used to optically separate donor and acceptor fluorescence emission. Thus, it is typically difficult to directly convert measured FRET values into physical distances within the sample. In many cases, however, this problem can be circumvented by calibrating the experimental system with a set of samples with varying donor–acceptor distances that are known by design. In the nucleosome example, observed FRET values can be correlated to the octamer position quantitatively by calibration with a nucleosome ruler (Fig. 3.3C). To generate such a ruler, several ruler nucleosome constructs are prepared that differ from the construct shown in Fig. 3.1B only in the varying length of the exit side DNA linker that is terminally labeled with Cy5. Each ruler nucleosome therefore has a distinct but known bp distance between the Cy5 dye and the edge of the nucleosome. For each ruler nucleosome construct, a FRET histogram is obtained, and the peak position of the population with a single proximal donor dye is determined as a function of the bp distance between the Cy5 acceptor dye and the edge of the nucleosome. The calibration ruler obtained this way enables a direct correlation of measured FRET values with the bp position of the histone octamer on the DNA.

Careful analysis of single-molecule FRET time traces enables the determination of key kinetic parameters as a function of experimental conditions

---

begin and end time points are indicated for the first intermediate FRET plateau. (C) Initial FRET value (before remodeling) as a function of the exit linker DNA length (n bp). The black line depicts a linear fit to the data with a slope of -0.051 ± 0.002. The deviation from the expected nonlinear dependence is caused by the flexible linkers connecting the dyes to the nucleosome or the DNA. For this ruler, data from nucleosomes with a single Cy3 dye on the proximal H2A subunit are used. (D) FRET distribution of the pauses. (E) Distribution of $t_{p,1}$, the duration of the first translocation pause. (See Color Insert.)

such as enzyme and substrate concentrations, buffer conditions, or temperature. FRET changes in single-molecule time traces sometimes proceed with pauses or plateaus at distinct FRET states. Distinct FRET states identified in time traces typically correspond to intermediate states that can provide mechanistic insight into the system under study. For example, a typical nucleosome remodeling trace (Fig. 3.3B) shows that translocation by ACF does not proceed at a constant rate. Instead, the FRET time trace displays periods of gradual decrease interrupted by translocation pauses at intermediate FRET values. Once characteristic intermediate FRET states have been identified at the single-trace level, they can be further analyzed by compiling information from many traces into a histogram. In the nucleosome remodeling example, such a histogram can be constructed the following way: First, a traces file is read by a Matlab script, and individual FRET time traces are displayed on screen. In traces that exhibit remodeling, the start and end time points of each intermediate FRET state are identified manually or via an automated algorithm (Fig. 3.3B). Next, an average plateau FRET value is calculated as the mean FRET value in the defined time interval. Average FRET plateau values from many traces and typically several remodeling experiments are then used to construct a histogram that reveals the frequency with which a given FRET plateau value occurs (Fig. 3.3D). In our example, the histogram displays three well-separated peaks, and their center positions are obtained by fitting the histogram with three Gaussians. By comparing with the calibration curve shown in Fig. 3.3C, we found that these three peaks correspond to nucleosome translocation of 7 bp (for the first peak) and 3–4 bp (for the subsequent two peaks).

The duration of characteristic phases in the time traces, such as dwell times of these FRET plateaus or the duration of the transition time between these plateaus, can be analyzed in a very similar way. In this case, start and end time points of the phases under study are determined, and a histogram is constructed from the interval durations. For example, Fig. 3.3E shows a dwell time histogram for the duration of the first pause $(t_{p,1})$, as indicated in Fig. 3.3B. The shape of the dwell time distribution often can be used to infer kinetic properties of the transition. For example, if the distribution follows a single-exponential decay, the corresponding transition likely has only one rate-limiting step; conversely, a lagged exponential implies more than one rate-limiting step.

In systems with more complex dynamic transitions between multiple distinct FRET states, the identification of these states can be facilitated by applying a hidden Markov modeling routine (HaMMy) (McKinney, Joo,

& Ha, 2006). In this approach, the time-binned FRET trace is modeled as a Markov process of state-to-state transitions that all display single-exponential decay kinetics.

The precise details of single-molecule FRET analysis depend on the system under study, and the analysis tools described here are intended to provide a starting point for further analysis tailored to the specific questions being addressed.

6. SUMMARY

During the past decade, single-molecule FRET has been established as a powerful biophysical technique for the mechanistic study of a wide range of biological systems. Here we provide a detailed description of how to set up, carry out, and analyze single-molecule FRET measurements, with a particular emphasis on experiments that require a rapid buffer exchange during data acquisition. As an illustrative example, we describe the observation of nucleosome translocation by single-molecule FRET. While some of the sections regarding sample preparation are specific to this particular example, most procedures and protocols described here are readily adapted to a variety of experimental systems.

ACKNOWLEDGMENTS

We thank William Hwang and Bryan Harada for critical reading of the manuscript. This work is in part supported by the Howard Hughes Medical Institute. X. Z. is a Howard Hughes Medical Institute Investigator. S. D. is a Merck fellow of the Jane Coffin Childs Memorial Fund for Medical Research.

REFERENCES

Amitani, I., Baskin, R. J., & Kowalczykowski, S. C. (2006). Visualization of Rad54, a chromatin remodeling protein, translocating on single DNA molecules. *Molecular Cell, 23*, 143–148.
Axelrod, D. (2003). Total internal reflection fluorescence microscopy in cell biology. *Biophotonics, Part B, 361*, 1–33.
Baruah, H., Puthenveetil, S., Choi, Y. A., Shah, S., & Ting, A. Y. (2008). An engineered aryl azide ligase for site-specific mapping of protein-protein interactions through photo-cross-linking. *Angewandte Chemie International Edition, 47*, 7018–7021.
Becker, P. B., & Horz, W. (2002). ATP-dependent nucleosome remodeling. *Annual Review of Biochemistry, 71*, 247–273.
Benesch, R. E., & Benesch, R. (1953). Enzymatic removal of oxygen for polarography and related methods. *Science, 118*, 447–448.
Blosser, T. R., Yang, J. G., Stone, M. D., Narlikar, G. J., & Zhuang, X. W. (2009). Dynamics of nucleosome remodelling by individual ACF complexes. *Nature, 462*, 1022–1027.

Clapier, C. R., & Cairns, B. R. (2009). The biology of chromatin remodeling complexes. *Annual Review of Biochemistry, 78,* 273–304.
Dawson, P. E., Muir, T. W., Clarklewis, I., & Kent, S. B. H. (1994). Synthesis of proteins by native chemical ligation. *Science, 266,* 776–779.
Dyer, P. N., Edayathumangalam, R. S., White, C. L., Bao, Y. H., Chakravarthy, S., Muthurajan, U. M., et al. (2004). Reconstitution of nucleosome core particles from recombinant histones and DNA. *Methods in Enzymology, 375,* 23–44.
Fernandez-Suarez, M., Baruah, H., Martinez-Hernandez, L., Xie, K. T., Baskin, J. M., Bertozzi, C. R., et al. (2007). Redirecting lipoic acid ligase for cell surface protein labeling with small-molecule probes. *Nature Biotechnology, 25,* 1483–1487.
Fernandez-Suarez, M., & Ting, A. Y. (2008). Fluorescent probes for super-resolution imaging in living cells. *Nature Reviews Molecular Cell Biology, 9,* 929–943.
Flaus, A., & Owen-Hughes, T. (2004). Mechanisms for ATP-dependent chromatin remodelling: Farewell to the tuna-can octamer? *Current Opinion in Genetics & Development, 14,* 165–173.
Forster, T. (1949). Experimentelle und theoretische Untersuchung des zwischenmolekularen Ubergangs von Elektronenanregungsenergie. *Zeitschrift für Naturforschung A, 4,* 321–327.
Griffin, B. A., Adams, S. R., & Tsien, R. Y. (1998). Specific covalent labeling of recombinant protein molecules inside live cells. *Science, 281,* 269–272.
Ha, T. (2001). Single-molecule fluorescence resonance energy transfer. *Methods, 25,* 78–86.
Ha, T., Enderle, T., Ogletree, D. F., Chemla, D. S., Selvin, P. R., & Weiss, S. (1996). Probing the interaction between two single molecules: Fluorescence resonance energy transfer between a single donor and a single acceptor. *Proceedings of the National Academy of Sciences of the United States of America, 93,* 6264–6268.
Ha, T., Rasnik, I., Cheng, W., Babcock, H. P., Gauss, G. H., Lohman, T. M., et al. (2002). Initiation and re-initiation of DNA unwinding by the Escherichia coli Rep helicase. *Nature, 419,* 638–641.
Hamiche, A., Sandaltzopoulos, R., Gdula, D. A., & Wu, C. (1999). ATP-dependent histone octamer sliding mediated by the chromatin remodeling complex NURF. *Cell, 97,* 833–842.
Heyduk, T. (2002). Measuring protein conformational changes by FRET/LRET. *Current Opinion in Biotechnology, 13,* 292–296.
Ito, T., Bulger, M., Pazin, M. J., Kobayashi, R., & Kadonaga, J. T. (1997). ACF, an ISWI−containing and ATP−utilizing chromatin assembly and remodeling factor. *Cell, 90,* 145–155.
Kapanidis, A. N., Ebright, Y. W., & Ebright, R. H. (2001). Site-specific incorporation of fluorescent probes into protein: Hexahistidine-tag-mediated fluorescent labeling with (Ni2+: Nitrilotriacetic acid)(n)-fluorochrome conjugates. *Journal of the American Chemical Society, 123,* 12123–12125.
Kassabov, S. R., Henry, N. M., Zofall, M., Tsukiyama, T., & Bartholomew, B. (2002). High-resolution mapping of changes in histone-DNA contacts of nucleosomes remodeled by ISW2. *Molecular and Cellular Biology, 22,* 7524–7534.
Kornberg, R. D. (1974). Chromatin structure—Repeating unit of histones and DNA. *Science, 184,* 868–871.
Kornberg, R. D., & Lorch, Y. (1999). Twenty-five years of the nucleosome, fundamental particle of the eukaryote chromosome. *Cell, 98,* 285–294.
Langst, G., Bonte, E. J., Corona, D. F. V., & Becker, P. B. (1999). Nucleosome movement by CHRAC and ISWI without disruption or trans-displacement of the histone octamer. *Cell, 97,* 843–852.
Lee, K. M., & Narlikar, G. (2001). Assembly of nucleosomal templates by salt dialysis. *Current Protocols in Molecular Biology,* Chapter 21: Unit 21.6.

Li, G., Levitus, M., Bustamante, C., & Widom, J. (2005). Rapid spontaneous accessibility of nucleosomal DNA. *Nature Structural & Molecular Biology, 12*, 46–53.
Lia, G., Praly, E., Ferreira, H., Stockdale, C., Tse-Dinh, Y. C., Dunlap, D., et al. (2006). Direct observation of DNA distortion by the RSC complex. *Molecular Cell, 21*, 417–425.
Luger, K., Rechsteiner, T. J., & Richmond, T. J. (1999). Preparation of nucleosome core particle from recombinant histones. *Chromatin, 304*, 3–19.
McKinney, S. A., Joo, C., & Ha, T. (2006). Analysis of single-molecule FRET trajectories using hidden Markov modeling. *Biophysical Journal, 91*, 1941–1951.
Mekler, V., Kortkhonjia, E., Mukhopadhyay, J., Knight, J., Revyakin, A., Kapanidis, A. N., et al. (2002). Structural organization of bacterial RNA polymerase holoenzyme and the RNA polymerase-promoter open complex. *Cell, 108*, 599–614.
Mendel, D., Cornish, V. W., & Schultz, P. G. (1995). Site-directed mutagenesis with an expanded genetic-code. *Annual Review of Biophysics and Biomolecular Structure, 24*, 435–462.
Miller, L. W., Cai, Y. F., Sheetz, M. P., & Cornish, V. W. (2005). In vivo protein labeling with trimethoprim conjugates: A flexible chemical tag. *Nature Methods, 2*, 255–257.
Muir, T. W., Sondhi, D., & Cole, P. A. (1998). Expressed protein ligation: A general method for protein engineering. *Proceedings of the National Academy of Sciences of the United States of America, 95*, 6705–6710.
Narlikar, G. J., Fan, H. Y., & Kingston, R. E. (2002). Cooperation between complexes that regulate chromatin structure and transcription. *Cell, 108*, 475–487.
Nimonkar, A. V., Amitani, I., Baskin, R. J., & Kowalczykowski, S. C. (2007). Single molecule imaging of Tid1/Rdh54, a Rad54 homolog that translocates on duplex DNA and can disrupt joint molecules. *Journal of Biological Chemistry, 282*, 30776–30784.
Popp, M. W., Antos, J. M., Grotenbreg, G. M., Spooner, E., & Ploegh, H. L. (2007). Sortagging: A versatile method for protein labeling. *Nature Chemical Biology, 3*, 707–708.
Prasad, T. K., Robertson, R. B., Visnapuu, M. L., Chi, P., Sung, P., & Greene, E. C. (2007). A DNA–translocating Snf2 molecular motor: Saccharomyces cerevisiae Rdh54 displays processive translocation and extrudes DNA loops. *Journal of Molecular Biology, 369*, 940–953.
Rasnik, I., McKinney, S. A., & Ha, T. (2006). Nonblinking and longlasting single-molecule fluorescence imaging. *Nature Methods, 3*, 891–893.
Rice, S., Lin, A. W., Safer, D., Hart, C. L., Naber, N., Carragher, B. O., et al. (1999). A structural change in the kinesin motor protein that drives motility. *Nature, 402*, 778–784.
Roy, R., Hohng, S., & Ha, T. (2008). A practical guide to single-molecule FRET. *Nature Methods, 5*, 507–516.
Shahian, T., & Narlikar, G. J. (2012). Analysis of changes in nucleosome conformation using fluorescence resonance energy transfer. *Methods in Molecular Biology, 833*, 337–349.
Shundrovsky, A., Smith, C. L., Lis, J. T., Peterson, C. L., & Wang, M. D. (2006). Probing SWI/SNF remodeling of the nucleosome by unzipping single DNA molecules. *Nature Structural & Molecular Biology, 13*, 549–554.
Sirinakis, G., Clapier, C. R., Gao, Y., Viswanathan, R., Cairns, B. R., & Zhang, Y. (2011). The RSC chromatin remodelling ATPase translocates DNA with high force and small step size. *The EMBO Journal, 30*, 2364–2372.
Smith, C. L., & Peterson, C. L. (2005). ATP-dependent chromatin remodeling. *Current Topics in Developmental Biology, 65*(65), 115–148.
Stryer, L., & Haugland, R. P. (1967). Energy transfer—A spectroscopic ruler. *Proceedings of the National Academy of Sciences of the United States of America, 58*, 719–726.
Tanaka, T., Yamamoto, T., Tsukiji, S., & Nagamune, T. (2008). Site-specific protein modification on living cells catalyzed by Sortase. *ChemBioChem, 9*, 802–807.
Tsukiyama, T., Palmer, J., Landel, C. C., Shiloach, J., & Wu, C. (1999). Characterization of the imitation switch subfamily of ATP-dependent chromatin-remodeling factors in Saccharomyces cerevisiae. *Genes & Development, 13*, 686–697.

Tsukiyama, T., & Wu, C. (1997). Chromatin remodeling and transcription. *Current Opinion in Genetics & Development, 7*, 182–191.
Yang, J. G., Madrid, T. S., Sevastopoulos, E., & Narlikar, G. J. (2006). The chromatin-remodeling enzyme ACF is an ATP-dependent DNA length sensor that regulates nucleosome spacing. *Nature Structural & Molecular Biology, 13*, 1078–1083.
Zhang, Y., Smith, C. L., Saha, A., Grill, S. W., Mihardja, S., Smith, S. B., et al. (2006). DNA translocation and loop formation mechanism of chromatin remodeling by SWI/SNF and RSC. *Molecular Cell, 24*, 559–568.
Zhuang, X., Bartley, L. E., Babcock, H. P., Russell, R., Ha, T., Herschlag, D., et al. (2000). A single-molecule study of RNA catalysis and folding. *Science, 288*, 2048–2051.

SECTION 2

Higher Order Chromatin Interactions

CHAPTER FOUR

4C Technology: Protocols and Data Analysis

Harmen J.G. van de Werken, Paula J.P. de Vree, Erik Splinter, Sjoerd J.B. Holwerda, Petra Klous, Elzo de Wit, Wouter de Laat[1]

Hubrecht Institute-KNAW and University Medical Center Utrecht, Utrecht, The Netherlands
[1]Corresponding author: e-mail address: w.delaat@hubrecht.eu

Contents

1. Introduction	90
2. 4C Template Preparation	92
2.1 Principles of 4C technology	92
2.2 Cross-linking and cell lysis	94
2.3 RE digestion of cross-linked chromatin	96
2.4 Ligation of cross-linked DNA fragments	97
2.5 Reversal of cross-links	98
2.6 Trimming the size of DNA circles: Secondary RE digestion and ligation	99
3. 4C-Seq Primer Design and PCR	100
3.1 Primer design 4C sequencing	100
3.2 PCR of 4C template	101
4. High-Throughput Sequencing of 4C PCR Products	103
5. Data Analysis	103
5.1 Mapping of 4C-seq reads to genome	103
5.2 Statistics and quality control of the 4C-seq	104
5.3 Data analysis in *cis*	107
5.4 *Trans* analysis	110
Acknowledgments	110
References	110

Abstract

Chromosome conformation capture (3C) technology and its genome-wide derivatives have revolutionized our knowledge on chromatin folding and nuclear organization. 4C-seq Technology combines 3C principles with high-throughput sequencing (4C-seq) to enable for unbiased genome-wide screens for DNA contacts made by single genomic sites of interest. Here, we discuss in detail the design, application, and data analysis of 4C-seq experiments. Based on many hundreds of different 4C-seq experiments, we define criteria to assess data quality and show how different restriction enzymes and cross-linking conditions affect results. We describe in detail the mapping strategy of 4C-seq reads and show advanced strategies for data analysis.

1. INTRODUCTION

Grasping the size of our genome in relation to that of the cell nucleus, which contains it, is completely impossible. The three billion base pairs of nucleotides that form the genome, typically indicated by the letters G, A, T, and C, spell out a sentence of 9000 km if using the average times roman font size 12. In reality, even the best microscope cannot visualize single nucleotides, let alone read the code. Still, if you would stretch the genome, it would measure an impressive 2 m. This 2 m of DNA is typically stored in a nucleus with a diameter of roughly 10 µm. Adding up all DNA in your head can therefore form a rope that sizes 40,000 times the distance from the earth to the moon.

While incomprehensible, these dimensions do teach us that a tremendous degree of packing, bending, and twisting is required to fit the genome in a cell nucleus. Until a decade ago, we had little understanding of the details of the shape of our genome. The development of chromosome conformation capture (3C) technology (Dekker, Rippe, Dekker, & Kleckner, 2002), however, changed our ability to study genome topology, and consequently altered our view on genome function and nuclear organization. 3C Experiments demonstrated that remote mammalian enhancers physically loop to the genes they control (Tolhuis, Palstra, Splinter, Grosveld, & de Laat, 2002). In addition, it has been shown that transcription factors are responsible for the formation of chromatin loops between genes and regulatory DNA sequences (Drissen et al., 2004; Spilianakis & Flavell, 2004), and that other proteins exist that bring together genomic sites, which are not genes (Hadjur et al., 2009; Splinter et al., 2006).

3C Technology is referred to as a one-versus-one method as it analyzes interactions between selected pairs of DNA sites. The strategy relies on formaldehyde cross-linking of proteins to proteins and to DNA, the subsequent digestion of cross-linked DNA by restriction enzymes (REs), the ligation of cross-linked DNA fragments, and quantitative frequency assessment of selected ligation junctions by PCR. If two distal sites on the linear chromosome form more ligation junctions with each other than with intervening sequences, a chromatin loop is demonstrated to exist between these sites *in vivo* (Dekker, 2006; Simonis, Kooren, & de Laat, 2007).

The availability of 3C technology triggered the further development of genome-scale variants thereof. These include 4C technology (one-versus-all) (Simonis et al., 2006; Zhao et al., 2006), 5C technology (many-versus-many) (Dostie et al., 2006), ChIA-PET (an approach combining chromatin immunoprecipitation (ChIP) and 3C) (Fullwood et al., 2009), and Hi-C (all-versus-all) (Lieberman-Aiden et al., 2009). Each strategy has unique advantages and disadvantages, and the method of choice relies on the specific research question asked (de Wit & de Laat, 2012; van Steensel & Dekker, 2010). Collectively, they have been quickly adopted by the large community studying nuclear organization, gene regulation, and DNA replication, leading to an ever-increasing body of work showing how the shape of the genome affects its function (Ong & Corces, 2011; Splinter et al., 2011).

Here, we will focus on 4C technology, a technology designed to study in a detailed and unbiased manner the DNA contacts made across the genome by a given genomic site of interest. 4C Technology has been applied to demonstrate that individual gene loci can be engaged in many long-range DNA contacts with loci elsewhere on the same chromosome and on other chromosomes (Simonis et al., 2006). It confirmed at a much higher resolution, observations made earlier by microscopy that active and inactive chromatin separately in the nucleus (Noordermeer, Leleu, et al., 2011; Simonis et al., 2006; Splinter et al., 2011). In *Drosophila*, 4C was used to demonstrate that genes are bound by polycomb group (PcG) proteins, and are far apart on the chromosome, frequently meet in nuclear space (Bantignies et al., 2011; Tolhuis et al., 2011). Long-range interactions among conserved noncoding sequences were identified by 4C (Robyr et al., 2011), and it was applied to demonstrate that regulatory sequences cannot autonomously decide where to go in the nucleus; for their overall spatial positioning, they instead rely on their chromosomal context (Hakim et al., 2011; Noordermeer, de Wit, et al., 2011). 4C Technology and variants thereof have not only been applied to identify contacts between larger genomic regions, but now also started to be used for the identification of more local and defined interactions between regulatory sequences (Lower et al., 2009; Montavon et al., 2011; Soler et al., 2010). A robust standard protocol for this still seems lacking though. Finally, 4C not only allows the detection of three-dimensional DNA contacts, it also, and primarily, picks up sequences that are close in space because they happen to be proximal on the linear chromosome template. This realization has lead to an unexpected application of 4C technology in molecular diagnostics, as a robust strategy for the identification and fine-mapping of balanced and unbalanced chromosomal rearrangements near sites of interest. In leukemia, for example, 4C already enabled the detection of multiple novel translocations and inversions and the discovery of several new oncogenes (Homminga et al., 2011; Simonis et al., 2009).

Based on many hundreds of different 4C experiments, most of them unpublished, we will explain how to design, perform, and analyze 4C experiments, and how to judge the quality.

2. 4C TEMPLATE PREPARATION

2.1. Principles of 4C technology

4C technology enables the identification of all regions in the genome that contact a genomic site of interest. We refer to such sites of interest as "viewpoints" or "baits," while contacting regions that are cross-linked and ligated to the viewpoint are called "captures." An outline of 4C technology is given in Fig. 4.1. In brief, cells are treated with formaldehyde, which cross-links proteins to proteins and to DNA. Cross-linked chromatin is subsequently digested with a primary RE that creates "hairballs" (or aggregates) of cross-linked DNA. Next, chromatin is diluted and religated to fuse the ends of DNA fragments present in the same "hairball." Large circles containing multiple cross-linked restriction fragments are usually the outcome of this first ligation step. A second round of digestion and ligation is performed to trim these large circles and make them suitable for PCR. The trimming of the DNA circles is carried out after the removal of cross-links by heating, DNA is further digested by a second RE (usually a four cutter) that digest the RE fragments from the primary RE into smaller fragments-ends (fragends). These fragends are religated under diluted conditions to create much smaller DNA circles, which contain mostly a primary ligation junction. Inverse PCR primers that are specific for the viewpoint and have 5′ adapter overhangs compatible with Illumina sequencing (Fig. 4.2) are then used for amplifying captures that are ligated to the viewpoint. Subsequently, high-throughput sequencing is used to read the captured regions.

The size of viewpoints and captures is defined by the primary REs used. Both ends of a fragment can ligate to other fragments present in a cross-linked chromatin aggregate; the junctions formed provide a means to identify cross-linked pairs of fragments. A given captured fragment can be identified from both ends. Similarly, for a viewpoint fragment, it does

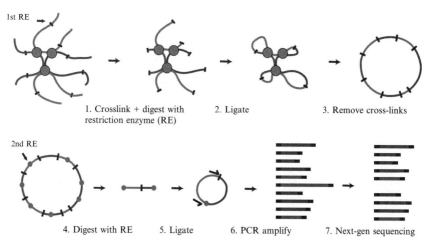

Figure 4.1 4C technology. Outline of the strategy is provided. After cross-linking by formaldehyde and digestion with the first restriction enzyme (1st RE), "hairballs" of cross-linked DNA are created (1). Chromatin is diluted and religated to fuse the ends of DNA fragments present in the same "hairball" (2). The ultimate outcome of this ligation event is large DNA circles encompassing multiple restriction fragments. Cross-links are removed by heating (3); DNA is digested by a second RE (usually a four cutter) (4) and religated under diluted conditions to create small DNA circles, most of which carry a primary ligation junction. Inverse PCR primers specific for the fragment of interest ("viewpoint," "bait") and carrying 5′ adapter overhangs for next-gen sequencing allow amplification of all its captured sequences followed by high-throughput sequencing. (See Color Insert.)

not matter which side is chosen for the analysis of ligation junctions. Another aspect considering is that 4C and other 3C technologies are based on the analysis of populations of cells. Diploid cells generally have two copies of each fragment, including the viewpoint that each can only fuse with one capture. A single cell therefore contributes maximally two captures per viewpoint. For 4C-seq to produce a reproducible contact profile (i.e., collection of captures), analysis should be directed to many cells. We routinely analyze the equivalent of 0.5 million cells, representing roughly 1 million ligation junctions. We start our sample preparation, however, with more cells, preferably 10 million, as the procedure is rather laborious and these amounts best guarantee reproducible DNA yields in the end. For correlating traits such as expression status to DNA contact profiles, relatively homogeneous cell populations need to be analyzed.

Below a detailed protocol will be given and, where relevant, background information will be provided.

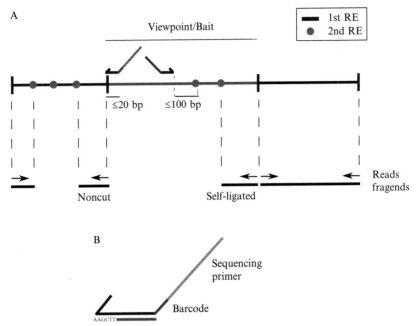

Figure 4.2 Details of the 4C PCR strategy. (A) First (read) primer is designed on top of the primary restriction site; second primer is designed within 100 base pairs of neighboring secondary restriction site on the viewpoint fragment (red). Green and blue: neighboring restriction fragments. Black horizontal bars indicate the amplifiable genomic parts in 4C (so-called fragends). Horizontal arrows show start and direction of sequencing reads. (B) Details of first primer, designed on top of a *Hin*dIII restriction site. Barcode (2–3 nucleotides) allows pooling of multiple 4C experiments from the same viewpoint in a single sequencing Illumina lane. (See Color Insert.)

2.2. Cross-linking and cell lysis

Background: Formaldehyde is able to cross-link protein to protein and protein to DNA and acts within 3 Å. Active and inactive chromatin regions both readily find corresponding chromatin regions elsewhere in the genome in 4C and Hi-C experiments (Lieberman-Aiden et al., 2009), suggesting that there is no dramatic bias in cross-link ability between these distinct states of chromatin. It is known, however, that regulatory DNA sequences are relatively difficult to cross-link: formaldehyde-assisted isolation of regulatory elements uses this property to identify these sites (Giresi, Kim, McDaniell, Iyer, & Lieb, 2007).

4C protocol: start

Our protocol has been optimized for 10 million cells and a *Hin*dIII/ *Dpn*II digestion, but can be performed with other REs as well. Before starting the 4C protocol, prepare a lysis buffer fresh and store it on ice.

10 ml lysis buffer consists of:
- 500 µl of 1 M Tris pH 7.5 (50 mM Tris)
- 300 µl of 5 M NaCl (150 mM NaCl)
- 100 µl of 0.5 M EDTA (5 mM EDTA)
- 250 µl of 20% NP-40 (0.5% NP-40)
- 100 µl Triton X-100 (1% TX-100) 200 µl of 50 × protease inhibitors.

1. Count the cells and centrifuge for 5 min at 1100 rounds per minute (RPM) (280 g) at room temperature (RT).
2. Discard the supernatant and resuspend the pellet in 5 ml phosphate buffered saline (PBS)/10% fetal calf serum (FCS) (1 × 10^7 cells).
3. Add 5 ml of 4% formaldehyde in PBS/10% FCS and incubate (2% final concentration), while tumbling, for 10 min at RT.
4. Add 1.425 ml of 1 M glycine and put the tubes immediately on ice to quench the cross-linking reaction. Continue immediately with step 5.
5. Centrifuge for 8 min at 1300 RPM (400 g) at 4 °C and remove all the supernatant.
6. Resuspend the pellet in 5 ml cold lysis buffer and incubate for 10 min on ice.
7. Determine if cell lysis is complete: Mix 3 µl of cells with 3 µl of methyl green–pyronin staining on a microscope slide and cover with a coverslip. View staining under a microscope. Cytoplasm stains pink and the nuclei/DNA stains blue/green. If cell lysis is not complete, use a douncer to increase efficiency.
8. Centrifuge for 5 min at 1800 RPM (750 g) at 4 °C and remove the supernatant.
9. Cells can be stored at −80 °C:
    9.1. Take up the cell pellets in PBS and divide over 1.5-ml safe lock tubes, make aliquots of 1 × 10^7 cells per tube.
    9.2. Centrifuge for 2 min at 2400 RPM (600 g) at 4 °C.
    9.3. Take off the supernatant; freeze the pellets in liquid nitrogen and store at −80 °C.
10. Take up the pellet in 440 µl Milli-Q and continue with step 11.

2.3. RE digestion of cross-linked chromatin

Background: The first digestion is performed on cross-linked material: this digest defines the sizes and position of fragments between which contacts (cross-links) are analyzed. Most REs are not able to properly digest cross-linked material. We use the six base pair-cutting enzymes *Hin*dIII, *Eco*RI, *Bgl*II or the more frequently cutting four base pair-recognizing enzymes *Dpn*II, *Nla*III, and *Csp*6I (none blocked by CpG methylation), as these in our hands digest cross-linked chromatin well. The four bp-cutters are, in theory, able to generate a higher resolution 4C-seq profile because these enzymes cut every 256 bp instead of every 4 kb, as six bp-cutters do. This increases the power of statistical analysis as 16 × more fragments can be analyzed per genomic distance and directs analysis to fragments that are similarly sized as most regulatory DNA sequences (mostly considered to be several hundred base pairs). Theoretically, they, therefore, allow pinpointing local interactions between regulatory elements (van de Werken, in preparation).

To control for digestion efficiency, running digested and undigested DNA in parallel on an agarose gel should be applied routinely. A qPCR analysis with primers across multiple restriction sites can be used for precise calculation of digestion efficiency. The choice of the primary RE is also determined by the ability to design 4C primers around the genomic site of interest. Viewpoints preferably contain such site, or at least are close to it. We ask viewpoint fragments to be at least 500 bp, a size arbitrarily considered large enough to frequently be subject to cross-linking (see primer design).

4C protocol: continued (1)
11. Add 60 µl of 10 × RE buffer.
12. Place the tube at 37 °C and add 15 µl of 10% SDS.
13. Incubate for 1 h at 37 °C while shaking at 900 RPM.
14. Add 75 µl of 20% Triton X-100.
15. Incubate for 1 h at 37 °C while shaking at 900 RPM.
16. Take a 5-µl aliquot of the sample as the "undigested control."
17. Add 200 U RE (*Hin*dIII) and incubate for 4 h at 37 °C while shaking at 900 RPM.
18. Add 200 U RE and incubate (*Hin*dIII) o/n at 37 °C while shaking at 900 RPM.
19. Add 200 U RE and incubate (*Hin*dIII) for 4 h at 37 °C while shaking at 900 RPM.
20. Take a 5-µl aliquot of the sample as the "digested control."
21. Determine the digestion efficiency:

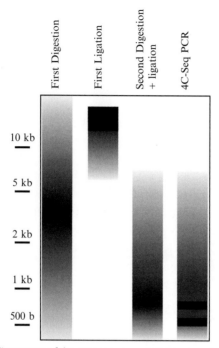

Figure 4.3 Intermediate steps of the 4C procedure. Artist impression of expected results obtained after the first digestion with a six cutter (cross-link reversed DNA aliquot; lane 1), first ligation (cross-link reversed DNA aliquot; lane 2), second digestion and ligation (lane 3), and PCR (lane 4). PCR often results in a smear of DNA with two abundant products, being the products of noncut DNA and the self-ligated viewpoint fragment.

21.1. Add 90 μl of 10 mM Tris–HCl (pH 7.5) to the 5-μl samples from steps 16 and 20.
21.2. Add 5 μl Prot K (10 mg/ml) and incubate for 1 h at 65 °C.
21.3. Load ~20 μl on a 0.6% agarose gel.
21.4. If digestion is of good quality, proceed with step 22. If digestion is of poor quality, repeat steps 17–21.

DNA fragments collected after the first digestion can be separated and viewed on an agarose gel (Fig. 4.3, lane 1).

2.4. Ligation of cross-linked DNA fragments

Background: Most schematic illustrations of 3C or 4C methodologies depict the result of cross-linking and digestion as two bridged fragments, but in fact the cross-linked aggregates consist of many more fragments (Fig. 4.1). Many ligation junctions can therefore be made in a single aggregate, resulting in high molecular weight DNA after ligation (Fig. 4.3). Although most of these ligation products are expected to be circles and therefore potential templates for 4C PCR, PCR is dramatically less efficient on these large circles. Further processing of the DNA is therefore recommended.

4C protocol: continued (2)
22. Inactivate the enzyme (if necessary) by incubating for 20 min at 65 °C.
23. Transfer the samples to a 50-ml Falcon tube.
24. Add 700 µl of 10× ligation buffer.
25. Add Milli-Q to 7 ml.
26. Add 10 µl ligase (Roche; 5 U/µl) and incubate o/n at 16 °C.
27. Take a 100-µl aliquot of the sample as the "ligation control."
28. Determine the ligation efficiency:
    28.1. Add 5 µl Prot K (10 mg/ml) and incubate for 1 h at 65 °C.
    28.2. Load ~20 µl on a 0.6% agarose gel.
    28.3. If ligation is of good quality, proceed with step 29. If ligation is poor of quality, add some fresh ATP and repeat steps 26–28.

DNA fragments collected after the first ligation can be separated, viewed, and judged on an agarose gel (Fig. 4.3, lane 2).

2.5. Reversal of cross-links

4C protocol: continued (3)
29. Add 30 µl Prot K (10 mg/ml) and de-cross-link o/n at 65 °C.
30. Add 30 µl RNAse A (10 mg/ml) and incubate for 45 min at 37 °C.
31. Add 7 ml phenol–chloroform and mix vigorously.
32. Centrifuge for 15 min at 3750 RPM (3300 g) at RT.
33. Transfer the water phase to a new 50-ml tube and add:
    33.1. 7 ml Milli-Q.
    33.2. 1.5 ml of 2 M NaAC pH 5.6.
    33.3. 35 ml of 100% ethanol.
34. Mix and incubate at −80 °C until the sample is completely frozen.
35. Spin for 20 min at 9000 RPM (8300 g) at 4 °C.
36. Remove the supernatant and add 10 ml of cold 70% ethanol.
37. Centrifuge for 15 min at 3750 RPM (3300 g) at 4 °C.
38. Remove the supernatant and briefly dry the pellet at RT.
39. Dissolve the pellet in 150 µl of 10 mM Tris–HCl pH 7.5 at 37 °C.
40. Continue with step 41 or store samples at −20 °C.

2.6. Trimming the size of DNA circles: Secondary RE digestion and ligation

Background: The second digestion of the template is required to trim the large circles and isolate ligation junctions in smaller fragments that can be PCR amplified. A frequent four bp-cutter is therefore used as the secondary enzyme. The RE of choice needs to be insensitive to CpG methylation, preferably create cohesive ("sticky") ends and recognizing sites with 50% GC content, to ensure a more homogeneous digestion across the genome. We mostly use *Dpn*II, *Nla*III, or *Csp*6I (*Cvi*QI). The actual choice between these enzymes then relies on the viewpoint fragment and the ability of the secondary enzyme to generate a viewpoint "fragend" for which primers can be designed (see primer design).

4C protocol: continued (4)

41. To 150 μl of single digestion/ligation "3C" sample ($\sim 1 \times 10^7$ cells), add:
    41.1. 50 μl of 10× *Dpn*II restriction buffer.
    41.2. 50 U *Dpn*II enzyme.
    41.3. Milli-Q to 500 μl.
42. Incubate o/n at 37 °C.
43. Take a 5-μl aliquot of the sample as the "digestion control."
44. Determine the digestion efficiency:
    44.1. Add 95 μl of 10 m*M* Tris (pH 7.5) to the 5 μl sample from step 43.
    44.2. Load ~ 20 μl on a 0.6% agarose gel.
    44.3. If digestion is of good quality, proceed with step 45. If digestion is of poor quality, repeat steps 41.2, 42–44.
45. Inactivate the enzyme by incubating at 65 °C for 25 min.
46. Transfer the sample to a 50-ml tube and add:
    46.1. 1.4 ml of 10× ligation buffer.
    46.2. 20 μl ligase (100 U).
    46.3. *x* ml Milli-Q to 14 ml.
47. Ligate o/n at 16 °C and add:
    47.1. 1.4 ml of 2 *M* NaAC (pH 5.6).
    47.2. 14 μl glycogen.
    47.3. 35 ml 100% ethanol.
48. Store at -80 °C until completely frozen.
49. Spin for 20 min at 9000 RPM (8300 *g*) at 4 °C.
50. Remove the supernatant and add 15 ml of cold 70% ethanol.
51. Spin for 15 min at 3750 RPM (3300 *g*) at 20 °C.
52. Remove the supernatant and briefly dry the pellet at RT.

53. Dissolve the pellet in 150 μl of 10 mM Tris pH 7.5 at 37 °C.
54. Purify the samples with the QIAquick PCR purification kit.
    Use three columns per sample, maximum binding capacity is 10 μg DNA per column. Elute the columns with 50 μl of 10 mM Tris–HCl pH 7.5 and pool the samples.
55. Measure the concentration with NanoDrop.
56. Store the samples at −20 °C or continue with step 57.

DNA fragments collected after the second ligation (4C template) can be separated and viewed on an agarose gel (Fig. 4.3, lane 3).

The purified DNA sample is now ready for 4C PCR amplification.

3. 4C-Seq PRIMER DESIGN AND PCR

3.1. Primer design 4C sequencing

The RE fragends captured by the viewpoint are amplified by an inverse PCR. The primers are designed outward on the viewpoint (Fig. 4.2A). To circumvent the need for further library preparation steps necessary for Illumina sequencing, the primers are designed with 5′ overhangs encoding the Illumina single-end sequence adapter P5 and P7. As a result of this strategy, each read from the sequencer first shows the PCR primer sequence (i.e., the part complementary to the viewpoint) and then the sequence captured by the viewpoint. We design the reading primer (with P5 adapter) on top of the first RE recognition site (Fig. 4.2B), thus maximizing the number of captured fragend bases per read. Directing reading to the first RE site is also important as it prevents analysis of random ligations that can occur during the second ligation step. The reading primer can be extended with a barcode (Fig. 4.2B). A barcode is useful when multiple 4C experiments are using the same viewpoint and are loaded on a single Illumina lane. That is, when 4C experiments with a single viewpoint are generated under different conditions or with different cell types or tissues, we recommend a barcode between 2 and 3 nt, which is enough to distinguish the different experiment and has no dramatic effect to the mappability of the Illumina reads. The nonreading primer is designed with a length between 18 and 27 bp and as close to, but maximal 100 bp away from, the second RE recognition site. This procedure minimizes the PCR bias that can occur for long PCR products. Primer3 (Rozen & Skaletsky, 2000) can be used to design the primer pairs and the result can be checked for uniqueness with megablast (Zhang, Schwartz, Wagner, & Miller, 2000) using the settings (-p 88.88 -W 12 -e 1 -F T) or any another alignment tool to the genome of interest. The reading primer should be unique while the other primer can have maximum three perfect hits to the genome. Moreover, the primers should have less than 30 hits with more than 16/18 (88.88%) identity, aiming not to amplify repeats. As mentioned above, primers are exclusively designed for viewpoints that are at least 500 bp. For six cutters, we prefer viewpoints larger than 1500 bp, which guarantee best that they are involved in cross-links. Moreover, the viewpoint fragend size should not be lower than 300 bp; otherwise, it is hard to form a circle during the second ligation step (Rippe, von Hippel, & Langowski, 1995) with captured small fragends.

3.2. PCR of 4C template

The PCR DNA polymerase should preferably have little bias for length of the product and abundance. We use the Expand Long Template Polymerase (Roche) which meets the criteria best (Simonis et al., 2006). After testing the primers for specificity and functionality on a 4C template, the primers are used on sufficient number of ligation products to produce a complex 4C-seq library. To analyze long-range *cis* and *trans* interactions, many ligation junctions need to be amplified, as most of the ligations occur close to its viewpoint ($\sim$1 Mb). One million ligation events can be amplified with 16 × 200 ng DNA. We use such amounts to create a 4C-seq library that is complex enough to generate reproducible results.

4C protocol: continued (5)

57. Determine the linear range of amplification by performing a PCR using template dilutions of 12.5, 25, 50, and 100 ng 4C template. A typical 25 μl PCR consists of:
- 2.5 μl of 10 × PCR buffer 1 (supplied with the Expand Long Template Polymerase).
- 0.5 μl dNTP (10 mM).
- 35 pmol forward primer.
- 35 pmol reverse primer.
- 0.35 μl Expand Long Template Polymerase (Roche, #11759060001).
- x μl Milli-Q to a total volume of 25 μl.

A typical 4C-PCR program: 2′ 94 °C, 10″ 94 °C, 1′ 55 °C, 3′ 68 °C, 29 × repeat, 5′ 68 °C, and ∞ 12 °C. The concentration of primers used in a 4C-PCR is typically three times higher than a regular PCR as this often facilitates the efficiency of amplification.

58. Separate 15 μl PCR product on a 1.5% agarose gel and quantify to assess linear amplification and template quality.
59. Determine the functionality of the adapter primers by comparing them with the "short" primers, which do not have the Illumina adapters attached, from step 57. *Note*: the volume of the adapter primers is corrected for their length difference by using 4.5 μl and 3 μl of a 1/7 diluted 1 μg/μl stock solution of the $\sim$75 nt P5 primer and the $\sim$40 nt P7 primer. The adapter primers should cause a shift in PCR product size, which is visible when separated and compared on a 1.5% agarose gel.
60. When satisfied about the quality and quantity of the PCR product generated using the adapter primers, 4C template for sequencing is prepared as follows:
    – 80 μl of 10× PCR buffer 1.
    – 16 μl dNTP (10 mM).
    – 1.12 nmol 75 nt sequencing P5 primer (24 μl reading primer of a 1 μg/μl 75 nt primer stock).
    – 1.12 nmol 40 nt reverse P7 primer (16 μl reverse primer of a 1 μg/μl 40 nt primer stock).
    – Typically 3.2 μg 4C template.
    – 11.2 μl Expand Long Template Polymerase.
    – Milli-Q water till 800 μl total.
    – Mix and separate into 16 reactions of 50 μl before running the PCR.
61. Collect and pool the 16 reactions. Purify the sample using the High Pure PCR Product Purification Kit (Roche, #11732676001), which effectively separates between the nonused adapter primers ($\sim$75 nt; $\sim$40 nt) and the PCR product (>120 nt). Use minimal two columns per 16 reactions.
62. Determine the sample quantity and purity using the NanoDrop spectrophotometer. Typically, the yield resides between 10 and 20 μg with $A_{260}/A_{280}\sim1.85$ and $A_{260}/A_{230}>1.5$. Sample purity is an important control to prevent complications during the sequencing procedure. If absorption ratios deviate, repurification is advised.
63. Quality is determined by separation of 300 ng purified PCR product on a 1.5% agarose gel.
64. Combine 4C PCR products of different experiments in preferred ratios for sequencing.

DNA fragments collected after the 4C-seq PCR can be separated and viewed on an agarose gel (Fig. 4.3, lane 4) and are ready for Illumina sequencing.

4. HIGH-THROUGHPUT SEQUENCING OF 4C PCR PRODUCTS

Illumina's Genome Analyzer II and HiSeq 2000 can both be used to sequence the 4C-seq libraries. The current HiSeq 2000 generates more than 100 million 4C-seq reads, filtered for high quality, in a single lane. However, 4C-seq can generate specific sequencing problems. One major concern is the possibility of an identical nucleotide in all reads at the same position or cycle. This occurs when all the reading primers are of the same length and ending all with the same restriction site sequence, like GATC (*DpnII*). We recommend mixing experiments with different (primary) restriction digests and/or reading primer sequence length. When this is not possible, mix the 4C-seq PCR products up to 30% with foreign DNA, such as bacteriophage phi X 174 DNA. Using a reading primer length of 20 bp, an Illumina run creating 36-mer reads is enough to generate a good quality 4C-genome map. The mappability in 4C experiments increases only slightly when reading longer sequences and is therefore often not necessary.

5. DATA ANALYSIS

5.1. Mapping of 4C-seq reads to genome

The mapping of 4C-seq reads onto a reference genome is different compared to that of other next generation sequencing applications. We have developed a custom analysis pipeline written in perl to process the 4C-seq data. The first step of mapping the data is to bin the reads according to the reading primers and barcodes used in each lane. Usually, more than 90% (median 96% N: 49 lanes) of the reads that passed the Illumina quality filter can be binned. Each bin represents a single 4C-seq experiment from a single viewpoint with a unique reading primer and, if used, a barcode. We allow a single mismatch in the reading primer and a strict criterion for the barcode where no mismatch is allowed. The binned sequences are mapped to an *in silico* library of fragends that is generated based on the REs used for the 4C template preparation. We do not take the quality score of the Illumina sequencer into account, but do not allow any mismatch in the fragment-end. Our strategy allows for rapid mapping. Mapping algorithms used for other next generation sequencing techniques can also be used at this stage of 4C-seq mapping though. For analysis, we mostly focus on the unique fragends only. After properly mapping the 4C-seq Illumina reads, we generate wiggle track (WIG) and files in GFF formats for visualizing (UCSC browser (Kent et al., 2002)) and analysis purposes (see below).

5.2. Statistics and quality control of the 4C-seq

Mapping 4C-seq reads to the reduced 4C-seq genome results in a list of raw read counts per fragend. We use a combination of criteria to roughly estimate the quality of the 4C-seq genomic map. First, the total number of reads per experiment is relevant. Second, the number of mapped reads in *cis* over the total number of mapped reads (*cis*/overall ratio) is indicative of the quality, and third, the percentage of covered fragends within windows around the viewpoint is considered. For six cutters, we analyze coverage within 1 Mb on each side of the viewpoint. For four cutters, we do this in a window of 0.2 Mb (100 kb on each side). We are aiming for more than 1 million reads for a single 4C-seq experiment, which give, in general, high-quality data (Fig. 4.4). As can be seen, most experiments have very high local coverage, indicative of the high complexity of the sample. When coverage is low, caution must be taken in the interpretation of the data. The third criterion, *cis*/total ratio, will often be a reflection of the cross-link efficiency,

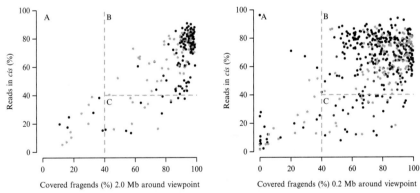

Figure 4.4 4C quality assessment. Percentage of total number of reads mapped in *cis* (i.e., to the chromosome containing the viewpoint) is plotted versus local coverage around the viewpoint for 4C experiments performed with a six (left, $n=225$) and four (right, $n=467$) cutter as primary restriction enzyme. Colors refer to the total number of reads obtained per experiment, being <300,000 (black), 300,000–1,000,000 (gray), 1–3 M (blue), and >3 M (red). In a high-quality 4C experiment, the great majority of reads map in *cis* and nearly all fragends 1 Mb (left, six cutter) and 0.1 Mb (right, four cutter) on either side of the viewpoint are read more than once. Note that reads from self-ligation and nondigested products were removed from the analysis. (See Color Insert.)

since many reads from other chromosomes are likely identified because of random ligations. We experienced *cis*/overall ratio of >40% acceptable and lower one as poor quality experiments (Fig. 4.4). Figure 4.4 depicts the three criteria and is segmented into three areas A, B, and C for which B represents the experiments with high quality that we preferably analyze.

Two prominent PCR products are often found, being the noncut fragend and the self-ligated fragend (Figs. 4.2 and 4.3). They are often visible as distinct PCR products on gel and appear the most common reads after sequencing. The noncut fragend is generated by the fact that the first RE was not able to cut the site that is analyzed by the reading primer of the viewpoint and is, therefore, a reflection of the digestion efficiency in cross-linked chromatin. Judged from the percentage of reads that originate from this product (Fig. 4.5), on average 20% of the DNA is not digested in chromatin that was cross-linked by 2% formaldehyde for 10 min. The self-ligation occurs when both ends of the viewpoint directly ligate to each other, something that will often happen when the viewpoint is not cross-linked to other fragments. The amount of self-ligated fragends (median: 0.5% of the reads across the 912 experiments) may therefore give an indication of the cross-linking efficiency with the viewpoint of interest. Figure 4.5A shows the abundance of these two products across many different 4C-seq experiments, performed with different primary REs. Provided the complexity is sufficiently high (Fig. 4.4), even contact profiles of experiments that appear rather poorly digested (i.e., lots of reads from the nondigested viewpoint) can still be meaningful and interpreted. We have also analyzed seven viewpoints under different cross-linking conditions (0.5, 1, and 2% formaldehyde for 10 min each). As shown in Fig. 4.5B, cross-linking stringency has impact on the self-ligation and noncut fragends. The median of the self-ligations goes down with increased stringency, in agreement with the idea that then the viewpoint is more likely to be cross-linked to another fragment that competes for ligation to the viewpoint ends. The proportion of noncut fragend reads goes up with higher formaldehyde concentration, which is also as expected, since more cross-linking hampers DNA RE digestion.

To assess the read distribution of 4C experiments, we divided fragends (without the self-ligation and the noncut fragends) in bins that cover differently sized genomic regions at increasing distance from the viewpoints (Fig. 4.6). The analysis shows that in order to capture the same amount of fragments, one has to 10-fold increase the size of bins that are further apart

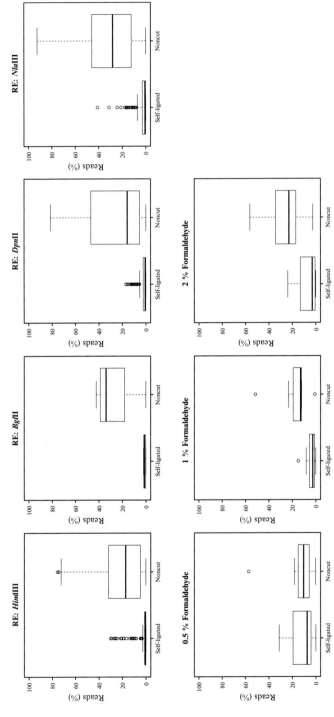

Figure 4.5 Self-ligated and nondigested 4C products. (A) Box plots showing the distribution of self-ligated and noncut reads (as their percentage of total number of mapped reads), as measured across experiments applying 2% formaldehyde to fix chromatin and using *HindIII* (n = 187), *BglII* (n = 7), *DpnII* (n = 196), or *NlaIII* (n = 200) as the primary restriction enzyme. Boxes indicate median and 25–75 percentile range, plus outliers. (B) Same, but for experiments fixing chromatin with 0.5%, 1%, or 2% formaldehyde (and digesting with either *DpnII* or *NlaIII* as RE1; n = 7).

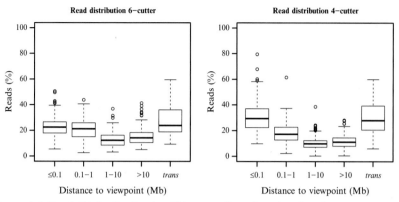

Figure 4.6 Read distribution in *cis* and in *trans*. Box plots showing the percentage of total mappable reads for bins representing sequences in *cis* within 100 kb, 0.1–1 Mb, 1–10 Mb, >10 Mb from viewpoint, and sequences in *trans* (on all other chromosomes) in high-quality experiments. Left panel: six cutters ($n=194$), right panel: four cutters ($n=396$). Note that reads from self-ligation and nondigested products were removed from the analysis.

from the viewpoint. The analysis also shows that four cutters more often capture the near *cis* regions that are within 0.1 Mb from the viewpoint, further arguing that they are most suited to analyze the local interactions between regulatory DNA sequences.

5.3. Data analysis in *cis*

The first step of 4C-seq data analysis is to divide the analysis of *cis* and *trans* data. Here we use, as an example, a viewpoint on *β-major*, a *β*-globin gene which is highly expressed in fetal liver. The template was digested with the REs *Hin*dIII and *Dpn*II. From the data, the noncut fragend and self-ligated fragend were removed. Subsequently, the raw data were displayed, using our preferred statistical programming language R (R Development Core Team, 2010) (Fig. 4.7A). The data-plot shows a tremendous peak around the viewpoint. This peak consist of captures that are only close to the viewpoint (<100 kb). The interactions at a larger distance are, however, not visible. When using a *y*-axis limit of 99.9% quantile (Fig. 4.7B), a clearer interacting *cis* profile appears, with high signals grouping together. Note that each uniquely captured fragend represents at least a single independent ligation event and that in each cell a maximum of two ligations with the viewpoint can occur. Capturing multiple neighboring fragends strongly indicates that the locus interacts with the viewpoint in multiple cells. In addition, the PCR technique can introduce bias for fragend properties such as size and

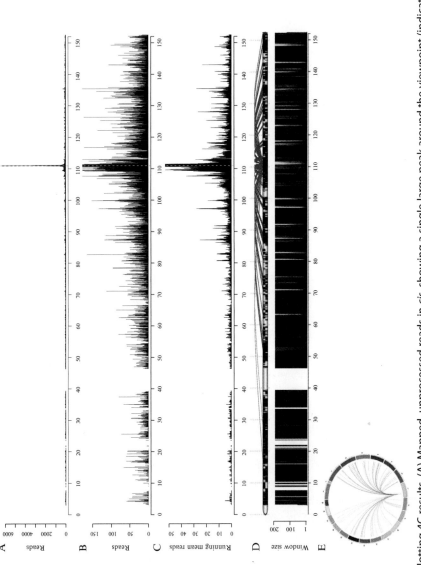

Figure 4.7 Plotting 4C results. (A) Mapped, unprocessed reads in *cis*, showing a single large peak around the viewpoint (indicated by a gray dashed line). Note that reads from self-ligation and nondigested products were removed from the analysis. (B) Same, but with y-axis scaled to show the many other reads in *cis*. (C) Same data processed by a running mean sliding window approach (31 fragends window) to allow better viewing of specific chromosomal contacts. (D) High-level analysis using a domainogram to show significance scores of contacts across a range of chromosomal window sizes. On top an arachnogram shows the contacts for a given window size. (E) Circo plot visualizing *trans* interactions

GC content and, as such, individual fragends, at large distance from viewpoint, should not be interpreted quantitatively. Therefore, we advice to apply a running window approach to smoothen the data. Running window approaches are robust indicators of domain interactions and were very useful analyzing the microarray 4C data (Simonis et al., 2006). After applying a running window approach to the raw β-*major* data (running median or running mean with an arbitrary window size of 31 that corresponds to $\sim$75 kb), interacting domains are popping up along the *cis* chromosome (Fig. 4.7C). Moreover, the running window approach shows that the height of 4C-seq signal has an inverse relationship with the distance to the viewpoint. We developed a 4C-algorithm (Splinter et al., 2011) to correct for this bias and visualize the results in a domainogram (de Wit, Braunschweig, Greil, Bussemaker, & van Steensel, 2008) (Fig. 4.7D). The 4C-algorithm first binarizes the data (captured and noncaptured fragends are set to one and zero, respectively). The rational for binarization is that 4C-seq signals are, as stated above, hard to interpret quantitatively and with binarization correct for PCR artifacts. Binarization is especially useful when analyzing long distance and *trans* reads because these reads are assumed to be generated by a single ligation event and differences in read count are the result of variations in PCR. After binarization, a window with a size from 1 to 200 is running along the chromosome, accompanied with a background window around the other window of 3001 fragends. For each window, a direct one-tailed binomial test can be applied, which test if the number of occurrences in first window (with size of the window as number of trails and with a expected probability based on proportion of captures in the background window) is greater than expected. An alternative strategy is to calculate the z-score first (Splinter et al., 2011). Both strategies give a p-value that is $-\log_{10}$ transformed to maximum of 10 (p-value of 1×10^{-10}) and color coded from black via red to yellow (Fig. 4.7D). We have empirically defined a background window of 3001 and a running window of 100 as the most optimal window size for selecting significantly contacting regions.

To correct for multiple testing, we employ a false discovery rate (FDR) to permuted ($N \geq 100$) *cis* datasets. regions are interacting if FDR $<$ 0.01. The interacting regions can be visualized by genome browsers (UCSC or IGV (Robinson et al., 2011)), after transforming the data to a proper format such as GFF, BED, or WIG. Figure 4.7D shows an arachnogram of the interacting regions that are detected in experiment with β-*major* as viewpoint. The domainogram, arachnogram, and to a lesser extent the running window graph give similar interacting regions, which indicate the robustness of the analyses.

5.4. Trans analysis

For the *trans* data analysis we have to be even more cautious than the *cis* data analysis, since many *trans* captures are random-ligation events of the first or second ligation step. In general, the *trans* data analysis is very similar to the *cis* data analysis. However, since the *trans* data have no viewpoint on their chromosome, there is no bias in the distribution of the 4C-signal along the chromosomes. Therefore, the 4C-domainogram algorithm can be applied without a running background correction. We are using a single background probability depending on the number of unique fragends in each *trans*-chromosome. Because *trans* fragends are less frequently captured than *cis* fragends, the running window is set to 500 fragends for the FDR analysis (q-value < 0.01). For example, in the β-*major* experiment, the *cis* long-range coverage (> 2 Mb from viewpoint) is 21% (while the *trans* coverage is 4%). The discrete data can be visualized with the excellent Circos tool (Krzywinski et al., 2009) (Fig. 4.7E), or with *trans* domainograms and in a quantitative manner with genome browsers.

ACKNOWLEDGMENTS

This work was financially supported by grants from the Dutch Scientific Organization (NWO) (91204082 and 935170621) and a European Research Council Starting Grant (209700, "4C")

REFERENCES

Bantignies, F., Roure, V., Comet, I., Leblanc, B., Schuettengruber, B., Bonnet, J., et al. (2011). Polycomb-dependent regulatory contacts between distant Hox loci in Drosophila. *Cell*, *144*, 214–226.
de Wit, E., Braunschweig, U., Greil, F., Bussemaker, H. J., & van Steensel, B. (2008). Global chromatin domain organization of the Drosophila genome. *PLoS Genetics*, *4*, e1000045.
de Wit, E., & de Laat, W. (2012). A decade of 3C technologies: Insights into nuclear organization. *Genes & Development*, *26*, 11–24.
Dekker, J. (2006). The three 'C' s of chromosome conformation capture: Controls, controls, controls. *Nature Methods*, *3*, 17–21.
Dekker, J., Rippe, K., Dekker, M., & Kleckner, N. (2002). Capturing chromosome conformation. *Science*, *295*, 1306–1311.
Dostie, J., Richmond, T. A., Arnaout, R. A., Selzer, R. R., Lee, W. L., Honan, T. A., et al. (2006). Chromosome conformation capture carbon copy (5C): A massively parallel solution for mapping interactions between genomic elements. *Genome Research*, *16*, 1299–1309.
Drissen, R., Palstra, R. J., Gillemans, N., Splinter, E., Grosveld, F., Philipsen, S., et al. (2004). The active spatial organization of the beta-globin locus requires the transcription factor EKLF. *Genes & Development*, *18*, 2485–2490.
Fullwood, M. J., Liu, M. H., Pan, Y. F., Liu, J., Xu, H., Mohamed, Y. B., et al. (2009). An oestrogen-receptor-alpha-bound human chromatin interactome. *Nature*, *462*, 58–64.

Giresi, P. G., Kim, J., McDaniell, R. M., Iyer, V. R., & Lieb, J. D. (2007). FAIRE (formaldehyde-assisted isolation of regulatory elements) isolates active regulatory elements from human chromatin. *Genome Research, 17,* 877–885.

Hadjur, S., Williams, L. M., Ryan, N. K., Cobb, B. S., Sexton, T., Fraser, P., et al. (2009). Cohesins form chromosomal cis-interactions at the developmentally regulated IFNG locus. *Nature, 460,* 410–413.

Hakim, O., Sung, M. H., Voss, T. C., Splinter, E., John, S., Sabo, P. J., et al. (2011). Diverse gene reprogramming events occur in the same spatial clusters of distal regulatory elements. *Genome Research, 21,* 697–706.

Homminga, I., Pieters, R., Langerak, A. W., de Rooi, J. J., Stubbs, A., Verstegen, M., et al. (2011). Integrated transcript and genome analyses reveal NKX2-1 and MEF2C as potential oncogenes in T cell acute lymphoblastic leukemia. *Cancer Cell, 19,* 484–497.

Kent, W. J., Sugnet, C. W., Furey, T. S., Roskin, K. M., Pringle, T. H., Zahler, A. M., et al. (2002). The human genome browser at UCSC. *Genome Research, 12,* 996–1006.

Krzywinski, M., Schein, J., Birol, I., Connors, J., Gascoyne, R., Horsman, D., et al. (2009). Circos: An information aesthetic for comparative genomics. *Genome Research, 19,* 1639–1645.

Lieberman-Aiden, E., van Berkum, N. L., Williams, L., Imakaev, M., Ragoczy, T., Telling, A., et al. (2009). Comprehensive mapping of long-range interactions reveals folding principles of the human genome. *Science, 326,* 289–293.

Lower, K. M., Hughes, J. R., De Gobbi, M., Henderson, S., Viprakasit, V., Fisher, C., et al. (2009). Adventitious changes in long-range gene expression caused by polymorphic structural variation and promoter competition. *Proceedings of the National Academy of Sciences of the United States of America, 106,* 21771–21776.

Montavon, T., Soshnikova, N., Mascrez, B., Joye, E., Thevenet, L., Splinter, E., et al. (2011). A regulatory archipelago controls Hox genes transcription in digits. *Cell, 147,* 1132–1145.

Noordermeer, D., de Wit, E., Klous, P., van de Werken, H., Simonis, M., Lopez-Jones, M., et al. (2011). Variegated gene expression caused by cell-specific long-range DNA interactions. *Nature Cell Biology, 13,* 944–951.

Noordermeer, D., Leleu, M., Splinter, E., Rougemont, J., De Laat, W., & Duboule, D. (2011). The dynamic architecture of Hox gene clusters. *Science, 334,* 222–225.

Ong, C. T., & Corces, V. G. (2011). Enhancer function: New insights into the regulation of tissue-specific gene expression. *Nature Reviews. Genetics, 12,* 283–293.

R Development Core Team. (2010). R: A language and environment for statistical computing. Vienna, Austria: R Foundation for Statistical Computing.

Rippe, K., von Hippel, P. H., & Langowski, J. (1995). Action at a distance: DNA-looping and initiation of transcription. *Trends in Biochemical Sciences, 20,* 500–506.

Robinson, J. T., Thorvaldsdottir, H., Winckler, W., Guttman, M., Lander, E. S., Getz, G., et al. (2011). Integrative genomics viewer. *Nature Biotechnology, 29,* 24–26.

Robyr, D., Friedli, M., Gehrig, C., Arcangeli, M., Marin, M., Guipponi, M., et al. (2011). Chromosome conformation capture uncovers potential genome-wide interactions between human conserved non-coding sequences. *PloS One, 6,* e17634.

Rozen, S., & Skaletsky, H. (2000). Primer3 on the WWW for general users and for biologist programmers. *Methods in Molecular Biology, 132,* 365–386.

Simonis, M., Klous, P., Homminga, I., Galjaard, R. J., Rijkers, E. J., Grosveld, F., et al. (2009). High-resolution identification of balanced and complex chromosomal rearrangements by 4C technology. *Nature Methods, 6,* 837–842.

Simonis, M., Klous, P., Splinter, E., Moshkin, Y., Willemsen, R., de Wit, E., et al. (2006). Nuclear organization of active and inactive chromatin domains uncovered by chromosome conformation capture-on-chip (4C). *Nature Genetics, 38,* 1348–1354.

Simonis, M., Kooren, J., & de Laat, W. (2007). An evaluation of 3C-based methods to capture DNA interactions. *Nature Methods, 4*, 895–901.

Soler, E., Andrieu-Soler, C., de Boer, E., Bryne, J. C., Thongjuea, S., Stadhouders, R., et al. (2010). The genome-wide dynamics of the binding of Ldb1 complexes during erythroid differentiation. *Genes & Development, 24*, 277–289.

Spilianakis, C. G., & Flavell, R. A. (2004). Long-range intrachromosomal interactions in the T helper type 2 cytokine locus. *Nature Immunology, 5*, 1017–1027.

Splinter, E., de Wit, E., Nora, E. P., Klous, P., van de Werken, H. J., Zhu, Y., et al. (2011). The inactive X chromosome adopts a unique three-dimensional conformation that is dependent on Xist RNA. *Genes & Development, 25*, 1371–1383.

Splinter, E., Heath, H., Kooren, J., Palstra, R. J., Klous, P., Grosveld, F., et al. (2006). CTCF mediates long-range chromatin looping and local histone modification in the beta-globin locus. *Genes & Development, 20*, 2349–2354.

Tolhuis, B., Blom, M., Kerkhoven, R. M., Pagie, L., Teunissen, H., Nieuwland, M., et al. (2011). Interactions among polycomb domains are guided by chromosome architecture. *PLoS Genetics, 7*, e1001343.

Tolhuis, B., Palstra, R. J., Splinter, E., Grosveld, F., & de Laat, W. (2002). Looping and interaction between hypersensitive sites in the active beta-globin locus. *Molecular Cell, 10*, 1453–1465.

van Steensel, B., & Dekker, J. (2010). Genomics tools for unraveling chromosome architecture. *Nature Biotechnology, 28*, 1089–1095.

Zhang, Z., Schwartz, S., Wagner, L., & Miller, W. (2000). A greedy algorithm for aligning DNA sequences. *Journal of Computational Biology, 7*, 203–214.

Zhao, Z., Tavoosidana, G., Sjolinder, M., Gondor, A., Mariano, P., Wang, S., et al. (2006). Circular chromosome conformation capture (4C) uncovers extensive networks of epigenetically regulated intra- and interchromosomal interactions. *Nature Genetics, 38*, 1341–1347.

CHAPTER FIVE

A Torrent of Data: Mapping Chromatin Organization Using 5C and High-Throughput Sequencing

James Fraser, Sylvain D. Ethier, Hisashi Miura, Josée Dostie[1]
Department of Biochemistry and Goodman Cancer Research Center, McGill University, Montréal, Québec, Canada
[1]Corresponding author: email: josee.dostie@mcgill.ca

Contents

1. Introduction — 114
   1.1 Genome architecture — 115
   1.2 The 5C technology — 116
   1.3 Analyzing 5C libraries — 118
   1.4 The *Hox* clusters as model systems — 119
2. Preparing Torrent 5C Libraries — 121
   2.1 Torrent 5C primer design — 122
   2.2 Torrent 5C library preparation — 124
   2.3 Torrent 5C library amplification and purification — 126
   2.4 Quantifying Torrent 5C libraries for sequencing — 128
   2.5 Selecting the right amount of Torrent 5C library — 128
3. Overview of the Ion Torrent Sequencing Protocol — 129
4. Torrent 5C Data Processing — 131
   4.1 Sequencing quality control — 131
   4.2 Torrent 5C data mapping — 133
   4.3 Analysis formats — 135
   4.4 Data quality control and normalization — 137
5. Conclusion — 138
Acknowledgments — 138
References — 138

Abstract

The study of three-dimensional genome organization is an exciting research area, which has benefited from the rapid development of high-resolution molecular mapping techniques over the past decade. These methods are derived from the chromosome conformation capture (3C) technique and are each aimed at improving some aspect of 3C. All 3C technologies use formaldehyde fixation and proximity-based ligation to capture chromatin contacts in cell populations and consider *in vivo* spatial proximity more or less inversely proportional to the frequency of measured interactions. The 3C-carbon copy (5C) method is among the most quantitative of these approaches. 5C is extremely robust and can be used to study chromatin organization at various scales. Here, we present a modified 5C analysis protocol adapted for sequencing with an Ion Torrent Personal Genome Machine™ (PGM™). We explain how Torrent 5C libraries are produced and sequenced. We also describe the statistical and computational methods we developed to normalize and analyze raw Torrent 5C sequence data. The Torrent 5C protocol should facilitate the study of *in vivo* chromatin architecture at high resolution because it benefits from high accuracy, greater speed, low running costs, and the flexibility of in-house next-generation sequencing.

1. INTRODUCTION

Recent developments in genome mapping technologies have led to a better understanding of three-dimensional chromatin organization and its role in the regulation of genes (Ethier, Miura, & Dostie, 2012). The chromosome conformation capture (3C) and 3C-derived molecular methods have been instrumental in acquiring this new insight. By measuring formaldehyde-captured chromatin contacts across millions of cells, the 3C techniques can help identify fundamental characteristics of genome folding in addition to features that are more cell-type specific. These techniques include the 3C-carbon copy (5C) method which, unlike the original 3C method published in 2002 (Dekker, Rippe, Dekker, & Kleckner, 2002), can be used to interrogate chromosome structure *in vivo* over very large genomic intervals (Dostie et al., 2006). 5C is ideal for mapping the organization of chromatin fibers at high resolution and to monitor the physical connectivity between regulator DNA elements. This approach can be adapted to study various biological questions, is easily scalable, and can be conducted high throughput. 5C is also highly quantitative when using an optimal detection method. 5C products were previously measured on custom microarrays and by deep DNA sequencing, which may be prone to noise or prohibitively expensive, respectively (Bau et al., 2011; Fraser, Rousseau, Blanchette, & Dostie, 2010; Fraser et al., 2009). We report here how 5C can be conducted with the Ion Torrent Personal Genome Machine™ (PGM™) sequencer. The PGM™ is a benchtop sequencer that uses semiconductor technology to measure polymerase-driven base incorporation in real time. This new approach offers higher accuracy, greater speeds, lower running costs, and the flexibility of in-house next-generation sequencing. We describe how Torrent 5C libraries are prepared and sequenced and how Torrent 5C data are processed and analyzed. We introduce this new method by characterizing the three-dimensional organization of the human *HoxA* cluster during cellular differentiation. We have previously reported a similar analysis and compared our results with those already published. Below, we start by introducing key biological concepts relevant to this report.

1.1. Genome architecture

Like any other complex system, cells have developed sophisticated strategies selected over the course of evolution to respond to their environment. Complex response mechanisms can be found to regulate all essential cellular processes such as transcription. A major challenge in molecular biology is to deconvolute these control mechanisms and identify what represent "normal" or healthy responses to those found in disease. The human genome, for example, encodes DNA sequences that are recognized by transcription factors that either activate or repress the expression of genes. *In vivo*, our genome is packaged with histone and non-histone proteins in the form of chromatin, which serves at least two functions. First, protein binding can act as a physical barrier to restrict access of transcription factors to the DNA. The composition and physical properties of chromatin can be modified posttranslationally and how histone marks affect transcription is an important research area in the field of epigenetics. Chromatin also appears essential to reduce the length of chromosomes during interphase. An unfolded human genome, for instance, is estimated to measure approximately 2 meters in length and must be contained within micron size nuclei. DNA packaging thus raises interesting questions relating to the coregulation or sequential activation of genes and suggests that genome organization is not random.

Functional genome organization was originally supported by microscopy studies demonstrating that chromosomes occupy distinct nuclear territories termed "chromosome territories" (Cremer & Cremer, 2010). The fact that gene-rich chromosomes were found to preferentially localize to the nucleus center and gene-poor at the periphery and the observation that coregulated genes often colocalize also suggested an important relationship between genome structure and function (Pombo & Branco, 2007; Towbin, Meister, & Gasser, 2009). This relationship has since been confirmed at high resolution using 3C by uncovering long-range physical contacts between genes and regulatory DNA elements located within (*cis*) or between (*trans*) chromosomes (Pomerantz et al., 2009; Schmidt et al., 2010; Spilianakis & Flavell, 2004; Splinter et al., 2006; Tolhuis, Palstra,

Splinter, Grosveld, & de Laat, 2002). Because these interactions have been found genome-wide, the human genome is viewed as an elaborate network of physical contacts that are either structural or dynamic in nature. These observations together reveal a complex interplay between genome structure and function, where structure can impinge on function and vice versa. Genome packaging thus additionally plays the role of a higher-order mechanism that regulates transcription and other cellular processes.

The physical relationship between genes and control DNA elements can easily be probed with techniques such as 3C, 4C, and 5C where scale, resolution, and throughput mostly dictate the method of choice (Dekker et al., 2002; Dostie et al., 2006; Ling et al., 2006; Simonis et al., 2006; Würtele & Chartrand, 2006; Zhao et al., 2006). 3C-related methods such as Hi-C and TCC designed for genome-wide mapping have also begun to uncover basic principles of genome folding (Duan et al., 2010; Kalhor, Tjong, Jayathilaka, Alber, & Chen, 2011; Lieberman-Aiden et al., 2009). Since this chapter focuses on how 5C can be performed using an Ion PGM^{TM} sequencer, we outline below the underlying principles of this technique.

1.2. The 5C technology

The 5C technique is a direct derivative of the 3C method in that it simply measures chromatin contacts with ligation-mediated amplification instead of by semiquantitative PCR (Fig. 5.1). Thus, as in 3C, cells are first cross-linked with formaldehyde to capture protein–protein, protein–DNA, and protein–RNA interactions. Formaldehyde is usually used instead of other cross-linkers because its short cross-links are thought to capture closer and more relevant interactions. The chromatin is next digested overnight with a restriction enzyme and ligated under experimental conditions that favor intermolecular ligation of cross-linked DNA fragments. The cross-links are then reversed by heating, proteinase K digestion, and phenol–chloroform extraction, which yield the 3C library template used to generate 5C libraries. The protocol and other considerations of 3C library preparation have been described in detail elsewhere and are not presented here for this reason (Dekker et al., 2002; Dostie & Dekker, 2007; Dostie et al., 2006; Dostie, Zhan, & Dekker, 2007; Miele, Gheldof, Tabuchi, Dostie, & Dekker, 2006).

3C libraries are converted to 5C libraries in three simple steps. First, the 3C library is denatured and annealed to 5C primers overnight. The sequences of 5C primers correspond to the ends of restriction fragments such

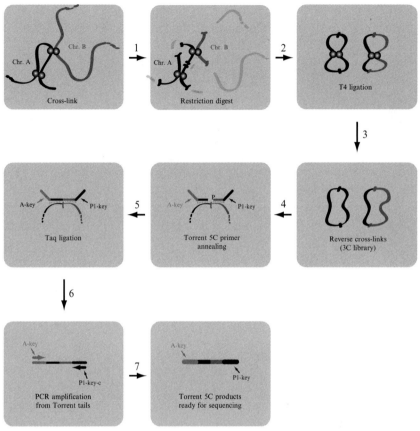

Figure 5.1 Production of Torrent 5C libraries. To prepare libraries, cells are first cross-linked with formaldehyde, digested with a restriction enzyme, and ligated with T4 DNA ligase. The cross-links are then reserved and the DNA purified by phenol–chloroform extraction. Resulting 3C libraries represent the template onto which Torrent 5C primers (forward and reverse) are annealed at high level of multiplexing. The annealed Torrent primers are next ligated at 3C junctions with Taq DNA ligase. The Torrent 5C libraries are finally amplified by PCR with primers that are complementary to the universal A-key and P1-key tails, and purified onto MinElute columns. The purified Torrent 5C products is the material used for EmPCR and sequencing. Spheres represent cross-linked proteins. Interacting DNA segments are shown as black and gray lines.

that primers will anneal pairwise at the junctions of 3C ligation products only if they exist in the 3C library. Annealing is followed by ligation with Taq DNA ligase, which can only ligate primers that are perfectly paired onto the same template strand when there are no gaps between them. Because 3C products derived from the ligation of opposite restriction fragment ends are usually characterized during 3C and 5C analysis (the so-called head-to-head or tail-to-tail 3C products; see above-cited publications for details), ligation can only occur between one forward and one reverse 5C primer. Also, because forward and reverse 5C primers predicted for each fragment are at the same restriction end and are complementary to each other, individual restriction fragments can only be probed with either a single forward or reverse 5C primer at a time. 5C primers can nonetheless be mixed at very high level of multiplexing such that thousands of different 5C products can be measured simultaneously during each experiment.

As discussed in Section 2.1, 5C primers are usually designed computationally with online tools by entering the region of interest and the restriction enzyme used to create 3C libraries (Fraser & Dostie, 2009; Lajoie & Dekker, 2009). These computer programs also require the selection of universal tail sequences. These sequences are different for forward and reverse 5C primers and are used for PCR amplification in the third step of 5C library preparation (Fig. 5.1). Purified 5C libraries can be analyzed on custom microarrays or by deep sequencing, and the choice of universal tail sequences differs based on the method of analysis as described below.

1.3. Analyzing 5C libraries

The choice of detection method is an important aspect of 5C analysis because it can impact the accuracy, speed, and cost of the study. 5C libraries were previously characterized successfully on custom microarrays and by deep sequencing (Bau et al., 2011; Dostie et al., 2006; Fraser et al., 2009). Although each approach has pros and cons, sequencing remains by far the preferred method for achieving the maximum quantitative potential of 5C. Microarray analysis requires the fluorescent labeling of 5C libraries with a labeled PCR primer during the amplification step and their hybridization on custom microarrays. This approach has been described in detail, and although very simple, low cost, and accessible, it can be prone to noise (Dostie et al., 2006; Fraser, Rousseau, Blanchette, et al., 2010; Fraser et al., 2009). Deep sequencing, on the other hand, although highly accurate, can be very expensive. Sequencing has nonetheless a number of additional advantages over microarray analysis. First, 5C experimental design can be scaled and modified much more easily with sequencing as compared to arrays. Because custom microarrays must feature all possible 5C products in a given experiment, what can be analyzed with a given design is restricted. Any changes in the region of interest or any modification to the 5C experimental design (enzyme selected, etc.) therefore requires a new array design, which will greatly increase the cost and time required to complete a study. These are not issues when using sequencing technologies since 5C products are simply read, mapped to a user-specified reference genome, and counted. The 5C primer pool can thus be modified and expanded at any time with sequencing. Moreover, as the primer pool increases in number, the number of potential 5C contacts may grow considerably—even exponentially—depending on the experimental design. Detection of rare contacts such as long-range interactions in complex 5C libraries is not a concern with sequencing because it is possible to simply sequence deeper. The only question remaining is how deep the sequencing should be. This is in contrast to array detection where the maximum number of features possible on each chip is limited.

A second advantage that makes sequencing more suitable for 5C analysis is the absence of the background signals present on microarrays. Array hybridization can bear a significant amount of nonspecific signal depending on the nature and complexity of the library, the amount of library hybridized, and the quality of the washes. Although this background can mostly be controlled for, hybridization of products with significant homology can be problematic. Although 5C sequencing still displays some degree of background, these signals are more clearly identifiable and can be removed from the list of mapped reads during analysis.

1.4. The *Hox* clusters as model systems

The *Hox* genes encode evolutionary conserved transcription factors with pivotal roles during development. In human, there are 39 *Hox* genes localized into 4 clusters termed *HoxA, B, C,* and *D*. *Hox* transcription factors are responsible for anterior–posterior body patterning, the formation of limbs and genitalia, and can contribute to cancer in adults when misregulated. We previously used the *HoxA* cluster to study the relationship between gene expression changes and chromatin architecture (Ferraiuolo et al., 2010; Fraser et al., 2009). In one study, we used the THP-1 cellular differentiation system to characterize the *HoxA* cluster by 3C and 5C (Fig. 5.2A). THP-1 cells are myelomonocytes derived from a 1-year-old infant male with acute myeloid leukemia. These cells are known to express the oncogenic MLL–AF9 fusion protein, which

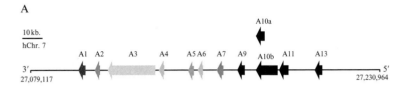

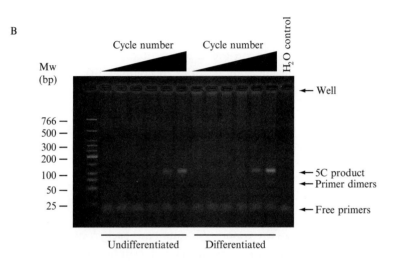

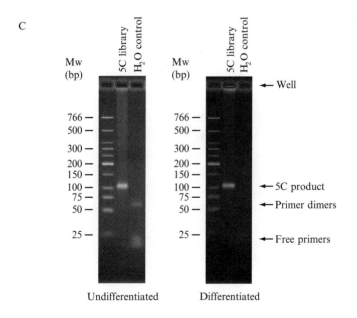

has been shown to upregulate *HoxA9* and *10* transcriptions, two oncogenes contributing to the aberrant proliferation of the cells (Ayton & Cleary, 2003; Kroon et al., 1998; Pession et al., 2003; Thorsteinsdottir et al., 1997). THP-1 cells can be terminally differentiated into a macrophage-like lineage with phorbol myristate acetate (PMA) and thus have been extensively used to study cellular differentiation (Suzuki et al., 2009). Differentiation with PMA is also associated with transcriptional repression of the *HoxA* 5′-end genes (*HoxA9, 10, 11,* and *13*). In these cells, we found that the *HoxA* cluster 5′-end is more compact and organized in a series of chromatin loops when the corresponding genes are repressed following differentiation. Here, we studied the *HoxA* cluster in the THP-1 differentiation model to develop and validate the Torrent 5C protocol because this model had already been characterized by 5C microarray. We now compare and contrast our new data with previously published results throughout the text to highlight the robustness of the Torrent 5C protocol.

2. PREPARING TORRENT 5C LIBRARIES

The first step in Torrent 5C analysis is the generation of 3C libraries. Others and we have previously published a number of procedures detailing how to prepare 3C libraries, test their quality by contact content profiling of gene deserts or other genomic regions, and determine the amount required to convert them into 5C libraries (Dostie & Dekker, 2007; Dostie et al., 2007; Ferraiuolo, Sanyal, Dekker, & Dostie, 2012; Fraser, Rousseau, Blanchette, et al., 2010). These protocols can be used to generate template 3C libraries for Torrent 5C library production. An important additional consideration when designing the 3C experiment for later analysis with the Ion Torrent sequencer is to avoid using restriction

Figure 5.2 Amplifying and quantifying purified *HoxA* Torrent 5C libraries. (A) Linear representation of the human *HoxA* cluster region used to optimize the Torrent 5C protocol. Genomic positions are in hg18. (B) Selecting the right PCR amplification cycle number. Five microliters of Torrent 5C products from 25 μl PCR reactions of 20, 22, 24, 26, 28, and 30 cycles were loaded onto a 2.5% agarose gel. The condition selected for purification was 28 cycles. (C) Quality control and quantification of purified *HoxA* Torrent 5C libraries. Torrent 5C products from purified 28 cycle reactions and from a water control were led on a 2.5% gel and quantified against a molecular weight marker with a gel quantification system.

enzymes cutting at palindromes containing two identical consecutive nucleotides. The reason for this will become apparent when discussing 5C data mapping in Section 4.2. A key difference between the Torrent 5C procedure and previously published 5C protocols is the use of different 5C primer universal tails. These tails are first used to PCR-amplify sufficient quantities of 5C libraries for purification and during the emulsion PCR (EmPCR) when performing clonal amplification on beads.

2.1. Torrent 5C primer design

3C libraries are composed of entire genomes in addition to the new ligation products derived from chromatin fragments once proximal in the nuclear space. When considering the polarity of restriction fragments, pairwise ligation products can be found in various configurations that include head-to-head, tail-to-tail, and head-to-tail. As mentioned above, head-to-head or tail-to-tail contacts are usually probed by 3C and 5C to avoid detecting uncut genomic products when measuring neighboring interactions. Thus, contacts are measured with 5C by ligating one forward and one reverse 5C primer annealed at 3C junctions. The forward and reverse 5C primers can be designed easily with online tools (Fraser & Dostie, 2009; Lajoie & Dekker, 2009; Lajoie, van Berkum, Sanyal, & Dekker, 2009). Primers are multiplexed and the complexity of the resulting libraries is dependent on the number of forward and reverse primers used. As with previous 5C libraries, Torrent 5C libraries are first purified to remove unincorporated primers, salt, and proteins. This step requires that libraries be first PCR-amplified to obtain sufficient material for purification. Amplification is performed with primers recognizing universal tail sequences on 5C primers. For Ion Torrent sequencing, all forward 5C primers should include the A-key Ion Torrent primer sequence at their $5'$-ends, and all reverse 5C primers must include the P1-key Ion Torrent primer sequence at their $3'$-ends. The sequences of these tails and of the corresponding PCR primers are shown in Table 5.1.

Multiplexing of different 5C libraries may be feasible depending on the complexity of Torrent 5C libraries, sequencing depth desired, and the type of Ion chip used. Different 5C libraries can be sequenced simultaneously onto a single chip simply by barcoding forward A-key PCR primers with the sequences shown in Table 5.1. These sequences are those used in the Ion Torrent barcoding Kit and allow multiplexing up to 16 samples (Torrent/Applications, 2011).

Table 5.1 Torrent primer sequences

Sequence name	DNA sequence (5′–3′)	Length (nt)
A-key sequence (forward 5C 5′ tail)	CCATCTCATCCCTGCGTGTCT CCGACTCAG-(5C-specific)	30
P1-key sequence (reverse 5C 3′ tail)	(5C-specific)-ATCACCGACTGCCCATAGAGAGG[a]	23
A-key PCR primer	CCATCTCATCCCTGCGTGTCTCC GACTCAG	30
P1-key-c PCR primer	CCTCTCTATGGGCAGTCGGTGAT	23
Barcoded A-key PCR primer	CCATCTCATCCCTGCGTGTCTCC GACTCAG-Barcode	40
BC IonXpress_001	CTAAGGTAAC	10
BC IonXpress_002	TAAGGAGAAC	10
BC IonXpress_003	AAGAGGATTC	10
BC IonXpress_004	TACCAAGATC	10
BC IonXpress_005	CAGAAGGAAC	10
BC IonXpress_006	CTGCAAGTTC	10
BC IonXpress_007	TTCGTGATTC	10
BC IonXpress_008	TTCCGATAAC	10
BC IonXpress_009	TGAGCGGAAC	10
BC IonXpress_010	CTGACCGAAC	10
BC IonXpress_011	TCCTCGAATC	10
BC IonXpress_012	TAGGTGGTTC	10
BC IonXpress_013	TCTAACGGAC	10
BC IonXpress_014	TTGGAGTGTC	10
BC IonXpress_015	TCTAGAGGTC	10
BC IonXpress_016	TCTGGATGAC	10

[a]Note that reverse 5C primers must be 5′-end phosphorylated.
[b]BC refers to barcode sequence.

2.2. Torrent 5C library preparation

Torrent 5C libraries must be prepared following a very specific procedure to reduce background. This background originates mainly from the 5C primers as they can yield template-independent and ligase-independent 5C products when improperly handled. This background can be virtually eliminated by diluting the primers with the procedure described below. We also describe below how annealing and ligation of Torrent 5C primers should be performed.

2.2.1 Diluting Torrent 5C primers

1. Resuspend lyophilized primers at 80 μM in 1× TE, pH 8.0, and heat at 65 °C for 15 min. Spin down stocks, aliquot, and store tubes at −80 °C.
2. From the 80 μM samples, prepare a second set of stocks at 20 μM in 1× TE, pH 8.0. Aliquot and store tubes at −80 °C.

A second set of stocks at lower concentration will decrease the risks of stock contamination and improve accuracy when primers are mixed for dilution. When specific primer pools are routinely used, 20 μM stocks can be mixed at high concentration to reduce the number of tubes at the dilution step. However, forward and reverse primers should be kept separate until the final dilution step (see below).

3. On the day of the experiment, thaw individual 20 μM primer stocks or stock primer pools on ice.
4. On ice, mix equal volumes of individual forward or reverse primer stocks in separate tubes and dilute each stock mix *in water* to a final concentration of 0.1 μM.

If forward and reverse stock pools are thawed, simply dilute them separately to 0.1 μM in water.

5. Dilute forward and reverse 5C primer pools individually *in water* by 10-fold serial dilution until a final concentration of 0.004 μM.
6. Immediately before use, mix equal volumes of forward and reverse primer pools to obtain a final concentration of 0.002 μM. Keep on ice.
7. Per library, mix 1.7 μl of diluted 5C primer pools with 1 μl of 10× annealing buffer 10× NEBuffer 4 (New England Biolabs inc.) on ice. Mix by pipetting and spin down samples. This is the 5C primer master mix.

Throw away any remaining primer dilution made in water. Freeze–thaw of primers stored in water alters their working concentration.

2.2.2 Annealing Torrent 5C primers onto 3C libraries

8. Dispense approximately 600 ng or 100,000 diploid genome equivalents of each 3C library in individual PCR tubes or plates on ice.

The amount of 3C library used should be estimated by titration as described previously (Dostie & Dekker, 2007; Dostie et al., 2007; Ferraiuolo et al., 2012; Fraser, Rousseau, Blanchette, et al., 2010).

A "no ligase" control for each 3C template and a "no template" control should also be prepared at this point. These samples are prepared by omitting the Taq DNA ligase in Section 2.2.3 or the 3C library, respectively. These controls verify that production of 5C products require the ligase activity and are formed at 3C junctions.

9. Add salmon testis DNA to each 3C sample for a total mass of 1.5 μg. Adjust the volume to 7.3 μl with ice-cold water. Mix by pipetting and briefly spin down.
10. Add 2.7 μl of 5C primer master mix to each sample. Mix by pipetting and centrifuge briefly.
11. Denature samples 5 min at 95 °C and ramp down 0.1 °C per second to preferred annealing temperature. Annealing temperature should be between 48 and 55 °C depending on the GC content of 5C primers in pools.

Slow cooling to desired temperature improves annealing efficiency of the 5C primers.

12. Incubate 16 h at 48–55 °C to anneal Torrent 5C primers at 3C template junctions.

2.2.3 Ligating Torrent 5C primers at 3C junctions

13. The next day, prepare at least 20 μl of ligation master mix for each 5C annealing reaction. The ligation master mix consists of 1 × Taq DNA ligase buffer and 0.25 μl (10 U) of Taq DNA ligase.
14. While keeping the samples at the selected annealing temperature, add 20 μl of ligation master mix to each 5C reaction. Gently mix by pipetting.

Prepare ligation master mix at room temperature immediately before use. Do not forget to prepare master mix for the "no ligase" controls by substituting the Taq with water.

15. Incubate samples at the annealing temperature for 1 h to ligate annealed primers.
16. Incubate samples at 75 °C for 10 min to terminate ligation reactions. Ligation products represent the 5C libraries.

The unamplified 5C libraries can be stored at $-20\ °C$ for later use.

2.3. Torrent 5C library amplification and purification

5C libraries must be PCR-amplified to obtain enough material for purification and sequencing. PCR amplification is performed with PCR primers complementary to the Torrent 5C tail sequences as shown in Table 5.2. We recommend first verifying the linearity of PCR amplification by testing a range of PCR cycle numbers to identify optimal conditions.

17. On ice, prepare at least five PCR reactions in tubes or in plates for each test and control 5C library.

A PCR "water control" containing everything but a 5C library should also be prepared to control for possible contamination.

18. Set up a thermal cycler with the PCR program shown in Table 5.3. We suggest testing a range of 22, 24, 26, 28, and 30 cycles as a starting point. The number of cycles will depend on the 5C design and the contact frequencies in 3C libraries.
19. Transfer 5 μl of each PCR reaction to fresh Eppendorf tubes. Add 2.5 μl of 4× agarose gel loading buffer (10% Ficoll, 0.15% xylene cyanol) and 3 μl of water to each tube and mix by pipetting.
20. Resolve PCR samples on a 2.5% agarose gel prepared with 0.5 × TBE. Include a molecular weight marker of known concentration to help estimate the amount of 5C products on gel.

Table 5.2 Composition of 5C PCR amplification reactions

Stock component	Volume (μl) per rxn[a]	Final per rxn
10× PCR buffer	2.5	1×
50 mM MgSO$_4$	2.0	4.0 mM
25 mM dNTPs	0.2	0.2 mM
100 ng/μl STD[b]	1.5	150 ng
5 μM A-key PCR primer	2.0	0.4 μM
5 μM P1-key-c PCR primer	2.0	0.4 μM
Taq DNA polymerase 5 U/μl[c]	0.2	1 U
H$_2$O	11.6	
5C library	3.0	

[a]Volumes are for 25 μl reactions.
[b]STD; salmon testis DNA stock diluted in water.
[c]If experiencing problems with primer dimers, AmpliTaq gold (Life Technologies) can be used instead of a regular Taq DNA polymerase.

Table 5.3 Torrent 5C library PCR amplification conditions

Number of cycles	Denature	Anneal[a]	Extend	Cool
1	95 °C			
	5 min			
X[b]	95 °C	60 °C	72 °C	
	30 s	30 s	30 s	
1	94 °C	60 °C	72 °C	
	30 s	30 s	8 min	
1				10 °C
				Infinity

[a]Annealing temperature is optimized for Torrent PCR primers (A-key PCR primer and P1-key-c PCR primer).
[b]Variable number of cycles.

21. At room temperature, stain gel 30 min with a solution containing 0.5× TBE and 2.0 μg/ml ethidium bromide. Rinse 5 min in 0.5× TBE.
22. Catalog results with a gel documentation system and quantify products in each lane. You will want to select a condition that yields at least 5 ng of material on this gel.

Figure 5.2B shows the results of a PCR amplification test on two Torrent *HoxA* 5C libraries. These libraries were prepared by mixing 57 Torrent 5C primers (29 forward and 28 reverse) with 3C libraries from undifferentiated and differentiated THP-1 cells. These libraries were used to optimize the Torrent 5C protocol and will be characterized further below (Fig. 5.5). The lowest possible cycle number should be chosen to minimize potential sequence duplication and amplification biases. The number of cycle tested here are 20, 22, 24, 26, 28, and 30. We selected cycle number 28 in these experiments and pooled 8 PCR reactions from that number for purification.

23. Select the lowest possible cycle number and amplify remaining 5C library with these conditions.
24. Pool PCR reactions and purify amplified 5C products on MinElute columns (Qiagen) as instructed by the manufacturer. Elute libraries once in 15 μl of the buffer provided (EB buffer).

2.4. Quantifying Torrent 5C libraries for sequencing

The selection of an optimal amount of DNA is an important step in preparing samples for sequencing on an Ion Torrent Personal Genome Machine^{TM} (PGM^{TM}). While too little DNA yields many "empty" Ion Sphere^{TM} particles (ISPs) and a low read count, too much DNA also results in low read counts by creating an abundance of unusable polyclonal ISPs (see Section 3). As a significant amount of genomic DNA is still present in purified 5C libraries, the concentration of purified 5C libraries should not be estimated by spectrophotometry or using PicoGreen® as these methods would overestimate the amount of actual 5C products present in samples. Although quantification by Bioanalyzer^{TM} analysis or real-time quantitative PCR is possible, we found that quantification on agarose gel is sufficient to approximate the amount of sample for sequencing.

25. After purification, resolve 5 μl of each amplified 5C library on a 2.5% agarose gel prepared with 0.5× TBE. Include a molecular weight marker of known concentration to help estimate the amount of 5C products on gel.
26. Stain gel 30 min at room temperature with a solution containing 0.5× TBE and 0.5 μg/ml ethidium bromide. Rinse 5 min with 0.5× TBE.
27. Use a gel documentation system to catalog results and quantify products in each lane. The amount of DNA required for sequencing is in the 0.05 ng range such that all you need here is to detect enough material to estimate sample concentrations using the molecular weight marker as reference.

Figure 5.2C shows the two *HoxA* 5C libraries produced with undifferentiated and differentiated THP-1 cells after purification on MinElute columns. We used these libraries for sequencing (Fig. 5.5).

2.5. Selecting the right amount of Torrent 5C library

The optimal number of DNA molecules might vary between Torrent 5C libraries and have to be optimized individually. Life Technologies^{TM} currently recommends using between 140×10^6 and 560×10^6 molecules per reaction to obtain between 10% and 25% positive (DNA-bound) ISPs. We have successfully used 478×10^6 molecules per reaction, which corresponds to the highest number of recommended molecules in the original Ion Xpress template protocol (v1.0). With this number of molecules, we reproducibly achieve final library reads between 60% and 75% of positive ISPs, which is after the filtering of polyclonal reads and other problematic sequences. The amount of purified 5C libraries is easily calculated as follows:

$$5C\,\text{library}\,(\text{ng}) = \left[\frac{\text{DNA molecules} \times \text{product length}(\text{bp}) \times 650(\text{Da/bp})}{6.022 \times 10^{23}(\text{Da/g})}\right] 1 \times 10^9$$

where "DNA molecules" is the number of DNA molecules per reaction recommended by Life Technologies™ which can be varied, and "product length" corresponds to the average length of 5C products in purified 5C Torrent libraries. The length of 5C products can vary depending on the length of 5C primers used and the presence of barcoded adaptors. The average length of our 5C products is approximately 120 bp, such that we normally use 0.05 ng of purified 5C library for sequencing. This amount of product is diluted in 18 μl of water before proceeding with the EmPCR as outlined below.

3. OVERVIEW OF THE ION TORRENT SEQUENCING PROTOCOL

The Ion PGM™ is the first sequencer that measures polymerase-driven base incorporation by detecting the release of hydrogen ions by the polymerase in real time. This technology uses semiconductor sensor chips that are scalable and that transform changes in pH into digital information. To sequence with a PGM™, DNA molecules must first be clonally amplified on beads by EmPCR (Fig. 5.3). This amplification step is necessary to magnify the changes in pH that occur when deoxynucleotides are incorporated. The procedure, which is outlined in steps 1–4 of Fig. 5.3, is carried out with the Ion Xpress™ Template Kit v2.0 mostly according to the manufacturer's instructions (Life Technologies™; Rothberg et al., 2011). Briefly, the purified 5C library (140–560 × 10⁶ molecules) is added to a PCR master mix containing ISPs and emulsified using the emulsion oil provided. This mixture is aliquoted into PCR plates and placed into a thermal cycler for EmPCR. After breaking the emulsion, the samples are washed and template-positive ISPs are enriched onto magnetic streptavidin beads. Template-positive ISPs are finally eluted and yield the Torrent 5C libraries. These libraries are ready for loading into ion chips for massively parallel sequencing and are sequenced with the Ion Sequencing Kit v2.0 exactly as recommended by the manufacturer (Life Technologies™).

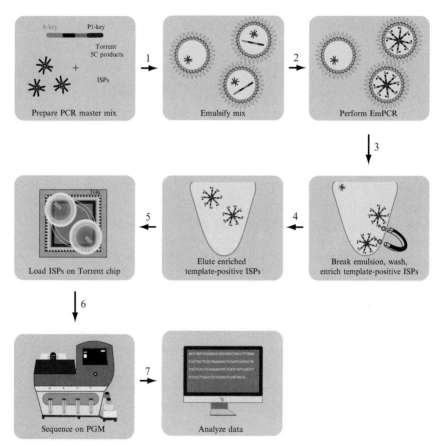

Figure 5.3 Overview of the Ion Torrent sequencing protocol.

Three different Ion chips, the Ion 314™, Ion 316™, and Ion 318™, are currently available for the Ion Torrent PGM™. The Ion 314™, 316™, and 318™ each feature 1.3, 6, and 11 million wells, respectively, and newer chips with increasingly more wells are expected to be released periodically. We have used the Ion 314™ chips for the *HoxA* data presented here and recovered at least 400,000 mappable reads from each of our libraries. An important question with any deep sequencing experiment is how many reads will be required to obtain an appropriate coverage across regions of interest. While too little 5C sequencing depth can increase the rate of false negatives, too much sequencing could become prohibitively expensive. A major factor contributing to the required 5C sequencing depth is the expected library complexity. More complex libraries will require more sequencing to maintain similar detection ranges. As described in Section 1.3, the number of detectable contacts can increase rapidly depending on the experimental design. The expected complexity of 5C libraries can be estimated simply by multiplying the number of forward and reverse 5C primers. For example, the *HoxA* cluster characterization reported here uses 29 forward and 28 reverse 5C primers, which can detect up to 812 different 5C products. With this design, an average read count of 100 reads per detectable contact should theoretically only require 81,200 reads if all contacts were present in equimolar ratio. However, because chromatin contacts exist in proportions that are inversely related to their original nuclear *in vivo* distances, the relative abundance of 5C products can only be defined empirically, and there is no easy way to calculate how deep sequencing should be. It should nonetheless be expected that 5C products derived from proximal DNA fragments, such as between neighbors, would be more highly represented in 5C libraries. The abundance of neighboring contacts can vary significantly depending on the region and its transcriptional state. These contacts can therefore make up a very large percentage of sequencing reads, and weaker interactions such as those derived from long-range contacts can be missed if sequencing is not deep enough. We were able to detect most but not all possible *HoxA* contacts with the experimental design described above. Thus, when the goal of an experiment is to detail chromatin architecture throughout a region, it is important to sequence until even contacts over large distances are represented.

4. TORRENT 5C DATA PROCESSING

Being able to correctly process and analyze the output from the Ion Torrent PGM^{TM} is a critical step. Raw sequence data must first be controlled for quality, each read mapped to a specific 5C contact, and the data collected in a coherent manner and normalized when necessary. The Ion Torrent sequencing software (Torrent Suite Software) automatically outputs a large number of metrics after each sequencing run that can help identify the source of problems if they arise. We discuss below important metrics specifically related to 5C analysis with an Ion Torrent PGM^{TM}, as well as the methods we have derived for the correct mapping and analysis of processed data. Additional aspects of the built-in analysis can be found in the Ion Torrent^{TM} Server guide (Torrent/Publications, 2012).

4.1. Sequencing quality control

The distribution of read lengths is an important predictor of the quality of the clonal bead amplification and the sequencing itself. Figure 5.4A shows a typical read length distribution obtained after a Torrent 5C sequencing run.

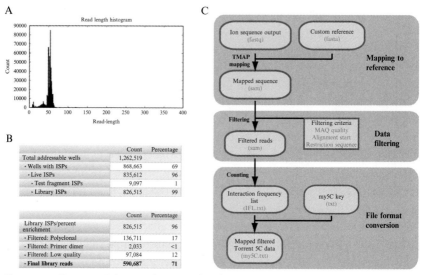

Figure 5.4 Torrent 5C data quality control and mapping workflow. (A) Example of a read length distribution histogram generated after each Ion Torrent sequencing run of 5C libraries. The number of reads is shown on the y-axis (count) and the x-axis shows the length of 5C inserts in the libraries (read length). (B) ISP diagnostic table explaining the difference between the total number of existing ion chip wells and the final number of mappable reads. (C) Workflow of Torrent 5C sequencing data analysis created on Galaxy.

This histogram should display read sizes corresponding mainly to the expected lengths of 5C product inserts, which will vary depending on the sizes of 5C primers used in the experiment. Regardless of differences in experimental design, the histogram peak from a given library should be centered around the same size visualized on the agarose gel minus the size of the A-key and P1-key adaptors, as the sequencing software removes them in the FASTQ output. As shown in Fig. 5.4A, the distribution of our *HoxA* 5C products is centered around 55–60 bp, which corresponds to what was observed on gel (Fig. 5.2C) when the forward (30 nt) and reverse (23 nt) 5C primer tails are subtracted. A number of sequences will also likely be observed at shorter lengths. These products might originate from sequences with poor sequencing quality that are prematurely truncated or might point to problems during the EmPCR. Truncated products might become problematic depending on their abundance and if their sizes are not sufficiently long to identify the fragments involved in pairwise interactions. When encountering this problem, we recommend examining the size distribution of the amplified libraries after the EmPCR step on agarose gel in order to distinguish between sequencing and amplification problems.

The Ion Torrent sequencing program also automatically generates an "ISP identification summary" which is a diagnostic of various ISP parameters relevant to the quality of the sequencing run (Fig. 5.4B). This summary explains the difference between the total possible number of reads on an Ion chip (total addressable wells) and the number of raw sequence reads in the output FASTQ file (final library reads). For instance, we can get approximately 600,000 reads from an Ion 314^{TM} chip when sequencing our *HoxA* 5C libraries (Fig. 5.4B). In the example shown here, we tend to lose chip capacity in three main areas. First, although the chip contains almost 1.3 million wells, not all wells become loaded with ISPs (69%). However, the wells that do contain ISPs tend to be "template-positive" (live ISPs) indicating a good bead enrichment during library preparation. The second source of capacity loss is the presence of polyclonal reads that are filtered by the sequencing program. Polyclonal reads are unavoidable and rise in number when too much library is used during the clonal amplification such that a single ISP is more likely to be populated with two or more different DNA molecules. To obtain the correct balance between maximizing the clonal ISPs and minimizing the number of polyclonal ISPs, it is important to use the correct amount of 5C library during the EmPCR. A third source of data loss is the filtering of low-quality sequences. "Low-quality" reads are filtered out of the final FASTQ file when a certain number of nucleotides within the sequence receive low-quality scores. This can be caused by problems with amplifying the DNA molecules onto the ISPs with the EmPCR or anything that affects the detection of bases by the Ion Torrent PGM^{TM}.

Another simple way to verify whether the experiment was successful is to estimate the percentage of sequence reads in the FASTQ file that contains the restriction site used to generate the 3C libraries. A good 5C sequencing experiment should contain at least 70% of sequences with a restriction site. Lower percentages might reflect an abundance of truncated reads or poor sequencing quality, as any incorrect base calls at the restriction site will exclude sequences.

4.2. Torrent 5C data mapping

High-quality data mapping is essential for quantitative detection of 5C contacts. Inaccurate mapping can increase background levels and create false positives or negatives. Software is currently available to map sequencing data against reference genomes. These include MAQ (mapping and assembly with quality) (Li, Ruan, & Durbin, 2008), BWA (Burrows-Wheeler Aligner)

(Li & Durbin, 2009), and Bowtie (Langmead, Trapnell, Pop, & Salzberg, 2009), which differ in speed and accuracy. The Torrent Analysis Suite includes the TMAP alignment software, which we used here to map our *HoxA* Torrent 5C data (Li & Durbin, 2009, 2010; Ning, Cox, & Mullikin, 2001; Smith & Waterman, 1981). TMAP makes use of the BWA and SSHA algorithms to handle the greater insertion and deletion frequency observed with the Ion PGM^{TM}. This software is also optimized to handle a broad range of sequence read lengths and reads that are more difficult to map due to low-quality scores. Compared to other sequencing technologies, the Ion PGM^{TM} is prone to certain types of errors that are mostly related to the detection method. For instance, stretches of identical nucleotides are not always reported accurately. As described in Section 3, the semiconductor chip electrically measures changes in pH during "normal" nucleotide incorporation in real time. Thus, if the polymerase encounters a region with identical consecutive deoxynucleotides, the electrical signal will increase proportionally. However, the signal variance also increases with increasing incorporations such that the software is sometimes unable to correctly determine the exact number of incorporations. This property is responsible for the greater insertion and deletion rate observed, especially with respect to sequences containing long stretches of the same nucleotide. It is for this reason that we recommend not using a restriction enzyme containing consecutive identical nucleotides, as the percentage of reads containing restriction sites will likely be slightly underestimated and not as reliable for quality control. Although in principle Torrent 5C data can be mapped with any software, the one selected should thus be efficient at handling gapped alignments.

To measure 5C contacts in our *HoxA* libraries, we mapped the Torrent sequencing data against a customized reference file with TMAP (Fig. 5.4C). The custom reference consists of a list of all possible products between forward and reverse 5C primers without the sequencing tails in FASTA format. This list replaces the reference genome normally used for mapping and increases mapping speed and accuracy as compared to individual alignment of each sequence halves against the human genome. This approach also simplifies data extraction since the abundance of a 5C contact directly corresponds to the number of times it is counted in the mapped sequence dataset. The mapped sequence output (SAM format) is then uploaded on our private instance of Galaxy and processed through a Torrent 5C data transformation pipeline (Fig. 5.4C) (Blankenberg et al., 2010; Giardine et al., 2005; Goecks, Nekrutenko, & Taylor, 2010). Through this pipeline, the mapped data are first filtered to remove low-quality reads (MAQ quality score of lower than 30), reads aligning more than two nucleotides away from the reference sequence start site, and reads which do not contain any restriction sites. The resulting filtered SAM file is then converted into an interaction frequency list (IFL) by counting the occurrence of each possible contact. This file can finally be transformed into a table of read counts compatible with the my5C visualization tool (Lajoie et al., 2009). This step requires a "my5C key" file of the regions characterized, which identifies the position of each forward and reverse 5C primer in a reference genome (e.g., hg19). The positions are used to label each pairwise interaction with genomic coordinates.

4.3. Analysis formats

We routinely use two file formats to analyze and visualize 5C data. A simple way to examine 5C data is in the form of an IFL with a spreadsheet program such as Excel. The IFL is a tab-delimited text file containing multiple columns and as many rows as there are possible contacts in the experimental design. Each column in the IFL represents a different feature, including the name of forward and reverse 5C primers, the number of read counts of the corresponding pairwise interaction, and the distance between them. This type of list is very useful when conducting basic analysis such as when examining the interaction between a given fragment and the remaining region of interest.

We also find the "my5C" format generated at the end of our 5C mapping pipeline very useful for data analysis. The my5C format is a text file containing a table of read counts that can be uploaded directly to the my5C viewer (Lajoie & Dekker, 2009; Lajoie et al., 2009) for visualization in heatmap form. Figure 5.5A shows the results of the *HoxA* cluster Torrent 5C analysis in undifferentiated and differentiated THP-1 cells. As we previously reported, more interactions could be detected throughout the *HoxA* cluster after differentiation with PMA as compared to the undifferentiated DMSO control. We tested the reproducibility of the Torrent 5C approach by generating three independent technical replicates for each THP-1 states. We found a very high correlation within the undifferentiated (Pearson's $r=0.999$) and the differentiated (Pearson's $r=0.999$) datasets, demonstrating the reproducibility of the PGM^{TM}. The my5C heatmaps are linked to the UCSC genome browser and facilitates the examination of the three-dimensional contacts in the

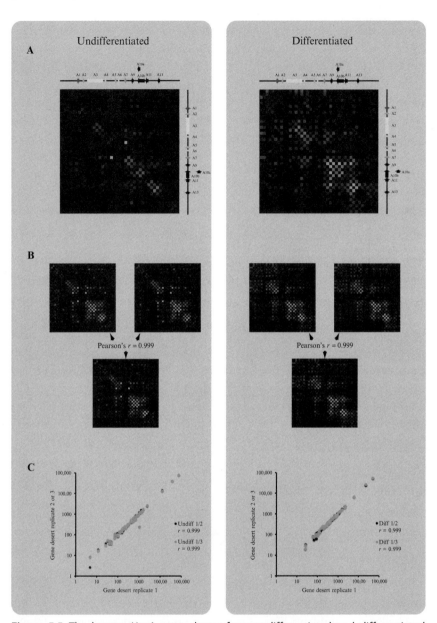

Figure 5.5 The human *HoxA* gene cluster from undifferentiated and differentiated THP-1 cells analyzed by Ion Torrent sequencing. (A) Heatmap representation of the *HoxA* gene cluster Torrent 5C data in undifferentiated (left) and differentiated (right) THP-1 cells. (B) Heatmap representation of 3 *HoxA* Torrent 5C technical replicates. The replicates are from a different set of 5C libraries than shown in (A) and were each processed for Torrent 5C sequencing three times. (C) Correlation of the gene desert data between the three technical replicates after normalization.

context of the two-dimensional annotations present within the browser. The my5C Web site also includes a suite of very useful online analysis tools, including moving window analysis, smoothing, fixed point graphs, and differential analysis between dataset pairs (Lajoie et al., 2009).

4.4. Data quality control and normalization
4.4.1 Control with a gene desert region
The quality of 5C datasets can be verified by examining the interaction profile of a control region. The control region selected should bear a predictable chromatin architecture that preferentially remains unchanged between datasets because this region should also be used to calculate a normalization factor (see below). We recommend using gene desert regions for quality control because their interaction profiles are usually very predictable and constant. Interaction frequencies usually decrease with increasing distance in these regions and reach a minimum at approximately 12–20 kb under the 3C experimental conditions we previously described (Dostie & Dekker, 2007; Dostie et al., 2007; Fraser, Rousseau, Blanchette, et al., 2010).

4.4.2 Normalization between different datasets
In addition to its value in the quality control of 5C data, the gene desert region is also useful to normalize between datasets. This normalization approach assumes that the chromatin organization of the region selected does not significantly change in different cell types or states. Although any invariable genomic domain could in principle serve as normalization reference, gene desert regions are particularly attractive because the absence of transcription units makes them more likely to remain constant across samples. Normalization with an internal reference controls for differences at many levels including variations in cutting efficiency, the amount of 3C template used to generate Torrent 5C libraries, ligation efficiency, differences in the total number of read counts, and many more parameters. In the event that the entire domain stays unchanged, all points of the interaction profile can be used to calculate a normalization factor that will be applied to one of the datasets. When long-range interactions tend to fluctuate because of their low interaction frequencies, we recommend using only neighboring interactions or interactions separated by <8 kb. Normalization factors are defined by first calculating the $\log_{10}$ ratio of each corresponding pairwise interaction frequencies. These ratios are then averaged and returned to the original base 10 notations. The value derived from this calculation is

the normalization factor used to either divide or multiply one of the datasets. Direct comparison between regions of interests in different datasets becomes possible after this step.

Genomic regions where chromatin structure can be predicted are interesting control tools because they measure a number of parameters that reflect the overall quality of 5C experiments. Normalization can, however, also be achieved by transforming read counts into $\log_{10}$ values and calculating Z-scores for each contact. This method, which is available on the my5C browser, was previously used to normalize 5C data from a genomic region containing the alpha-globin gene domain (Bau et al., 2011). A similar approach was also used for 5C data normalization in a study characterizing the three-dimensional architecture of a *Caulobacter* genome (Umbarger et al., 2011).

5. CONCLUSION

The relationship between three-dimensional chromatin organization and genome function remains poorly understood. This research field has greatly benefitted from a number of important technological advances over the past decade, which includes the development of 3C technologies. The 5C technique is among the most quantitative of these approaches and is ideally suited to study high-resolution genome architecture on a megabase scale. The method we describe here combines 5C and deep sequencing with the Ion Torrent PGM^{TM} to improve the accuracy, speed, and convenience of the approach. We believe that this protocol will make high-resolution, high-throughput chromatin structure analysis accessible and affordable to everyone.

ACKNOWLEDGMENTS

We thank members of our laboratory for critical reading of this chapter. We thank the Canadian Institutes of Health Research (CIHR; MOP-86716) and the Canadian Cancer Society Research Institute (CCSRI; 019252) for funding support. We are also grateful to the CIHR for graduate scholarships awarded to J. F. (Frederick Banting and Charles Best Canada Graduate Scholarship) and S. D. E. (Systems Biology Training Program). J. D. is a CIHR New Investigator and FRSQ Research Scholar (Fonds de la Recherche en Santé du Québec).

REFERENCES

Ayton, P. M., & Cleary, M. L. (2003). Transformation of myeloid progenitors by MLL oncoproteins is dependent on Hoxa7 and Hoxa9. *Genes & Development, 17*, 2298–2307.

Bau, D., Sanyal, A., Lajoie, B. R., Capriotti, E., Byron, M., Lawrence, J. B., et al. (2011). The three-dimensional folding of the alpha-globin gene domain reveals formation of chromatin globules. *Nature Structural and Molecular Biology, 18*, 107–114.

Blankenberg, D., Vonkuster, G., Coraor, N., Ananda, G., Lazarus, R., Mangan, M., et al. (2010). Galaxy: A web-based genome analysis tool for experimentalists. *Current Protocols in Molecular Biology*, Chapter 19, Unit 19.10.11-21.

Cremer, T., & Cremer, M. (2010). Chromosome territories. *Cold Spring Harbor Perspectives in Biology*, 2, a003889.

Dekker, J., Rippe, K., Dekker, M., & Kleckner, N. (2002). Capturing chromosome conformation. *Science*, 295, 1306–1311.

Dostie, J., & Dekker, J. (2007). Mapping networks of physical interactions between genomic elements using 5C technology. *Nature Protocols*, 2, 988–1002.

Dostie, J., Richmond, T. A., Arnaout, R. A., Selzer, R. R., Lee, W. L., Honan, T. A., et al. (2006). Chromosome Conformation Capture Carbon Copy (5C): A massively parallel solution for mapping interactions between genomic elements. *Genome Research*, 16, 1299–1309.

Dostie, J., Zhan, Y., & Dekker, J. (2007). Chromosome conformation capture carbon copy technology. *Current Protocols in Molecular Biology*, Chapter 21, Unit 21.14.

Duan, Z., Andronescu, M., Schutz, K., McIlwain, S., Kim, Y. J., Lee, C., et al. (2010). A three-dimensional model of the yeast genome. *Nature*, 465, 363–367.

Ethier, S. D., Miura, H., & Dostie, J. (2012). Discovering genome regulation with 3C and 3C-related technologies. *Biochimica et Biophysica Acta*, 1819(5), 401–410.

Ferraiuolo, M. A., Rousseau, M., Miyamoto, C., Shenker, S., Wang, X. Q., Nadler, M., et al. (2010). The three-dimensional architecture of Hox cluster silencing. *Nucleic Acids Research*, 38, 7472–7484.

Fraser, J., & Dostie, J. (2009). http://Dostielab.biochem.mcgill.ca.

Fraser, J., Rousseau, M., Blanchette, M., & Dostie, J. (2010). Computing chromosome conformation. *Methods in Molecular Biology*, 674, 251–268.

Fraser, J., Rousseau, M., Shenker, S., Ferraiuolo, M. A., Hayashizaki, Y., Blanchette, M., et al. (2009). Chromatin conformation signatures of cellular differentiation. *Genome Biology*, 10, R37.

Giardine, B., Riemer, C., Hardison, R. C., Burhans, R., Elnitski, L., Shah, P., et al. (2005). Galaxy: A platform for interactive large-scale genome analysis. *Genome Research*, 15, 1451–1455.

Goecks, J., Nekrutenko, A., & Taylor, J. (2010). Galaxy: A comprehensive approach for supporting accessible, reproducible, and transparent computational research in the life sciences. *Genome Biology*, 11, R86.

Kalhor, R., Tjong, H., Jayathilaka, N., Alber, F., & Chen, L. (2011). Genome architectures revealed by tethered chromosome conformation capture and population-based modeling. *Nature Biotechnology*, 30, 90–98.

Kroon, E., Krosl, J., Thorsteinsdottir, U., Baban, S., Buchberg, A. M., & Sauvageau, G. (1998). Hoxa9 transforms primary bone marrow cells through specific collaboration with Meis1a but not Pbx1b. *The EMBO Journal*, 17, 3714–3725.

Lajoie, B.R, & Dekker, J. (2009). http://my5c.umassmed.edu/welcome/welcome.php.

Lajoie, B. R., van Berkum, N. L., Sanyal, A., & Dekker, J. (2009). My5C: Web tools for chromosome conformation capture studies. *Nature Methods*, 6, 690–691.

Langmead, B., Trapnell, C., Pop, M., & Salzberg, S. L. (2009). Ultrafast and memory-efficient alignment of short DNA sequences to the human genome. *Genome Biology*, 10, R25.

Li, H., & Durbin, R. (2009). Fast and accurate short read alignment with Burrows-Wheeler transform. *Bioinformatics*, 25, 1754–1760.

Li, H., & Durbin, R. (2010). Fast and accurate long-read alignment with Burrows-Wheeler transform. *Bioinformatics*, 26, 589–595.

Li, H., Ruan, J., & Durbin, R. (2008). Mapping short DNA sequencing reads and calling variants using mapping quality scores. *Genome Research*, 18, 1851–1858.

Lieberman-Aiden, E., van Berkum, N. L., Williams, L., Imakaev, M., Ragoczy, T., Telling, A., et al. (2009). Comprehensive mapping of long-range interactions reveals folding principles of the human genome. *Science, 326,* 289–293.
Ling, J. Q., Li, T., Hu, J. F., Vu, T. H., Chen, H. L., Qiu, X. W., et al. (2006). CTCF mediates interchromosomal colocalization between Igf2/H19 and Wsb1/Nf1. *Science, 312,* 269–272.
Miele, A., Gheldof, N., Tabuchi, T. M., Dostie, J., & Dekker, J. (2006). Mapping chromatin interactions by Chromosome Conformation Capture (3C). In F. M. Ausubel, R. Brent, R. E. Kingston, D. D. Moore, J. G. Seidman & J. A. Smith, et al. *Current protocols in molecular biology* (pp. 21.11.20–21.11.21). Hoboken, NJ: John Wiley & Sons.
Ning, Z., Cox, A. J., & Mullikin, J. C. (2001). SSAHA: A fast search method for large DNA databases. *Genome Research, 11,* 1725–1729.
Pession, A., Martino, V., Tonelli, R., Beltramini, C., Locatelli, F., Biserni, G., et al. (2003). MLL-AF9 oncogene expression affects cell growth but not terminal differentiation and is downregulated during monocyte-macrophage maturation in AML-M5 THP-1 cells. *Oncogene, 22,* 8671–8676.
Pombo, A., & Branco, M. R. (2007). Functional organisation of the genome during interphase. *Current Opinion in Genetics and Development, 17,* 451–455.
Pomerantz, M. M., Ahmadiyeh, N., Jia, L., Herman, P., Verzi, M. P., Doddapaneni, H., et al. (2009). The 8q24 cancer risk variant rs6983267 shows long-range interaction with MYC in colorectal cancer. *Nature Genetics, 41,* 882–884.
Rothberg, J. M., Hinz, W., Rearick, T. M., Schultz, J., Mileski, W., Davey, M., et al. (2011). An integrated semiconductor device enabling non-optical genome sequencing. *Nature, 475,* 348–352.
Schmidt, D., Schwalie, P. C., Ross-Innes, C. S., Hurtado, A., Brown, G. D., Carroll, J. S., et al. (2010). A CTCF-independent role for cohesin in tissue-specific transcription. *Genome Research, 20,* 578–588.
Simonis, M., Klous, P., Splinter, E., Moshkin, Y., Willemsen, R., de Wit, E., et al. (2006). Nuclear organization of active and inactive chromatin domains uncovered by chromosome conformation capture-on-chip (4C). *Nature Genetics, 38,* 1348–1354.
Smith, T. F., & Waterman, M. S. (1981). Identification of common molecular subsequences. *Journal of Molecular Biology, 147,* 195–197.
Spilianakis, C. G., & Flavell, R. A. (2004). Long-range intrachromosomal interactions in the T helper type 2 cytokine locus. *Nature Immunology, 5,* 1017–1027.
Splinter, E., Heath, H., Kooren, J., Palstra, R. J., Klous, P., Grosveld, F., et al. (2006). CTCF mediates long-range chromatin looping and local histone modification in the beta-globin locus. *Genes & Development, 20,* 2349–2354.
Suzuki, H., Forrest, A. R., van Nimwegen, E., Daub, C. O., Balwierz, P. J., Irvine, K. M., et al. (2009). The transcriptional network that controls growth arrest and differentiation in a human myeloid leukemia cell line. *Nature Genetics, 41,* 553–562.
Thorsteinsdottir, U., Sauvageau, G., Hough, M. R., Dragowska, W., Lansdorp, P. M., Lawrence, H. J., et al. (1997). Overexpression of HOXA10 in murine hematopoietic cells perturbs both myeloid and lymphoid differentiation and leads to acute myeloid leukemia. *Molecular and Cellular Biology, 17,* 495–505.
Tolhuis, B., Palstra, R. J., Splinter, E., Grosveld, F., & de Laat, W. (2002). Looping and Interaction between hypersensitive sites in the active beta-globin locus. *Molecular Cell, 10,* 1453–1465.
Torrent/Applications, I. (2011). http://www.iontorrent.com/applications/.
Torrent/Applications, I. (2012). http://www.invitrogen.com/site/us/en/home/Products-and-Services/Applications/Sequencing/Semiconductor-Sequencing/Publications.html.

Towbin, B. D., Meister, P., & Gasser, S. M. (2009). The nuclear envelope—A scaffold for silencing? *Current Opinion in Genetics and Development, 19*, 180–186.

Umbarger, M. A., Toro, E., Wright, M. A., Porreca, G. J., Bau, D., Hong, S. H., et al. (2011). The three-dimensional architecture of a bacterial genome and its alteration by genetic perturbation. *Molecular Cell, 44*, 252–264.

Würtele, H., & Chartrand, P. (2006). Genome-wide scanning of HoxB1-associated loci in mouse ES cells using an open-ended Chromosome Conformation Capture methodology. *Chromosome Research, 14*, 477–495.

Zhao, Z., Tavoosidana, G., Sjolinder, M., Gondor, A., Mariano, P., Wang, S., et al. (2006). Circular chromosome conformation capture (4C) uncovers extensive networks of epigenetically regulated intra- and interchromosomal interactions. *Nature Genetics, 38*, 1341–1347.

SECTION 3

Genome wide Analyses of Chromatin and Transcripts

CHAPTER SIX

Genome-Wide Mapping of Nucleosomes in Yeast Using Paired-End Sequencing

Hope A. Cole, Bruce H. Howard, David J. Clark[1]

Program in Genomics of Differentiation, *Eunice Kennedy Shriver* National Institute of Child Health and Human Development, National Institutes of Health, Bethesda, Maryland, USA
[1]Corresponding author: e-mail address: clarkda@mail.nih.gov

Contents

1. Introduction   146
2. Preparation of Nucleosome Core Particles from Yeast   148
   2.1 MNase digestion of yeast nuclei   149
   2.2 Gel purification of nucleosome core particle DNA   150
3. Preparation of Core Particle DNA for Sequencing   152
   3.1 Repair of purified mononucleosomal DNA   152
   3.2 Quality control for purified mononucleosomal DNA   153
   3.3 Modification of the ends of the nucleosomal DNA prior to adapter ligation 154
   3.4 Ligation of paired-end adapters to nucleosomal DNA   155
   3.5 PCR amplification of adapter-ligated nucleosomal DNA   157
4. Paired-End Sequencing   158
5. Bioinformatic Analysis of Nucleosome Sequences   159
   5.1 Data files   159
   5.2 Length distribution histograms   160
   5.3 Nucleosome occupancy maps   161
   5.4 Nucleosome position maps   164
6. Some General Experimental Considerations   165
   6.1 Sequence bias of MNase   165
   6.2 Formaldehyde cross-linking of chromatin prior to mapping   166
   6.3 Analysis of all MNase digestion products   166
Acknowledgments   166
References   167

Abstract

The DNA of eukaryotic cells is packaged into chromatin by histone proteins, which play a central role in regulating access to genetic information. The nucleosome core is the basic structural unit of chromatin: it is composed of an octamer of the four major core histones (two molecules each of H2A, H2B, H3, and H4), around which are wrapped ~ 1.75 negative superhelical turns of DNA, a total of 145–147 bp. Nucleosome cores are regularly spaced along the DNA *in vivo*, separated by linker DNA. Nucleosomes are compact structures capable of blocking access to the DNA that they contain. For example, they may prevent the binding of transcription factors to their cognate sites. It is therefore very important to obtain quantitative information on the positions of nucleosomes with respect to regulatory regions *in vivo*. The advent of high-throughput sequencing methods has revolutionized this field. We describe the use and advantages of paired-end sequencing to map nucleosomal DNA obtained by micrococcal nuclease digestion of budding yeast nuclei. This approach provides high-quality genome-wide nucleosome occupancy and position maps.

1. INTRODUCTION

Eukaryotic DNA is organized into chromatin in order to pack it efficiently into the cell nucleus. The basic structural unit of chromatin is the nucleosome core, which is composed of an octamer of the four major core histones (two molecules each of H2A, H2B, H3, and H4), around which are wrapped $\sim$1.75 negative superhelical turns of DNA, a total of 145–147 bp (Luger, Mäder, Richmond, Sargent, & Richmond, 1997). Nucleosome cores are regularly spaced along the DNA *in vivo*, separated by "linker" DNA. The spacing of nucleosomes ("repeat length") is characteristic of the cell type; in budding yeast it is $\sim$165 bp, corresponding to a linker of 15–20 bp.

The nucleosome is a very compact structure, severely limiting access to the DNA within it. For example, the inclusion of the binding site for a specific regulatory factor within a nucleosome is usually sufficient to reduce or prevent the binding of that factor to its site (e.g., Imbalzano, Kwon, Green, & Kingston, 1994). There are some exceptions, termed "pioneer factors" (Zaret et al., 2008). Thus, the precise positioning of nucleosomes with respect to DNA sequence *in vivo* is of great interest. Various types of chromatin remodeling enzymes affect the position of a nucleosome with respect to regulatory sites, or the degree to which the DNA within the nucleosome is accessible. These include ATP-dependent remodeling complexes which can move a nucleosome from one position to another, or effect conformational changes in nucleosomes (reviewed by Cairns, 2005). In addition, histone-modifying complexes affect histone–DNA interactions and mark nucleosomes for recruitment of additional regulatory factors, by catalyzing posttranslational modifications such as acetylation, methylation, and phosphorylation (reviewed by Gardner, Allis, & Strahl, 2011).

The nucleosome core can be isolated as a particle containing ~147 bp of DNA by digestion of chromatin with micrococcal nuclease (MNase). Early in digestion, MNase cleaves the linker DNA endonucleolytically to yield chromatin fragments with protruding linker DNA. Later in digestion, this linker DNA is trimmed off by MNase acting as an exonuclease, which eventually stops at the border of the nucleosome, resulting in a core particle (reviewed by Clark, 2010). Therefore, the position of a nucleosome can be defined by the DNA sequence within the core particle. Thus, a map of nucleosome positions along a DNA of interest can be obtained by determining the sequences of DNA isolated from nucleosome core particles.

There are two types of nucleosome map: the "occupancy" or "coverage" map and the "position" map (Segal & Widom, 2009). The occupancy map is a plot of the occurrence of a particular base pair in a nucleosome as a function of the chromosomal coordinate. The position map is a histogram of the observed occurrence of a nucleosome at each possible position defined by the midpoint of a 147-bp window (corresponding to the nucleosomal dyad) along the chromosome. If all cells have every nucleosome in the same position all the time, then the occupancy and position maps would be redundant. However, there are in fact multiple alternative positions for virtually all nucleosomes, and therefore the occupancy and position maps display different information (Cole, Howard, & Clark, 2011a).

Recently, two different methods have been adapted for genome-wide mapping of nucleosome positions using nucleosome core particle DNA: (1) hybridization of nucleosomal DNA to tiled oligonucleotide microarrays (Lee et al., 2007; Yuan et al., 2005) and (2) high-throughput sequencing of DNA from individual core particles (Albert et al., 2007; Field et al., 2008; Kaplan et al., 2009; Valouev et al., 2008; Zhang et al., 2009). If we use the strict definition of the position of a nucleosome—that it is defined by the DNA sequence within the core particle, it is clear that a hybridization method cannot yield positioning data, because it cannot determine the borders of individual nucleosomes. Instead, the microarray method gives only a nucleosome occupancy map, which can be thought of as averaged positioning or low-resolution mapping. It is a valuable method, but it is limited in scope. The next-generation sequencing methods constitute a major advance over the microarray method, because they provide both position and occupancy information; the position data from millions of individual nucleosome sequences can be used to build the occupancy map.

Most genome-wide nucleosome sequencing studies have utilized single-end sequencing of nucleosomal DNA, which provides the sequence for one end of each nucleosome (e.g., the first 40 nt). The position is then inferred, assuming that each sequence represents an intact 147-bp core particle. However, the accuracy of a position depends on how precisely the nucleosome core was trimmed by MNase; if there is residual linker DNA (DNA ≳ 147 bp), or if there is an internal cleavage of the nucleosome by MNase (DNA ≲ 147 bp), the position determined will be relatively inaccurate. This problem can be solved using 454-sequencing which can give read lengths greater than the size of the nucleosome (Field et al., 2008), but the number of reads generated by this technology is currently far lower than that generated by Illumina sequencing. Both problems are solved by using Illumina paired-end sequencing, in which both ends of the same DNA molecule are sequenced and then aligned to the genome to deduce the length of the sequenced fragment. Paired-end sequencing makes it possible to distinguish intact, fully trimmed, core particles from those that are too long or too short. The sequences of fully trimmed intact core particles can be used to define a set of accurate positions (Cole et al., 2011a; Cole, Howard, & Clark, 2011b). Paired-end sequencing of nucleosomal DNA has also been described by others (Hennikoff, Belsky, Krassovsky, MacAlpine, & Henikoff, 2011; Kent, Adams, Moorhouse, & Paszkiewicz, 2011).

Briefly, to derive the occupancy map using paired-end data, the deduced nucleosome sequences are aligned along the chromosome, and the number of times each coordinate base appears in a nucleosome sequence is summed along the chromosome. The position map is obtained by determining the midpoint coordinate of every aligned nucleosome sequence and counting the number of times the same midpoint occurs as a function of the chromosomal coordinate.

Below, we describe the experimental methods and analysis that we employ to measure nucleosome occupancy and positioning using genome-wide paired-end sequencing of *Saccharomyces cerevisiae* (Cole et al., 2011a, 2011b).

2. PREPARATION OF NUCLEOSOME CORE PARTICLES FROM YEAST

The aim of this protocol is to purify the DNA from intact, fully trimmed, nucleosome core particles. Nuclei are prepared from budding yeast cells and digested extensively with MNase. Mononucleosomal DNA is then gel-purified.

2.1. MNase digestion of yeast nuclei

Nuclei are purified as described previously (Kim, Shen, & Clark, 2004), with some modifications.

1. Prepare buffers and solutions as follows:

    Spheroplasting medium (SM) (growth medium containing 1 M sorbitol to stabilize spheroplasts) (500 ml): 10 g D-glucose, 3.35 g yeast nitrogen base (Difco), 0.41 g complete supplement mixture, and 91.1 g sorbitol (1 M).

    ST Buffer (500 ml): 91.1 g Sorbitol (1 M) and 25 ml of 1 M Tris–HCl pH 8.0 (50 mM).

    F-Buffer (500 ml): 90 g Ficoll-400 (18% w/v), 20 ml of 1 M K-phosphate pH 7.0 (40 mM), and 500 µl of 1 M MgCl$_2$ (1 mM). Adjust pH to 6.5. *Note*: Ficoll dissolves very slowly.

    FG Buffer (500 ml): 35 g Ficoll-400 (7% w/v), 100 ml glycerol (20%), 20 ml of 1 M K-phosphate pH 7.0 (40 mM), and 500 µl of 1 M MgCl$_2$ (1 mM). Adjust pH to 6.5.

    MNase: Dissolve MNase (Worthington) to 10 U/µl in 5 mM Na-phosphate pH 7.0, 0.025 mM CaCl$_2$. Store in aliquots at -80 °C.

    MNase Digestion Buffer (10 ml): 100 µl of 1 M HEPES pH 7.5 (10 mM), 70 µl of 5 M NaCl (35 mM), 5 µl of 1 M MgCl$_2$ (0.5 mM), and 50 µl of CaCl$_2$ (0.5 mM).

    MNase Stop Solution (1 ml): 130 µl of 0.5 M Na-EDTA (65 mM), 360 µl of 20% SDS (7%), and 510 µl water.

2. Thaw a frozen yeast pellet derived from 250 ml culture at OD$_{600}$ between 0.5 and 0.8 in 12.5 ml SM buffer containing 20 mM 2-mercaptoethanol (2-ME) by warming to 30 °C for 20 min.

3. Dissolve 25,000 units of high-grade lyticase (Sigma L-5263) in 3 ml SM buffer with 20 mM 2-ME and warm to 30 °C. Follow spheroplasting by measuring the OD$_{600}$ of cells dispersed in 1% SDS: use 30 µl cells in 1 ml of 1% SDS at zero-time. Add the lyticase to the cells and continue incubating at 30 °C, with occasional gentle agitation to prevent the cells from settling. Measure the OD$_{600}$ of 37.5 µl cells in 1 ml of 1% SDS after 5, 10, and 20 min (using 37.5 µl cells for time-points after zero-time accounts for the dilution due to the enzyme).

4. Collect spheroplasts when the OD$_{600}$ reaches ~10% of the starting value, usually 20–30 min, by centrifugation (7500 rpm, 5 min, Sorvall SA600 rotor, 4 °C). Wash once with 25 ml ST buffer and collect again by centrifugation (7500 rpm, 5 min, Sorvall SA600 rotor, 4 °C).

5. Lyse spheroplasts by vigorous resuspension in 20 ml F-Buffer containing 5 mM 2-ME and protease inhibitors (Roche EDTA-free tablet 05056489001). Then layer the resuspension on top of 15 ml FG Buffer containing 5 mM 2-ME and protease inhibitors as above in a high-speed polycarbonate tube with lid and spin (12,500 rpm, 20 min, Sorvall SA600 rotor, 4 °C).
6. Drain the nuclear pellet and place on ice. Resuspend the nuclei in 2.4 ml MNase Digestion Buffer containing 5 mM 2-ME and protease inhibitors as above, and divide into six aliquots of 400 μl in 1.7-ml microcentrifuge tubes.
7. Warm the aliquots of nuclei to room temperature for 2 min and then add increasing amounts of MNase (10 U/μl) to each tube and mix thoroughly, for example, 3, 6, 12, 24, 48, and 96 μl MNase. Incubate the digests at room temperature for 3 min. Stop the reaction by adding 50 μl MNase Stop Solution to each digestion, mixing, and leaving at room temperature for 5 min.
8. Add 25 μl of 20% SDS to each tube and mix (final SDS ~1.8% and final volume ~500 μl). Then add 125 μl of 5 M potassium acetate (~1 M) and mix.
9. Extract the DNA by adding 625 μl chloroform, mixing, and spinning in a microcentrifuge (5 min at top speed).
10. Repeat extraction: Transfer the supernatants to 1.7-ml tubes. Add 625 μl chloroform, mix, and spin down in microcentrifuge (5 min at top speed).
11. Transfer the supernatants to new 1.7-ml tubes and add 0.7 vol. isopropanol (490 μl) to precipitate the DNA. Mix and leave at −20 °C for 1 h. Pellet the DNA by spinning in a microcentrifuge (30 min at top speed).
12. Discard the supernatant. Wash the pellet once with 1 ml of 70% EtOH and spin in a microcentrifuge (5 min at top speed).
13. Discard the supernatant and dissolve each pellet in 50 μl of 10 mM Tris–HCl pH 8.0, 1 mM EDTA with 0.1 mg/ml RNase. Leave the samples overnight at room temperature. Store the DNA at -20 °C.

2.2. Gel purification of nucleosome core particle DNA

The MNase-digested DNA samples are analyzed in a gel to determine the extent of digestion. Mononucleosomal DNA is gel-purified from appropriately digested samples.

1. Analyze 5 µl of each DNA sample in a 2% (w/v) agarose gel (use agarose appropriate for small DNA fragments for maximum band resolution). Consider omitting bromophenol blue from the sample buffer, because the dye tracks with mononucleosomal DNA unless the gel is run far enough to separate the two. *Note*: For a DNA size marker, we use an *Msp*I digest of pBR322 (NEB N3032), which is particularly useful because it has marker bands at 147 and 160 bp.
2. Decide which MNase samples are appropriate for gel purification. Most of the DNA samples from the MNase titration should be predominantly mononucleosomal; an example is shown in Fig. 6.1. There are always some undigested dinucleosomes. Use the 160 bp and 147 bp bands in the pBR322 *Msp*I marker to assess the extent of mononucleosome digestion and trimming: The sample treated with the least MNase contains a strong mononucleosomal band migrating with or slightly above the 160 bp marker band; this corresponds to incompletely trimmed nucleosomes with protruding linker DNA that has not been trimmed off by MNase. In samples treated with more MNase, this band shifts to shorter

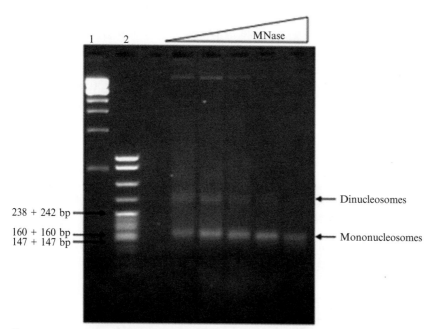

Figure 6.1 Titration of yeast nuclei with MNase. DNA purified from nuclei digested with increasing amounts of MNase was analyzed in a 2% agarose gel stained with ethidium bromide. DNA size markers: 1-kb DNA ladder (1); pBR322 *Msp*I digest (2).

sizes, toward the 147 bp marker band, as the mononucleosomes are trimmed to core particles. In the samples treated with the highest amounts of MNase, faint bands appear at <147 bp, which result from internal degradation of core particles. It is clear that a balance has to be struck between too little MNase digestion (significant dinucleosomes and oligonucleosomes remaining and incomplete trimming of mononucleosomes) and too much digestion (internal cleavage of core particles by MNase). We gel-purify the mononucleosomal DNA from the three best samples in the MNase titration and decide which one to send for sequencing after more quality control steps (see below).

3. Load the remaining ~45 μl of the chosen samples in a 2% agarose gel made using a preparative comb. Use the 1-kb ladder and the pBR322 *Msp*I digest as DNA length markers. Run the gel to obtain good band separation.
4. Excise the mononucleosomal bands from each lane. Proceed rapidly to minimize exposure to ultraviolet light and consequent DNA damage.
5. Purify the DNA from the gel pieces using the freeze/squeeze method (BioRad 732-6166): incubate at -20 °C for 30 min before spinning at top speed in a microcentrifuge for 5 min. Transfer the flow-through to a 1.7-ml microfuge tube and measure the volume.
6. Precipitate the DNA: Add 1 μl of 20 mg/ml glycogen (Roche 10901393001) and one-ninth volume 5 M potassium acetate and mix. Add 0.7 vol. isopropanol, mix, and leave at -20 °C for 1 h. Pellet the DNA by spinning in a microcentrifuge at top speed for 30 min. Remove the supernatant and wash the pellet once with 1 ml of 70% EtOH; spin in the microcentrifuge at top speed for 10 min. Remove the supernatant and dry the pellet. Resuspend each pellet in 15 μl of 10 mM Tris–HCl pH 8.0, 0.1 mM Na-EDTA, and store at -20 °C.

3. PREPARATION OF CORE PARTICLE DNA FOR SEQUENCING

Purified mononucleosomal DNA is first treated with DNA repair enzymes, subjected to quality control, and then processed for paired-end sequencing as described in the protocol for paired-end sequencing provided by Illumina, with some minor modifications.

3.1. Repair of purified mononucleosomal DNA

Extensive MNase digestion can result in nicking of the DNA on the outer surface of the nucleosome. This can be tested for by analyzing the nucleosomal DNA in a denaturing urea gel. As a precaution, we treat our mononucleosomal DNA with a mixture of repair enzymes, which will seal nicks and also repair any DNA damage due to exposure to ultraviolet light during gel purification.

1. Retain 0.5 µl mononucleosomal DNA (unrepaired control).
2. Repair the rest of the DNA using the PreCR Repair Mix Kit (NEB M0309) with 100 µM dNTPs in a 50-µl reaction, as described by the manufacturer. Incubate for 1 h at 37 °C and place on ice.
3. Purify the repaired DNA using the QIAquick PCR Purification Kit (Qiagen 28104) and elute with 50 µl Qiagen Elution Buffer.
4. Measure the DNA concentration using A_{260}; a minimum concentration of $\sim$4 ng DNA/µl is required to continue. Store at -20 °C.

3.2. Quality control for purified mononucleosomal DNA

Determine the length of the core particle DNA in a high-resolution polyacrylamide gel and check the DNA for nicks in a denaturing gel. Since the amount of DNA available is limited, it is necessary to analyze radiolabeled DNA.

1. Radioactively end-label aliquots of unrepaired DNA (0.5 µl), repaired DNA (use 1.5 µl), and a 50-bp DNA ladder (NEB) as size marker, in a 15-µl reaction volume, including 10 µl γ^{32}P-ATP (6000 Ci/mmol) and 0.5 µl T4 polynucleotide kinase at 30 U/µl (Affymetrix 70031). Incubate at 37 °C for 1 h.
2. Separate labeled DNA from unincorporated label using a Microspin G25 column (GE Illustra 27-5325-01).
3. Measure the counts in each sample using a liquid scintillation counter.
4. Analysis of DNA size: Load equal counts of labeled mononucleosomal DNA in a 6% (19:1) polyacrylamide native gel. Run the gel at 150 V for 1.5 h and dry.
5. Analysis of nicking: Denature equal counts of unrepaired and repaired DNA in sample buffer containing 85% formamide by heating to 95 °C for 10 min followed by rapid cooling on ice. Load in a denaturing 8% (19:1) polyacrylamide gel containing 50% (w/v) urea ($\sim$8 M). The gel should be prewarmed for 15 min at constant current (50 mA). Run the gel at 50 mA for 2 h. Soak the gel in 250 ml of 15% MeOH, 5% acetic acid for 15 min to remove the urea. Dry the gel.
6. Expose the gels to film or to a phosphorimager screen.
7. Choose the best mononucleosomal sample to proceed with: The majority of the DNA ($\sim$75%) should run next to the 150-bp marker band. There should be few DNA fragments shorter than $\sim$145 bp in the native

gel, and very little nicking as indicated by smearing below the mononucleosome band in the denaturing gel. Examples are shown in Fig. 6.2.

3.3. Modification of the ends of the nucleosomal DNA prior to adapter ligation

Modify the ends of the DNA samples as described in the protocol supplied by the manufacturer (Illumina), such that they can be ligated to the paired-end adapter supplied by Illumina. The adapter has noncomplementary sequences at one end and is therefore single stranded at this end; the other end is double stranded, with a 3′-dT and a 5′-phosphate. Ligation requires the addition of a 5′-phosphate and a 3′-dA to the nucleosomal DNA.

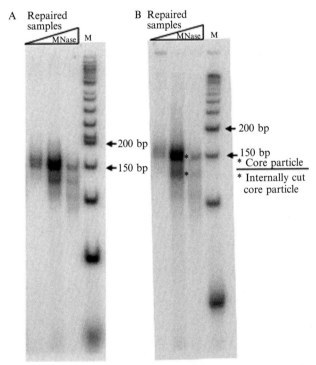

Figure 6.2 Quality control analysis of purified mononucleosomal DNA. Aliquots of gel-purified mononucleosomal DNA corresponding to three different MNase titration points were end-labeled with T4 polynucleotide kinase after treatment with DNA repair enzymes. (A) Analysis in a native 6% polyacrylamide gel. (B) Analysis in a denaturing 8% polyacrylamide gel containing 8 M urea. M = 50-bp DNA ladder. The second sample was chosen for sequencing.

1. Incubate the repaired core particle DNA in T4 polynucleotide kinase buffer with 1 mM ATP and 5 mM dithiothreitol in a total volume of 100 μl. Add 1 μl T4 polynucleotide kinase at 30 U/μl (Affymetrix 70031). Incubate at 37 °C for 1 h.
2. Purify the DNA using a QIAquick PCR purification column (Qiagen 28104) and elute with 30 μl Qiagen Elution Buffer.
3. Add dATP to a final concentration of 0.2 mM in a total volume of 50 μl and add 3 μl Klenow 3′–5′ exo- at 5 U/μl (NEB M0212). Incubate at 37 °C for 30 min.
4. Purify the DNA using a QIAquick PCR purification column and elute with 30 μl Qiagen Elution Buffer.

3.4. Ligation of paired-end adapters to nucleosomal DNA

This protocol for ligation of the paired-end adapter to nucleosomal DNA is modified from the Illumina protocol provided with the Paired-End Sample Prep Kit (Illumina PE-102-1001). The adapter sequence is under copyright but is available from Illumina. The adapter sequence can be modified with a "barcode" so that multiple samples can be mixed in the same sequencing lane.

3.4.1 Barcoded paired-end adapters

The increase in the number of reads obtained by high-throughput sequencing means that adequate coverage of small genomes, such as that of yeast, can still be obtained even if several samples are mixed together in the same sequencing lane. Barcode sequences are used to distinguish one sample from the others. A barcode is a short sequence tag (e.g., 5 bp) that is inserted at the double-stranded end of the adapter, just prior to the 3′-dT overhang. The same PCR primers can be used for all samples in the subsequent PCR amplification step. The beginning of each sequencing read will include the barcode and, consequently, each read pair can be assigned to the correct sample.

3.4.2 Ligation

It may be necessary to titrate the nucleosomal DNA with the paired-end adapter to optimize the amount of nucleosomal DNA with an adapter ligated to both ends. The manufacturer suggests a ratio of 10 adapters to one molecule of target DNA. We have found that 10–100 adapters per target DNA molecule work well for nucleosomal DNA.
1. Ligation: Mix 10 μl nucleosomal DNA, 3.5 μl of the paired-end adapter at 15 μM, and 5 μl Quick T4 DNA ligase (NEB Quick Ligation Kit

M2200) in a total volume of 50 µl at room temperature for 15 min and then place on ice.
2. Purify the DNA using a QIAquick PCR column (Qiagen 28104) and elute with 30 µl Qiagen Elution Buffer.

3.4.3 Purification of ligated DNA

After ligation, the nucleosomal DNA should have an adapter ligated to both ends. The correct product is separated from the other ligation products by gel purification. It is particularly important to remove self-ligated adapters (adapter dimers) which, if present, can dominate the PCR.

1. Load the ligation products in sample buffer lacking bromophenol blue onto a 6% (19:1) native polyacrylamide gel. Skip a lane between the sample and the markers (a 50-bp DNA ladder and an *Msp*I digest of pBR322; both from NEB). Run the gel: 150 V for 2 h. Stain the gel with ethidium bromide (1 µg/ml).
2. Excise the correct band from the gel (Fig. 6.3) as quickly as possible to limit exposure to ultraviolet light. *Important note*: The correct band,

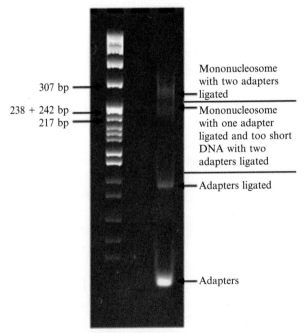

Figure 6.3 Ligation of paired-end adapters to mononucleosomal DNA. Mononucleosomal DNA was ligated to the paired-end adapters; the products were analyzed in a native 6% polyacrylamide gel stained with ethidium bromide. DNA size marker: pBR322 *Msp*I digest.

which should be ~220 bp, corresponding to two adapters ligated to nucleosomal DNA of ~150 bp, runs anomalously slowly in the gel because of the single-stranded adapter termini.
3. Transfer the gel piece to a 1.7-ml tube, crush it into small fragments, and soak the pieces in 200 µl TE with 0.1% SDS overnight on a tube rotator at room temperature. The SDS helps to prevent the gel pieces sticking to one another.
4. Freeze the gel pieces in a freeze–squeeze spin column (BioRad 732-6166) for 30 min at -20 °C. Spin at top speed in the microcentrifuge for 5 min and then transfer the flow-through to a 1.7-ml tube. Measure the volume.
5. Purify the DNA using a QIAquick PCR purification column (Qiagen 28104) and elute with 30 µl Qiagen Elution Buffer.

3.5. PCR amplification of adapter-ligated nucleosomal DNA

The adapter-ligated nucleosomal DNA is amplified by PCR prior to sequencing. The PCR protocol is as suggested by Illumina, using their paired-end PCR primers, PE1.0 (58 nt) and PE2.0 (61 nt); their sequences are available from Illumina. We use only 14 or 15 cycles to minimize possible bias due to PCR amplification. It is a good idea to check that the PCR is in the linear range by measuring the amount of DNA obtained after different numbers of cycles.
1. Use 1.4 µl adapter-ligated nucleosomal DNA in a 70-µl reaction with the Illumina paired-end PCR primers at 0.5 µM and 35 µl of 2× Phusion Master Mix (NEB M0531). Use 14 or 15 cycles. The PCR conditions are as described in the Illumina protocol.
2. Purify the DNA using a QIAquick PCR purification column (Qiagen 28104) and elute with 50 µl Qiagen Elution Buffer.
3. Determine the DNA concentration by measuring A_{260}. Store at −20 °C.
4. Analyze 5 µl of the amplified DNA in a 3% agarose gel. The amplified DNA should be ~280 bp in length, since the PCR primers extend past the ends of the adapters (Fig. 6.4A).
5. Purify the amplified DNA by excising the band from a 2% agarose gel with preparative wells (this step removes any amplified adapter dimers). Measure the weight of the gel slice. Extract the DNA using a QIAquick Gel Extraction Kit (Qiagen 28704) and elute with 50 µl Qiagen Elution Buffer.
6. Determine the DNA concentration by measuring the A_{260}. At least 10 µl of DNA at ~10 ng/µl is required.

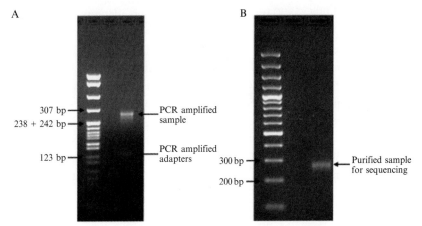

Figure 6.4 PCR amplification of adapter-ligated mononucleosomal DNA. (A) Analysis of PCR products in a 3% agarose gel stained with ethidium bromide. DNA size marker: pBR322 *Msp*I digest. This particular sample contains significant amounts of amplified ligated adapters (no insert); these were removed by subsequent gel purification of the required PCR product. (B) Gel-purified mononucleosomal PCR product: final analysis in a 2% agarose gel prior to sequencing. Size marker: 100-bp DNA ladder.

7. Confirm the quality of the purified DNA by analysis in a 2% agarose or a 6% polyacrylamide gel (Fig. 6.4B).

The nucleosomal DNA is now ready for paired-end sequencing.

4. PAIRED-END SEQUENCING

Our first sequencing reactions were performed at the Tufts University Core Facility (Boston, MA) using a Solexa Genome Analyzer GAII sequencer. They provided sorted ELAND alignments. More recently, Dr. Jun Zhu's Laboratory at the National Heart, Lung and Blood Institute (NHLBI) at the NIH (Bethesda, MD) have performed our sequencing reactions using their Illumina HiSeq machine. They aligned the sequences to the *S. cerevisiae* genome using a Burrows–Wheeler aligner program (Li & Durbin, 2009). The data are supplied in the form of .sam files (Li et al., 2009).

When we first began paired-end sequencing, a sequencing lane in a Solexa Genome Analyzer sequencer yielded ~15 million correctly aligned paired reads. Both reads of the pair were 40 nt. However, the technology is evolving rapidly: the new HiSeq sequencers produce >100 million sequences per lane, with read lengths of up to 100 nt. Given that our data for ~15 million *S. cerevisiae* nucleosomes are highly reproducible (Cole et al., 2011a, 2011b), one HiSeq lane is excess capacity for yeast. Consequently, it is reasonable to barcode and mix four different yeast samples and apply them to a single HiSeq sequencing lane. Since the cost is about the same per lane, the savings can be very large.

The increased number of sequences obtained from a single lane is also making sequencing nucleosomes from organisms with much larger genomes a reasonable proposition (Li, Schug, Tuteja, White, & Kaestner, 2011). The human and mouse genomes are ~ 200 times larger than the yeast genome and so ~ 200 times more sequences are required to obtain the same data quality, or ~ 3 billion sequences, equivalent to ~ 15 lanes. Taking into account that we have found that ~ 7 million sequences are enough to give excellent data for yeast and that some HiSeq lanes can yield 200 million sequences, it seems that just 3 or 4 lanes might give adequate coverage of the human or mouse genome at reasonable expense (the experiment must be repeated, of course). A secondary consideration is that, even though such genome-wide sequencing experiments are expensive, bioinformatic analysis of the data takes a long time and is inexpensive; therefore, the cost is spread over quite a long period.

5. BIOINFORMATIC ANALYSIS OF NUCLEOSOME SEQUENCES

5.1. Data files

The paired-end data are usually supplied as sorted files in which the paired reads have been aligned to the genome. The paired reads are the same length and may be up to 100 nt long; chromosome number and coordinates for both ends of the aligned reads are given. Read sequences are provided together with a quality score that indicates the likelihood that each base-call is correct. Quality scores may follow either the original Sanger convention or the convention adopted in ELAND alignments. In the former case, decoding may be accomplished by a perl script statement such as "integerScore = ord(phredSymbol) - 33," whereas in the latter case, the corresponding statement should be "integerScore = ord(phredSymbol) - 64." A cautionary point for Sanger-encoded scores is that the symbol "@" is the initial letter in the read ID line, but can also occur in the first position in the Phred score. If realignment is to be done, it may be desirable to prefilter the reads, excluding those with mismatches to the reference genome or quality scores less than 15. If Bowtie is used for realignment, the option for .sam format output can be designated. Since the .sam file format is now the most commonly used, there are useful tools for processing .sam files freely available on the Web (Li et al., 2009).

5.2. Length distribution histograms

The length of the nucleosomal DNA fragment is given by the difference between the 5′-coordinate nucleotides of the paired reads. A script designed to collect these lengths can be used to obtain the length distribution histogram of the sample at single-nucleotide resolution (Fig. 6.5). This is important for quality control and also indicates the fraction of intact, fully trimmed core particles. We have found that there is some variation in the length distributions of our samples even though they appear to be very similar in a gel. This is because the mononucleosome band is typically relatively broad and contains core particles and nucleosomes with an additional 10 bp of protruding linker DNA (sometimes two bands are distinguishable). All of our samples show a maximum peak at core particle size, but the length distribution around this peak varies somewhat; some samples are skewed toward the slightly larger particles while others are skewed toward subnucleosomal fragments, reflecting different degrees of digestion by MNase. It is probably important to compare samples with similar size distributions.

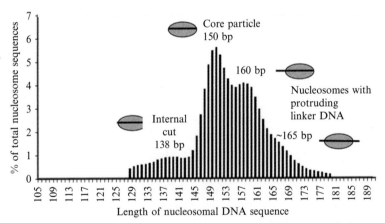

Figure 6.5 Length distribution of a typical mononucleosomal DNA sample. Histogram of the lengths of all nucleosome sequences obtained for a typical sample (∼12 million in this case). The fraction of sequences of a given length is expressed as a percentage of the total. The numbers indicate local peak maxima; the diagrams indicate the nucleosome structure corresponding to each peak. The scale is single-nucleotide resolution. *Adapted from Cole et al. (2011a).*

5.3. Nucleosome occupancy maps

The occupancy map is obtained using an algorithm designed to count the number of nucleosome sequences containing the chromosomal coordinate in question. Our scripts are published as Supplementary Information for Cole et al. (2011a). First, we divided the single alignment file into 16 files, corresponding to all of the paired reads aligning to each of the 16 chromosomes of budding yeast. This was done to reduce search times. The algorithm asks for the chromosome number (e.g., chromosome X) and the coordinates of interest (e.g., your favorite gene (*YFG1*) with some flanking sequences). Then all nucleosome sequences containing any part of the region of interest are found, and the occurrence of each coordinate nucleotide in the target region is counted.

For example, the occupancy of the first nucleotide in the ATG start codon of *YFG1*, the A at coordinate +1, is given by the number of nucleosome sequences containing A1. Even though A1 is likely to be in a "+1" nucleosome (i.e., just downstream of a nucleosome-depleted region), there will be sets of almost identical nucleosome sequences corresponding to several alternative nucleosome positions containing A1. Any nucleosome sequence containing A1 contributes to the occupancy measurement, that is, the relative probability of A1 being in a nucleosome.

For example, in a hypothetical data set (Fig. 6.6A), there are 18 nucleosome sequences containing A1: 3 identical sequences corresponding to nucleosome position 1, 10 sequences corresponding to position 2, and 5 sequences corresponding to position 3. Thus, the occupancy count for A1 is 18, but only 3 for the nucleotides in position 1 that do not overlap position 2, and 5 for those in position 3 sequences which do not overlap position 2. The sum of the occupancies for coordinate nucleotides in each of the three sets of sequences results in the occupancy peak shown in Fig. 6.6C. In practice, heterogeneity in the lengths of sequences derived from the same nucleosome will result in smoothing of the corners of the peaks, as illustrated.

We have found that occupancy data are remarkably similar for different experiments, with very close agreement between biological repeats. The quality of the data is such that it is unnecessary to apply any mathematical smoothing to occupancy plots; the number of sequences containing nucleotide "*n*" can be plotted directly against chromosomal coordinate, *n*. To compare two sets of data, it is necessary only to normalize for the total number of sequences in each data set. Thus, occupancy data require minimal

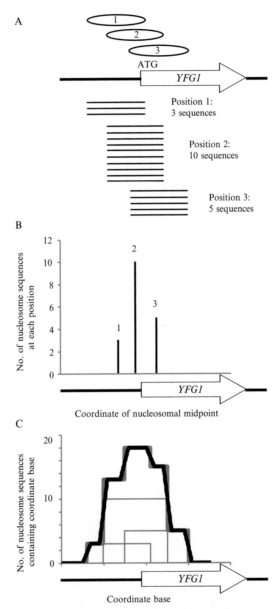

Figure 6.6 Derivation of nucleosome occupancy and position maps. An illustration using a hypothetical data set. (A) The data contain 18 nucleosome sequences (all 150 bp in length) that include the A1 nucleotide in the start codon of the *YFG1* gene. These 18 sequences correspond to three different overlapping positions. (B) Nucleosome position map: the peaks represent the midpoints of the three different positions. (C) Nucleosome occupancy map: the "square" peaks (thin gray lines) represent the occupancy values for the three different positions, which are summed to obtain the occupancy map (thick gray line); this line has been smoothed to simulate the effect of length heterogeneity (black line).

mathematical manipulation. An example of an occupancy map is shown in Fig. 6.7A. We find also that restricting the occupancy plot to nucleosome sequences close to core particle length has only a small effect on the plot, generally making the troughs between nucleosome peaks somewhat more

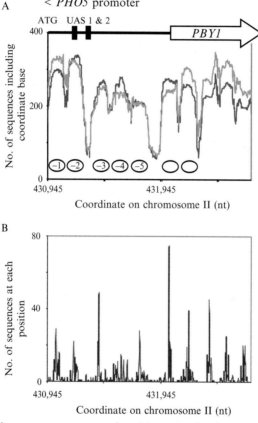

Figure 6.7 Nucleosome occupancy and position maps for the yeast *PHO5* promoter. (A) Occupancy map for the *PHO5* promoter and upstream *PBY1* gene on chromosome II. Two traces are shown: control cells (dark gray) and cells treated with 3-aminotriazole (3AT) (light gray). To normalize for different total numbers of nucleosome sequences in the two samples, the 3AT data were multiplied by 1.27. The data were not manipulated in any other way. Since *PHO5* and *PBY1* are unaffected by 3AT, we expect the two samples to be identical; the reproducibility is very high. Note the nucleosome-depleted regions corresponding to UAS2 (which contains transcription factor binding sites) and the *PBY1* promoter. The ovals represent the *PHO5* nucleosomes mapped by indirect end labeling (Svaren & Hörz, 1997). (B) Position map for nucleosomes in control cells. Note that each occupancy peak in (A) actually represents a cluster of alternative nucleosome midpoints in (B) (as in Fig. 6.6), rather than a single midpoint. Thus, the ovals in (A) actually represent average positions. *Adapted from Cole et al. (2011a).*

prominent (Cole et al., 2011a). We attribute this to removal of nucleosome sequences with protruding linker DNA (>147 bp).

A comparison of two samples can be facilitated using a Web-viewer program to visualize the data and to move along a chromosomal occupancy map.

5.4. Nucleosome position maps

The position map is a more detailed description of the data underlying the occupancy map. Since the position of a nucleosome is defined by the DNA sequence it contains, the nucleosome sequence data are nucleosome position data. Consequently, a position map is just a count of the number of times each particular nucleosome sequence is present in the data set, plotted against the chromosomal coordinate, n, which here represents the midpoint of the nucleosome sequence (dyad axis). An example is given in Fig. 6.6B: nucleosome position 1 is represented by a peak at its midpoint with a count of 3; similarly, nucleosomes 2 and 3 are represented by counts of 10 and 5, respectively, at their midpoints.

The major complication in constructing position maps is the fact that not all nucleosome sequences represent intact, fully trimmed core particles; some are too long (protruding linker DNA) and some are too short (internal cleavage) (Fig. 6.5). Therefore, we select the nucleosome sequences of the correct length (145–150 bp), typically representing about one-third of the data set. These are very accurate positions and could be used by themselves to plot the position map.

However, this approach typically eliminates about two-thirds of the nucleosome sequences. To include these data, we assume that they represent the same nucleosomes as in the accurate position set. The midpoint of each sequence is calculated and assigned to the count for the nearest accurate position. If the nearest accurate position midpoint is more than 10 bp away, the nucleosome sequence is discarded (Cole et al., 2011a). This approach results in inclusion of >90% of the sequences. The expanded data set, which has the same set of positions as the accurate set, but with increased counts, is then plotted to obtain a quantitative position map.

Analysis of a typical position map indicates that there are clusters of nucleosome midpoints that are closer together than the size of a nucleosome and consequently represent nucleosomes which overlap one another (Figs. 6.6 and 6.7B). Since nucleosomes cannot physically overlap one another on the same DNA molecule and still protect $\sim$147 bp, it follows that these midpoints must represent alternative positions. Thus, a nucleosome adopts a slightly different position in different cells (our yeast cells are haploid). In other words, a nucleosome peak in the occupancy map actually represents a cluster of alternative positions that reflect heterogeneity in the cell population. There is often a dominant position within a cluster, indicating that most cells have the nucleosome at this particular position. This "nucleosome position cluster" organization appears to be a general feature of yeast chromatin (Cole et al., 2011a; Kim, McLaughlin, Lindstrom, Tsukiyama, & Clark, 2006; Shen, Leblanc, Alfieri, & Clark, 2001) and perhaps also of mouse and human chromatin (Fragoso, John, Roberts, & Hager, 1995). We believe that it reflects a dynamic chromatin structure in living cells (Cole, Nagarajavel, & Clark, 2012). Note that the specialized centromeric nucleosome is a clear exception, with perfect positioning—the same position is adopted in every cell (Cole et al., 2011b).

6. SOME GENERAL EXPERIMENTAL CONSIDERATIONS
6.1. Sequence bias of MNase

MNase has a significant sequence bias and some have made efforts to correct for this bias. However, we would argue first, that correction is inappropriate, and second, that in any case, it is not possible to correct for the bias in a convincing way. The bias is apparent if digestion of protein-free DNA is halted at an early stage and the cut sites are examined (Chung et al., 2010; Dingwall, Lomonossoff, & Laskey, 1981; Hörz & Altenburger, 1981). For example, many nucleosome mapping studies using the classical indirect end-labeling method show free DNA controls; these are not smears but display sequence-specific band patterns (e.g., Roth, Dean, & Simpson, 1990). However, the important point is that protein-free DNA is destroyed by MNase by the time core particles generated from digestion of chromatin. It takes 10–100 times more enzyme to digest chromatin than DNA. Thus, any protein-free DNA is long gone from the reaction. This is a protection experiment: nucleosome mapping depends on the fact that the core histone octamer protects ∼150 bp from digestion. Even strong MNase cleavage sites are protected from digestion if present in a nucleosome; an example is the D5 nucleosome on *HIS3* (Clark, 2010). The nucleosome core particle is eventually attacked by MNase, but this occurs at a still later stage of digestion (McGhee & Felsenfeld, 1983). A strong bias for dinucleotides containing A or T is apparent when the ends of nucleosomal DNA are examined (Field et al., 2008). However, this consensus site ((A/T)|(A/T)) is very common, especially in AT-rich genomes such as that of yeast. This problem has been discussed in detail (Clark, 2010; Kaplan, Hughes, Lieb, Widom, & Segal, 2010).

6.2. Formaldehyde cross-linking of chromatin prior to mapping

In many studies, formaldehyde is used to cross-link the nucleosomes to DNA prior to isolation of mononucleosomes (e.g., Albert et al., 2007; Fragoso et al., 1995). The aim is to prevent nucleosomes sliding along the DNA from one position to another during cell lysis and subsequent steps. This is a reasonable approach. However, we do not use formaldehyde mainly because fixation is incomplete and therefore unlikely to be fully effective in preventing sliding, if it occurs. Complete fixation with formaldehyde renders yeast cells refractory to any kind of chromatin isolation. There might be some disadvantages associated with formaldehyde treatment, such as the requirement for complete reversal of cross-links without denaturing nucleosomal DNA. In addition, most reactions with formaldehyde result in monofunctional modification rather than cross-linking; modified DNA bases and protein side chains might introduce other artifacts.

6.3. Analysis of all MNase digestion products

This protocol is focused on analyzing only mononucleosomal DNA. An alternative approach would be to prepare the entire MNase digest for sequencing; most of the gel purification steps for mononucleosomal DNA could be omitted (Hennikoff et al., 2011; Kent et al., 2011). This would allow sequencing and analysis of all the species present in the digest, including di- and trinucleosomes; all products can be identified after paired-end sequencing by their length. The critical step here would be to gel-purify adapter-nucleosome ligation products away from adapter dimers, which tend to dominate the PCR.

ACKNOWLEDGMENTS

The authors thank Dwaipayan Ganguli and V. Nagarajavel for helpful comments on the manuscript. We thank Kip Bodi and Michael Berne at the Tufts University Core Facility, and Ting Ni and Jun Zhu at NHLBI for paired-end sequencing. We thank James Iben (NICHD) for help with the bioinformatic analysis. This research was supported by the Intramural Research Program of the National Institutes of Health (NICHD).

REFERENCES

Albert, I., Mavrich, T. N., Tomsho, L. P., Qi, J., Zanton, S. J., Schuster, S. C., et al. (2007). Translational and rotational settings of H2A.Z nucleosomes across the *Saccharomyces cerevisiae* genome. *Nature, 446,* 572–576.

Cairns, B. R. (2005). Chromatin remodeling complexes: Strength in diversity, precision through specialization. *Current Opinion in Genetics & Development, 15,* 185–190.

Chung, H., Dunkel, I., Heise, F., Linke, C., Krobitsch, S., Ehrenhofer-Murray, A. E., et al. (2010). The effect of micrococcal nuclease digestion on nucleosome positioning data. *PLoS One, 5,* e15754.

Clark, D. J. (2010). Nucleosome positioning, nucleosome spacing, and the nucleosome code. *Journal of Biomolecular Structure & Dynamics, 27,* 781–793.

Cole, H. A., Howard, B. H., & Clark, D. J. (2011a). Activation-induced disruption of nucleosome position clusters on the coding regions of Gcn4-dependent genes extends into neighbouring genes. *Nucleic Acids Research, 39,* 9521–9535.

Cole, H. A., Howard, B. H., & Clark, D. J. (2011b). The centromeric nucleosome of budding yeast is perfectly positioned and covers the entire centromere. *Proceedings of the National Academy of Sciences of the United States of America, 108,* 12687–12692.

Cole, H. A., Nagarajavel, V., & Clark, D. J. (2012). Perfect and imperfect nucleosome positioning in yeast. *Biochimica et Biophysica Acta, 1819,* 639–643.

Dingwall, C., Lomonossoff, G. P., & Laskey, R. A. (1981). High sequence specificity of micrococcal nuclease. *Nucleic Acids Research, 9,* 2659–2673.

Field, Y., Kaplan, N., Fondufe-Mittendorf, Y., Moore, I. K., Sharon, E., Lubling, Y., et al. (2008). Distinct modes of regulation by chromatin encoded through nucleosome positioning signals. *PLoS Computational Biology, 4,* 1–25.

Fragoso, G., John, S., Roberts, M. S., & Hager, G. L. (1995). Nucleosome positioning on the MMTV LTR results from the frequency-biased occupancy of multiple frames. *Genes & Development, 9,* 1933–1947.

Gardner, K. E., Allis, C. D., & Strahl, B. D. (2011). Operating on chromatin: A colourful language where context matters. *Journal of Molecular Biology, 409,* 36–46.

Hennikoff, J. G., Belsky, J. A., Krassovsky, K., MacAlpine, D. M., & Henikoff, S. (2011). Epigenome characterization at single base pair resolution. *Proceedings of the National Academy of Sciences of the United States of America, 108,* 18318–18323.

Hörz, W., & Altenburger, W. (1981). Sequence-specific cleavage of DNA by micrococcal nuclease. *Nucleic Acids Research, 9,* 2643–2658.

Imbalzano, A. N., Kwon, H., Green, M. R., & Kingston, R. E. (1994). Facilitated binding of TATA-binding protein to nucleosomal DNA. *Nature, 370,* 481–485.

Kaplan, N., Hughes, T. R., Lieb, J. D., Widom, J., & Segal, E. (2010). Contribution of histone sequence preferences to nucleosome organization: Proposed definitions and methodology. *Genome Biology, 11,* 140.

Kaplan, N., Moore, I. K., Fondufe-Mittendorf, Y., Gossett, A. J., Tillo, D., Field, Y., et al. (2009). The DNA-encoded nucleosome organization of a eukaryotic genome. *Nature, 458,* 362–366.

Kent, N. A., Adams, S., Moorhouse, A., & Paszkiewicz, K. (2011). Chromatin particle spectrum analysis: A method for comparative chromatin structure analysis using paired-end mode next-generation DNA sequencing. *Nucleic Acids Research, 39,* e26.

Kim, Y., McLaughlin, N., Lindstrom, K., Tsukiyama, T., & Clark, D. J. (2006). Activation of *Saccharomyces cerevisiae HIS3* results in Gcn4p-dependent, SWI/SNF-dependent mobilization of nucleosomes over the entire gene. *Molecular and Cellular Biology, 26,* 8607–8622.

Kim, Y., Shen, C. H., & Clark, D. J. (2004). Purification and nucleosome mapping analysis of native yeast plasmid chromatin. *Methods, 33,* 59–67.

Lee, W., Tillo, D., Bray, N., Morse, R. H., Davis, R. W., Hughes, T. R., et al. (2007). A high-resolution atlas of nucleosome occupancy in yeast. *Nature Genetics*, *39*, 1235–1244.

Li, H., & Durbin, R. (2009). Fast and accurate short read alignment with Burrows-Wheeler transform. *Bioinformatics*, *25*, 1754–1760.

Li, H., Handsaker, B., Wysoker, A., Fennell, T., Ruan, J., Homer, N., et al. (2009). The sequence alignment/map format and SAMtools. *Bioinformatics*, *25*, 2078–2079.

Li, Z., Schug, J., Tuteja, G., White, P., & Kaestner, K. H. (2011). The nucleosome map of the mammalian liver. *Nature Structural & Molecular Biology*, *18*, 742–746.

Luger, K., Mäder, A. W., Richmond, R. K., Sargent, D. F., & Richmond, T. J. (1997). Crystal structure of the nucleosome core particle at 2.8Å resolution. *Nature*, *389*, 251–260.

McGhee, J. D., & Felsenfeld, G. (1983). Another potential artifact in the study of nucleosome phasing by chromatin digestion with micrococcal nuclease. *Cell*, *32*, 1205–1215.

Roth, S. Y., Dean, A., & Simpson, R. T. (1990). Yeast α2 repressor positions nucleosomes in TRP1/ARS1 chromatin. *Molecular and Cellular Biology*, *10*, 2247–2260.

Segal, E., & Widom, J. (2009). What controls nucleosome positions? *Trends in Genetics*, *25*, 335–343.

Shen, C. H., Leblanc, B. P., Alfieri, J. A., & Clark, D. J. (2001). Remodeling of yeast *CUP1* involves activator-dependent repositioning of nucleosomes over the entire gene and flanking sequences. *Molecular and Cellular Biology*, *21*, 534–547.

Svaren, J., & Hörz, W. (1997). Transcription factors *vs.* nucleosomes: Regulation of the PHO5 promoter in yeast. *Trends in Biochemical Sciences*, *22*, 93–97.

Valouev, A., Ichikawa, J., Tonthat, T., Stuart, J., Ranade, S., Peckham, H., et al. (2008). A high-resolution nucleosome position map of *C. elegans* reveals a lack of universal sequence-dictated positioning. *Genome Research*, *18*, 1051–1063.

Yuan, G., Liu, Y., Dion, M. F., Slack, M. D., Wu, L. F., Altschuler, S. J., et al. (2005). Genome-scale identification of nucleosome positions in *S. cerevisiae*. *Science*, *309*, 626–630.

Zaret, K. S., Wandzioch, E., Watts, J., Xu, J., Smale, S. T., & Sekiya, S. (2008). Pioneer factors, genetic competence and inductive signaling: Programming liver and pancreatic progenitors from the endoderm. *Cold Spring Harbor Symposia on Quantitative Biology*, *73*, 119–126.

Zhang, Y., Moqtaderi, Z., Rattner, B. P., Euskirchen, G., Snyder, M., Kadonaga, J. T., et al. (2009). Intrinsic histone-DNA interactions are not the major determinant of nucleosome positions *in vivo*. *Nature Structural & Molecular Biology*, *16*, 847–852.

CHAPTER SEVEN

Measuring Genome-Wide Nucleosome Turnover Using CATCH-IT

Sheila S. Teves[*,†], Roger B. Deal[*], Steven Henikoff[*,‡,1]
[*]Division of Basic Sciences, Fred Hutchinson Cancer Research Center, Seattle, Washington, USA
[†]Molecular and Cellular Biology Program, University of Washington, Seattle, Washington, USA
[‡]Howard Hughes Medical Institute, Seattle, Washington, USA
[1]Corresponding author: e-mail address: steveh@fhcrc.org

Contents

1. Introduction — 170
2. Covalent Attachment of Tagged Histones to Capture and Identify Turnover — 171
   2.1 Solutions and materials — 172
   2.2 Methionine-free growth medium — 173
   2.3 Aha labeling and biotin coupling — 173
   2.4 Chromatin fragmentation and extraction — 175
   2.5 Streptavidin affinity capture — 176
   2.6 DNA isolation — 177
3. Modified Solexa Library Preparation — 178
   3.1 Solutions and materials — 179
   3.2 Paired-end adapter and primers — 180
   3.3 End repair — 180
   3.4 Phenol extraction and column purification — 181
   3.5 A-tailing — 181
   3.6 Adapter ligation and AMPure bead purification — 182
   3.7 PCR amplification and final purification — 182
References — 184

Abstract

The dynamic interplay between DNA-binding proteins and nucleosomes underlies essential nuclear processes such as transcription, replication, and DNA repair. Manifestations of this interplay include the assembly, eviction, and replacement of nucleosomes. Hence, measurements of nucleosome turnover kinetics can lead to insights into the regulation of dynamic chromatin processes. In this chapter, we describe a genome-wide method for measuring nucleosome turnover that uses metabolic labeling followed by capture of newly synthesized histones, which we have termed Covalent Attachment of Tagged Histones to Capture and Identify Turnover (CATCH-IT). Although CATCH-IT can be used with any genome-wide mapping procedure, high-resolution profiling is attainable using paired-end sequencing of native chromatin. Our protocol also includes an efficient Solexa DNA sequencing library preparation protocol that can be used for single base-pair resolution mapping of both nucleosome and sub-nucleosomal particles. We not only describe the use of these protocols in the context of a Drosophila cell line but also provide the necessary changes for adaptation to other model systems.

1. INTRODUCTION

Nucleosomes have evolved to tightly package DNA in chromosomes and so must be mobilized to allow DNA-binding proteins to gain access to their binding sites and perform DNA-templated processes. For DNA to be accessible, nucleosomes must be displaced, partially unwrapped, or evicted, and measuring these dynamic events can provide mechanistic insights into the regulation of chromatin-dependent processes. We have recently developed a method that combines kinetic measurement of nucleosome dissociation and replacement (turnover) with genome-wide readout technologies that we termed *Covalent Attachment of Tagged Histones to Capture and Identify Turnover* (CATCH-IT) (Deal et al., 2010). In CATCH-IT, a methionine analog containing an azide moiety is incorporated into newly synthesized proteins, which can then be biotinylated through copper-catalyzed cycloaddition reaction with a biotin-alkyne substrate. Nucleosomes containing the newly synthesized biotinylated histones can be isolated using streptavidin-coated beads and the extracted DNA used to measure the degree of nucleosome turnover genome-wide.

The increasing affordability of short-read massively parallel sequencing potentially enables the study of epigenomic events at single base-pair resolution. Paired-end (PE) sequencing using Illumina's Solexa platform (Bentley, 2006) is becoming especially valuable for epigenomic mapping, as it allows precise determination of both fragment lengths and positions. However, current Solexa library preparation protocols were designed for genomics applications, where fragmentation by random shearing and size selection were intended to provide a uniform population of DNA templates for bulk sequencing and so include a gel-based size-selection step to exclude both large and small fragments. However, for many epigenomic mapping applications, random fragmentation and size selection is inappropriate, and the requirement for large amounts of starting material can be limiting. To address these issues, we have developed a modified Solexa library protocol that removes gel-based size selection and streamlines DNA cleanup to allow for sequencing of as little as ~10 ng of starting material with fewer manipulations (Henikoff, Belsky, Krassovsky, Macalpine, & Henikoff, 2011). Combining this modified protocol with chromatin-probing experiments, such as micrococcal nuclease (MNase) mapping (Henikoff et al., 2011; Kent, Adams, Moorhouse, & Paszkiewicz, 2011), salt fractionation (Henikoff, Henikoff, Sakai, Loeb, & Ahmad, 2009), native chromatin immunoprecipitation (Teves & Henikoff, 2011; Weber, Henikoff, & Henikoff, 2010), and as we discuss in this chapter, CATCH-IT (Deal et al., 2010) allows for single base-pair resolution analyses of chromatin-based processes.

In this chapter, we present detailed protocols and tips to perform CATCH-IT at high resolution using a Drosophila cell line and provide illustrative data of expected results. These protocols should be adaptable to cell lines of other organisms, and we highlight the steps where necessary changes should be made when performing CATCH-IT for other systems.

2. COVALENT ATTACHMENT OF TAGGED HISTONES TO CAPTURE AND IDENTIFY TURNOVER

Below, we provide a detailed protocol for nucleosome turnover analysis through metabolic labeling. The method relies on the depletion of methionine in growth medium followed by incorporation of the methionine analog azidohomoalanine (Aha) into newly synthesized proteins in place of methionine. Following nuclei isolation from Aha-labeled cells, newly synthesized nuclear proteins containing an Aha-azide moiety can be coupled to an alkyne-linked biotin tag through a copper-catalyzed cycloaddition reaction. Chromatin can then be fragmented down to mononucleosomes using MNase, which digests away unprotected DNA fragments. A standard salt extraction step provides the input material for affinity purification of newly synthesized chromatin using streptavidin beads. Because the turnover rate of H2A/H2B dimers is faster than that of the central (H3/H4)$_2$ tetramer (Rufiange, Jacques, Bhat, Robert, & Nourani, 2007; Thiriet & Hayes, 2005), the immobilized chromatin is washed with a urea-containing solution that strips off the H2A/H2B from nucleosomes and also removes virtually all other bound proteins. DNA extracted from the bead-bound material can then be isolated for genome-wide analysis.

2.1. Solutions and materials

1. Drosophila S2 cells maintained in log-phase growth
2. Shields and Sang M3 (SSM3-Met) Insect growth medium without methionine
3. Azidohomoalanine (Anaspec cat # 63669)
4. Methionine (Sigma-Aldrich # M9625)
5. Round-bottom 30 mL Corex tubes
6. Table-top centrifuge
7. 1× phosphate-buffered saline (PBS)
8. *TM2 buffer (10 mM Tris, pH 7.5, 2 mM MgCl$_2$)
9. 10% NP-40 (Sigma-Aldrich # 74385)
10. Refrigerated table-top centrifuge
11. *HB125 buffer (0.125 M sucrose, 15 mM Tris, pH 7.5, 15 mM NaCl, 40 mM KCl, 0.5 mM spermidine, 0.15 mM spermine)
12. Copper(II) sulfate pentahydrate (Sigma-Aldrich # C7631)
13. L-Ascorbic acid (Sigma-Aldrich # A5960)
14. Biotin-alkyne (Invitrogen # B10185)
15. End-over-end microcentrifuge tube rotator (ex. Labquake Shaker—Thermo Scientific)
16. 0.5 M EDTA
17. 1 M CaCl$_2$
18. 37 °C water bath
19. MNase 200 U powder, resuspended to 0.2 U/μL (Sigma-Aldrich # N3755)
20. *CSB 350 buffer (1× PBS, 213 mM NaCl (350 mM total), 2 mM EDTA, 0.1% Triton X-100)
21. Dynabeads M-280 streptavidin (Invitrogen # 112.05D)
22. Magnetic rack for microcentrifuge tubes
23. *Urea/NaCl wash buffer (4 M urea, 0.3 M NaCl, 20 mM Tris, pH 8, 1 mM EDTA)
24. 10% SDS
25. Phenol–chloroform–isoamyl alcohol (25:24:1, v/v/v)
26. Glycogen 20 mg/mL
27. 200 proof ethanol
28. TE, pH 8 (10 mM Tris, pH 8, 1 mM EDTA)
29. Complete-mini, protease inhibitor cocktail EDTA free (Roche # 1 830 170)

*Buffers are supplemented with complete-mini, protease inhibitor cocktail prior to use.

2.2. Methionine-free growth medium

In many systems, growth medium without methionine is commercially available. However, for insect cell culture systems, this is not the case. We therefore prepare methionine-free SSM3-Met insect medium for use in CATCH-IT experiments. Table 7.1 lists the components of M3 medium without methionine for 1 L of medium.

1. Add 1.05 g of Bis–Tris to 650 mL of H_2O and adjust pH to 6.8.
2. Combine all solids (salts, vitamins, amino acids, sugar) and grind into fine powder using a mortar and pestle.
3. Slowly add the ground powder to the Bis–Tris–H_2O while mixing using a magnetic stir bar.
4. After all solids have dissolved, readjust the pH to 6.8 and bring the final volume to 1 L. Filter sterilize, and store at 4 °C.

2.3. Aha labeling and biotin coupling

1. Grow S2 cells in two 75 cm^2 flasks to late log phase in rich medium. Remove the medium and replace with 7 mL of SSM3-Met medium. Place the flasks back in the incubator for 30 min to starve cells of methionine.
2. Remove the medium and replace with 7 mL of SSM3-Met medium supplemented with 4 mM Aha for one flask and 4 mM Met for the other flask. Place the flasks back in the incubator for the desired amount of time (20 min to several hours).
3. Harvest the cells from each treatment in 30 mL round-bottom Corex tubes. Spin at $1200 \times g$ for 3 min. Decant the medium and wash cells with 10 mL of 1 × PBS, spin again, and decant PBS.

Note: The following steps (4–6) are used to isolate nuclei from Drosophila S2 cells. To adapt this method to other systems, the final concentration of NP-40 must be optimized. Alternatively, other standard protocols for nuclei preparation can be substituted.

4. Resuspend the cells in 1 mL of ice-cold TM2 buffer and transfer to 1.5 mL microcentrifuge tubes. Place on ice for 3 min.
5. Add 60 μL of 10% NP-40 and vortex for 5 s at the low-medium setting. Place cells back on ice for 3 min, vortex one more time as before, and spin out nuclei at $100 \times g$ for 10 min at 4 °C.
6. Wash nuclei with 1 mL cold TM2 buffer. Pellet nuclei at $100 \times g$ for 10 min at 4 °C, decant supernatant, and resuspend gently in 200 μL of cold HB 125 buffer. If performing a time course, leave each successive sample on ice at this point until all are collected and then proceed to step 7 with all samples.

Table 7.1 Components of Methionine-free growth medium 1 L of medium

	(g/L)
Bis–Tris	1.05
$CaCl_2$	0.76
$MgSO_4$	2.15
$NaHCO_3$	0.42
Na_2HPO_4	0.88
Glucose	10
Choline chloride	0.05
Oxalacetic acid	0.25
Amino acids	
L-α-Alanine	1.5
L-β-Alanine	0.25
L-Arginine	0.5
L-Asparagine	0.3
L-Aspartic acid	0.3
L-Cysteine-HCl	0.2
L-Glutamic acid-K	7.88
L-Glutamic acid-Na	6.53
L-Glutamine	0.6
L-Glycine	0.5
L-Histidine	0.55
L-Isoleucine	0.25
L-Leucine	0.4
L-Lysine-HCl	0.85
L-Phenylalanine	0.25
L-Proline	0.4
L-Serine	0.35
L-Threonine	0.5
L-Tryptophan	0.1
L-Tyrosine-2Na	0.3601
L-Valine	0.4

7. Prepare reagents as follows for the cycloaddition coupling reaction:
   a. Weigh out 25 mg of $CuSO_4$ and 88 mg of ascorbic acid into 1.5 mL tubes and dissolve each in 1 mL of H_2O to give solutions of 100 mM $CuSO_4$ and 500 mM ascorbic acid. Combine 100 μL of $CuSO_4$ with 100 μL of ascorbic acid. The solution turns yellow as Cu^{2+} is reduced to Cu^+ by ascorbic acid.
   b. Prepare 20 μL of 2 mM biotin–PEO–alkyne by 1:10 dilution of 20 mM stock solution.
8. Place nuclei suspensions at room temperature and then add the following to each 200-μL sample, mixing well after each addition:
   5 μL of 2 mM biotin–PEO–alkyne (50 μM final concentration)
   4 μL of $CuSO_4$/ascorbic acid mixture (final concentration is 1 mM $CuSO_4$, 5 mM ascorbic acid)
9. Place nuclei suspensions on a microcentrifuge tube rotator at 4 °C for 30 min. Save 2 μL (1% of total) for Western analysis at the end of the procedure.
10. Pellet nuclei at $100 \times g$ for 5 min at 4 °C.
11. Remove supernatant thoroughly and resuspend nuclei gently in 200 μL of cold HB 125 buffer. Repeat steps 8 through 10 using freshly prepared cycloaddition reagents.

We tested the labeling and coupling process using total cellular extracts from Met- or Aha-labeled cells. We performed Western blot analysis on aliquots saved from each step of the process and probed for biotin using α-streptavidin antibody (Fig. 7.1). We found that Aha- but not methionine-labeled cells have incorporated biotin into general cellular proteins. This test also showed that a second biotin coupling is required for saturation of all Aha-labeled proteins.

2.4. Chromatin fragmentation and extraction

1. Resuspend nuclei in 250 μL HB 125 with 1 mM EDTA and add $CaCl_2$ to 2 mM final concentration. Place tubes in a 37 °C water bath for 5 min.
2. Add 2 μL of the MNase solution to each tube in the water bath and mix gently by inversion and flicking several times. Continue digestion for 10 min with mixing each minute. This level of digestion should give mostly mononucleosomes (Fig. 7.2), but this should be optimized for each system by varying the concentration of MNase and/or the length of digestion time.
3. Add EDTA to 2 mM to stop the reaction and place the tubes on ice. Spin at $100 \times g$ for 10 min at 4 °C.

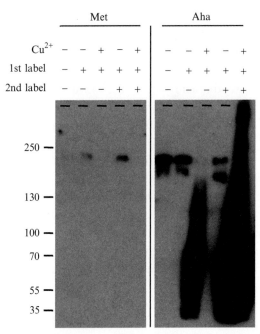

Figure 7.1 Incorporation of Aha into cellular proteins. Cells depleted of methionine were incubated in SSM3-Met medium supplemented with 4 m*M* Met or Aha for 20 min. Cells were then lysed in 1% SDS in 1× PBS, and the resulting total protein extract was subjected to biotin coupling as described either with or without Cu^+. Aliquots before treatment, after the first labeling, and after the second labeling were subjected to Western blot analysis using α-streptavidin to visualize biotinylated proteins.

4. Resuspend nuclei in 300 μL of CSB 350 and mix on a microcentrifuge tube rotator at 4 °C for at least 1 h to overnight.
5. Centrifuge nuclei at 100 × *g* for 10 min at 4 °C and save the supernatant (soluble chromatin). Clarify the soluble chromatin by centrifugation at full speed for 5 min and move to a fresh tube.
6. Remove a 10-μL aliquot of the soluble chromatin (Input) for DNA purification and 12 μL for Western analysis, coomassie gel, and protein concentration measurement. Resuspend the pellet in loading buffer and save for Western analysis.

2.5. Streptavidin affinity capture

1. To the soluble chromatin, add 25 μL (beads + buffer, same as original ratio as in stock slurry) of CSB-washed M280 streptavidin-coated Dynabeads. Rotate for 1.5 h at 4 °C.

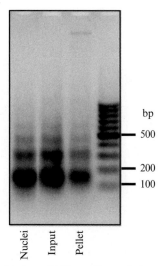

Figure 7.2 Nucleosome laddering of MNase-digested chromatin. After biotin coupling, intact nuclei were digested with 0.4 U MNase for 10 min and subjected to salt extraction. DNA from total nuclei, 350 mM salt-extracted chromatin, and pellet fractions were electrophoresed on a 1.5% agarose gel and visualized with ethidium bromide.

2. Place tubes on a Dynal magnet rack for several minutes, and then save the supernatant as "unbound" for Western analysis. Resuspend the beads in 700 μL of urea/NaCl wash buffer and rotate at 4 °C for 5 min to strip the H2A/H2B dimers from the chromatin.
3. Place tubes on the magnet rack, decant, and resuspend in 700 μL of CSB 350. Transfer the beads and buffer to a fresh tube and rotate at 4 °C for 5 min. Collect and decant the beads and proceed with DNA isolation or resuspend the beads in 20 μL CSB 350 and freeze at −20 °C.

Western analysis of total nuclei, input, unbound, and pellet fractions during the streptavidin affinity capture process (Fig. 7.3) shows lack of streptavidin signal in the unbound fraction compared to others, indicating an efficient capture of biotin-coupled, Aha-labeled proteins.

2.6. DNA isolation

1. Bring the volumes of input and beads to 200 μL with CSB 350 and add SDS to 0.5%. Add 1 μL of RNAse A and incubate at 37 °C for 10 min, and then add 1 μL of proteinase K and incubate at 70 °C for 10 min. Mix beads frequently during each enzyme digestion.

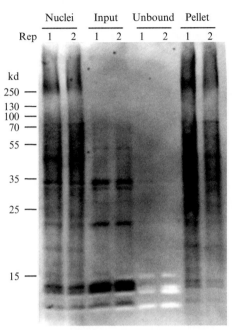

Figure 7.3 Efficient capture of biotinylated proteins. CATCH-IT was performed on Drosophila S2 cells in two replicates. Aliquots of total nuclei, input, unbound, and pellet fractions were subjected to a Western analysis as in Fig. 7.1.

2. Extract the DNA twice with phenol/chloroform and precipitate by adding 2.5 volumes of cold ethanol. Wash with 75% ethanol and resuspend in 20 μL of $0.1 \times$ TE, pH 8.
3. Quantify DNA either by using the PicoGreen fluorescence assay (Invitrogen) or by measuring the OD_{260} with a spectrophotometer such as Nanodrop.

3. MODIFIED SOLEXA LIBRARY PREPARATION

The following protocol for library uses the same enzymological steps as in the established Illumina Paired-End Sample Preparation protocol (Quail et al., 2008), with modifications to the cleanup and purification steps that allow for efficient recovery of fragments as small as 25 bp (Henikoff et al., 2011). In this process, the ends of the starting DNA material are made blunt and 5′-phosphorylated. An "A" nucleotide is added at the 3′-ends to prevent self-ligation and permit specific base-pairing of adapters with a 5′-phosphate and a 3′-T-overhang. The adapted samples are then amplified using primers with 3′-complementarity to the adapters, using a 60 °C extension step to minimize preferential loss of AT-rich fragments (Lopez-Barragan et al., 2010). In the Illumina protocol, the samples are size selected following the ligation of adapters both to remove unligated adapters and to isolate a specific subset of the samples for sequencing. This step is designed for samples that are randomly sheared or where single-end sequencing leads to fragment size ambiguity. However, if chromatin is enzymatically cleaved and PE sequenced, the fragment size distribution can reveal valuable features of the chromatin landscape. Therefore, we eliminated the gel-based size-selection step. Instead, we use Agencourt AMPure XP beads both to purify the ligated products from unligated primers and for post-PCR cleanup. Using the dilution factor specified by the manufacturer, the AMPure beads result in a strict size cutoff at 90–100 bp. Because the adapters add ~65 bp to the starting material, insert sizes as small as 25 bp will be present in the library. Another modification is that all QIAgen cleanup steps have been replaced with phenol/chloroform extractions to stop the reactions followed by spin column cleanup to purify the DNA. This modification, combined with the use of low-retention (siliconized) microcentrifuge tubes, minimizes the loss of DNA and allows for lower amounts of starting material to be used.

3.1. Solutions and materials

1. Low-retention 1.5 mL microcentrifuge tubes
2. 10× annealing buffer (0.5 M NaCl, 0.1 M Tris, pH 8, 0.01 M EDTA)
3. 10× T4 DNA ligase buffer with 10 mM ATP (New England Biolabs (NEB) # B0202S)
4. 40 mM dNTP (10 mM each)
5. T4 DNA polymerase 5 U/μL (Invitrogen # 100004994)
6. DNA polymerase I, large (Klenow) fragment 50 U/μL (NEB # M0210M)
7. T4 polynucleotide kinase (PNK) 10 U/μL (NEB # M0201L)
8. Phenol/chloroform/isoamyl (25:24:1)
9. Illustra MicroSpin S-300 HR columns (GE Healthcare # 27-5130-01)
10. SpeedVac
11. 10× NEB buffer 2 (NEB # B7002S)
12. 100 mM dATP (Invitrogen # 10216018)
13. Klenow fragment ($3' \rightarrow 5'$-exo-) 50 U/μL (NEB # M0212M)
14. Illumina PE adapter mix or independently synthesized adapter mix
15. T4 DNA ligase (Rapid) and 2× buffer (Enzymatics # L603-HC-L)
16. Agencourt AMPure XP magnetic beads (Agencourt # A63881)
17. Magnetic rack for microcentrifuge tubes

18. 70% ethanol
19. 0.1× TE (1 mM Tris, pH 8, 0.1 mM EDTA)
20. Phusion high-fidelity DNA polymerase with 5× HF buffer (Finnzymes # F-530L)
21. Illumina PE primers or independently synthesized primers
22. 10 mM dNTP (2.5 mM each)
23. Thermocycler

3.2. Paired-end adapter and primers

We have used both commercially available PE adapter mix and PCR primers from Illumina and custom-made adapter and primer oligos with comparable results. The following synthesized adapters and primers are compatible with the Illumina PE platform. Therefore, the protocol outlined below does not make a distinction between these options. The oligonucleotide sequences are as follows:

PE Adapter1. [Phosphate]GATCGGAAGAGCGGTTCAGCAGGAAT GCCGA★G

PE Adapter2. ACACTCTTTCCCTACACGACGCTCTTCCGATC★T

PE forward primer. AATGATACGGCGACCACCGAGATCTACAC TCTTTCCCTACACGACGCTCTTCCGATC ★T

PE reverse primer. CAAGCAGAAGACGGCATACGAGATCGGTC TCGGCATTCCTGCTGAACCGCTCTTCCGA TC★T

These oligonucleotides are PAGE purified. A phosphorothioate linkage before the last nucleotide as denoted in the sequence (★) is intended to prevent 3′-exonuclease activity during the ligation step (Quail et al., 2008).

Adapter oligonucleotides are dissolved to 100 μM in water, mixed in equimolar amounts, and annealed by addition of 10× annealing buffer to a final concentration of 50 mM NaCl, 10 mM Tris, pH 8, 1 mM EDTA. The mixed adapters are annealed by heating to 98 °C for 10 min in a thermocycler and slowly cooling (−1 °C/min) to 25 °C.

3.3. End repair

1. Measure the starting DNA concentration spectrophotometrically or by fluorescence. A total of 10–500 ng of enzymatically digested DNA can be used to prepare sequencing libraries. Bring the desired starting amount to 20 μL by addition of water or using a SpeedVac without heat to concentrate in a low-retention microcentrifuge tube.

2. Prepare the "End Repair" (ER) master mix by combining the following:
   a. 18.5 μL water
   b. 5 μL 10× T4 DNA ligase buffer
   c. 2 μL 40 mM dNTP
   d. 1.5 μL T4 DNA polymerase
   e. 0.5 μL Klenow fragment (diluted to 5 U/μL from 50 U/μL stock)
   f. 2.5 μL T4 PNK
3. If preparing multiple samples, combine a slight excess of each component of the ER master mix to compensate for pipetting losses during transfers. For example, to make libraries from eight different samples, combine 8.2 times each of the component in the master mix.
4. Add 30 μL of ER master mix to 20 μL of DNA (10–500 ng).
5. Incubate the sample in a 20 °C water bath for 30 min. A water bath can be as simple as an ice bucket filled with water that is measured to be 20 °C using a thermometer.

3.4. Phenol extraction and column purification

1. Extract the DNA with 50 μL of phenol/chloroform/isoamyl. Vortex to thoroughly mix, and separate the aqueous layer by centrifugation at maximum speed for 1 min using a table-top centrifuge. Carefully remove the organic layer from the bottom of the tube.
2. To prepare the MicroSpin S-300 HR column, snap off the bottom tip, slightly unscrew the cap, and place in tube holder. Spin buffer off for 1 min at 800 × g and replace the column into a new low-retention microcentrifuge tube.
3. Decant the aqueous layer from step 1 and drain the remaining organic layer by touching the pipette tip to the tube wall. Transfer the aqueous layer into the prepared MicroSpin S-300 HR column. Extract the DNA by centrifugation for 2 min at 800 × g. The resulting eluate will be about 50–60 μL in volume.

3.5. A-tailing

1. Reduce the eluate volume to 35 μL in a SpeedVac without heat for about 15 min.
2. Prepare the "A-tailing" (A-t) master mix by combining the following:
   a. 5 μL 10× NEB buffer 2
   b. 10 μL 1 mM dATP
   c. 0.3 μL Klenow exo- (50 U/μL)

3. Again, if preparing multiple samples, include a slight excess of each component as in the ER master mix procedure.
4. Add 15.3 μL of the A-t master mix to 35 μL of end-repaired DNA sample.
5. Incubate at 37 °C for 30 min.
6. Extract the DNA by phenol extraction and column purification (Section 3.3).

3.6. Adapter ligation and AMPure bead purification

1. Reduce the eluate volume to 18 μL in a SpeedVac without heat for about 30–45 min.
2. Add 1 μL of 1 mM PE adapter mix. This amount is best used for small amounts of starting material (10–50 ng of DNA). When using a large amount of starting material, adjust the adapter amount to give an estimated 10:1 adapter:starting DNA molar ratio.
3. Add 25 μL of 2× Rapid DNA ligase buffer (Enzymatics) and 5 μL of Rapid T4 DNA ligase. If preparing multiple samples, combine the buffer and ligase for a master mix and add 30 μL of this to the DNA-adapter mix.
4. Incubate the sample at 20 °C water bath for 15 min to ligate adapters.
5. To extract the DNA and remove excess adapters, add 90 μL of AMPure XP magnetic bead slurry to the sample, mix by pipetting 10 times, and hold at room temperature for 5 min.
6. Place the sample in a magnetic tube holder until the beads are cleared from the slurry, accumulate on the side (∼2 min), and aspirate off the solution.
7. While still on the magnet, wash the beads by adding 1 mL 70% ethanol. Aspirate the ethanol after 30 s. Repeat this wash one more time, carefully removing excess ethanol.
8. Allow the beads to dry for no more than 5 min. Remove the tube from the magnet and add 40 μL of 0.1× TE, pH 8. Mix thoroughly and replace the tube on the magnetic tube holder. Transfer the eluate into a new low-retention microcentrifuge tube. This now contains the DNA material with ligated PE adapters.

3.7. PCR amplification and final purification

1. A 20 μL PCR reaction volume is generally sufficient to produce enough products for PE sequencing using the Illumina platform. Prepare a "PCR" master mix by combining the following:

a. 4 μL 5 × Phusion buffer HF
   b. 1.6 μL 10 mM dNTP (2.5 mM each)
   c. 0.4 μL forward primer
   d. 0.4 μL reverse primer
   e. 0.2 μL Phusion HF Polymerase (Finnzymes)
   f. 8.4 μL H_2O
2. Add 15 μL of the PCR master mix to 5 μL of adapter-ligated material and proceed with PCR using the following cycling parameters:
   a. 98 °C for 30 s
   b. 12–18 cycles of:
      i. 98 °C for 10 s
      ii. 65 °C for 30 s
      iii. 60 °C for 30 s
   c. 60 °C for 5 min
   d. Hold at 8 °C
3. Clean up the PCR products by adding 36 μL of AMPure XP beads and following the purification method described above. Elute the sample with 40 μL of 0.1 × TE, pH 8. Check the library on a 2% agarose gel

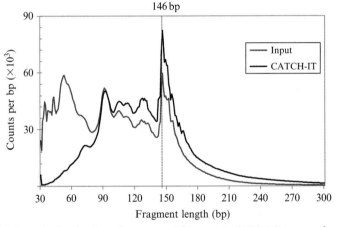

Figure 7.4 Length distribution of sequenced fragments. CATCH-IT was performed on Drosophila S2 cells, and a paired-end Solexa library was generated from the input and CATCH-IT material as described here. Paired-end sequencing for each sample was performed in a single lane of an Illumina Hi-Seq 2000 Instrument by the FHCRC Genomics Shared Resource (http://sharedresources.fhcrc.org/core-facilities/genomics), and fragments were mapped to the Drosophila genome using Novoalign alignment program (http://www.novocraft.com/main/index.php). The length distribution of all mapped fragments is shown here at base-pair resolution. The dominant nucleosomal peak is centered at 146 bp.

with ethidium bromide. The adapter and PCR primers add about 120 bp onto the starting DNA. Therefore, the distribution of the library should be shifted 120 bp higher than the starting DNA (Fig. 7.4).
4. Accurately measure the concentration of the library either by fluorescence-based assays such as the PicoGreen (Invitrogen) or by qPCR as outlined in Illumina protocols and dilute for application to the flow cell (2 μM for the Illumina Hi-Seq 2000).

REFERENCES

Bentley, D. R. (2006). Whole-genome re-sequencing. *Current Opinion in Genetics and Development, 16*, 545–552.

Deal, R. B., Henikoff, J. G., & Henikoff, S. (2010). Genome-wide kinetics of nucleosome turnover determined by metabolic labeling of histones. *Science, 328*, 1161–1164.

Henikoff, J. G., Belsky, J. A., Krassovsky, K., Macalpine, D. M., & Henikoff, S. (2011). Epigenome characterization at single base-pair resolution. *Proceedings of the National Academy of Sciences of the United States of America, 108*, 18318–18323.

Henikoff, S., Henikoff, J. G., Sakai, A., Loeb, G. B., & Ahmad, K. (2009). Genome-wide profiling of salt fractions maps physical properties of chromatin. *Genome Research, 19*, 460–469.

Kent, N. A., Adams, S., Moorhouse, A., & Paszkiewicz, K. (2011). Chromatin particle spectrum analysis: A method for comparative chromatin structure analysis using paired-end mode next-generation DNA sequencing. *Nucleic Acids Research, 39*, e26.

Lopez-Barragan, M. J., Quinones, M., Cui, K., Lemieux, J., Zhao, K., & Su, X. Z. (2010). Effect of PCR extension temperature on high-throughput sequencing. *Molecular and Biochemical Parasitology, 176*, 64–67.

Quail, M. A., Kozarewa, I., Smith, F., Scally, A., Stephens, P. J., Durbin, R., et al. (2008). A large genome center's improvements to the Illumina sequencing system. *Nature Methods, 5*, 1005–1010.

Rufiange, A., Jacques, P. E., Bhat, W., Robert, F., & Nourani, A. (2007). Genome-wide replication-independent histone H3 exchange occurs predominantly at promoters and implicates H3 K56 acetylation and Asf1. *Molecular Cell, 27*, 393–405.

Teves, S. S., & Henikoff, S. (2011). Heat shock reduces stalled RNA polymerase II and nucleosome turnover genome-wide. *Genes & Development, 25*, 2387–2397.

Thiriet, C., & Hayes, J. J. (2005). Replication-independent core histone dynamics at transcriptionally active loci in vivo. *Genes & Development, 19*, 677–682.

Weber, C. M., Henikoff, J. G., & Henikoff, S. (2010). H2A.Z nucleosomes enriched over active genes are homotypic. *Nature Structural and Molecular Biology, 17*, 1500–1507.

CHAPTER EIGHT

DNA Methyltransferase Accessibility Protocol for Individual Templates by Deep Sequencing

Russell P. Darst*, Nancy H. Nabilsi*, Carolina E. Pardo*, Alberto Riva[†], Michael P. Kladde*,[1]
*Department of Biochemistry and Molecular Biology, University of Florida Shands Cancer Center Program in Cancer Genetics, Epigenetics, and Tumor Virology, Gainesville, Florida, USA
[†]Department of Molecular Genetics and Microbiology, University of Florida Genetics Institute, Gainesville, Florida, USA
[1]Corresponding author: e-mail address: kladde@ufl.edu

Contents

1. Introduction	186
2. Materials	188
2.1 Reagents	188
2.2 Solutions	189
2.3 Supplies	190
3. Protocols	190
3.1 MAPit in nuclei	190
3.2 MAPit validation	192
3.3 Amplicon sequencing	192
3.4 Sequence analysis	195
Acknowledgments	202
References	202

Abstract

A single-molecule probe of chromatin structure can uncover dynamic chromatin states and rare epigenetic variants of biological importance that bulk measures of chromatin structure miss. In bisulfite genomic sequencing, each sequenced clone records the methylation status of multiple sites on an individual molecule of DNA. An exogenous DNA methyltransferase can thus be used to image nucleosomes and other protein–DNA complexes. In this chapter, we describe the adaptation of this technique, termed Methylation Accessibility Protocol for individual templates, to modern high-throughput sequencing, which both simplifies the workflow and extends its utility.

1. INTRODUCTION

Methylation Accessibility Protocol for individual templates (MAPit) assays chromatin structure on single molecules, as reviewed in Darst et al. (2012), Miranda, Kelly, Bouazoune, and Jones (2010), Pardo, Darst, Nabilsi, Delmas, and Kladde (2011), and Pondugula and Kladde (2008). The method combines DNA methyltransferase (DNMT) accessibility probing with bisulfite genomic sequencing. Exogenous DNMT probes can be used to footprint protein–DNA complexes, which block cytosine methylation (Kladde, Xu, & Simpson, 1996; Xu, Simpson, & Kladde, 1998). Chemical conversion of unmethylated cytosine to uracil and retention of methyl-5-cytosine (m^5C) upon bisulfite treatment then record methylation status of each residue as DNA sequence, as reviewed in Darst, Pardo, Ai, Brown, and Kladde (2010). Thus, by subsequent cloning, each sequence obtained records the chromatin landscape of a single parent molecule (Fatemi et al., 2005; Jessen, Hoose, Kilgore, & Kladde, 2006), as illustrated in Fig. 8.1. MAPit has been used to track chromatin transitions at single-molecule resolution *in vitro* (Bouazoune, Miranda, Jones, & Kingston, 2009; Dechassa et al., 2010), in isolated nuclei (Gal-Yam et al., 2006; Kilgore, Hoose, Gustafson, Porter, & Kladde, 2007; Lin et al., 2007), and *in vivo* (Jessen et al., 2006). MAPit can also probe protein–DNA interactions and endogenous DNA methylation simultaneously, using an enzyme cloned from a *Chlorella* virus, M.CviPI (Xu, Kladde, Van Etten, & Simpson, 1998a), which we have made commercially available. Whereas methylation in differentiated human cells is largely restricted to CG dinucleotides (Lister et al., 2009), M.CviPI methylates cytosine in GC dinucleotides exclusively. With M.CviPI, MAPit has been used to characterize chromatin diversity at a variety of loci (Andreu-Vieyra et al., 2011; Delmas et al., 2011; Han et al., 2011; Pardo, Carr, et al., 2011; Wolff et al., 2010).

To characterize variation in chromatin structure of a locus of interest across a population, enough sequences are needed to see every major variant and their intermediate states. We have found chromatin variants by MAPit that represented real biological phenomena and were only detectable when sequencing about 100 molecules (RPD, I. Haecker, CEP, A. Delmas, R. Renne, and MPK, unpublished observation). With modern high-throughput sequencing, one could study even rarer subjects within mixed populations, such as cancer stem cells or intermediates in chromatin remodeling. At the same time, direct deep sequencing of amplicons bypasses the need for cloning, streamlining the workflow. As for other chromatin analysis

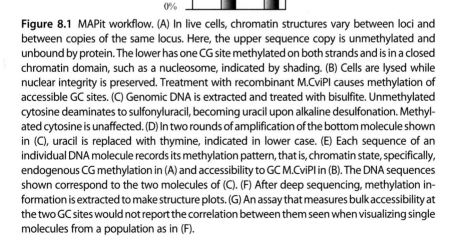

Figure 8.1 MAPit workflow. (A) In live cells, chromatin structures vary between loci and between copies of the same locus. Here, the upper sequence copy is unmethylated and unbound by protein. The lower has one CG site methylated on both strands and is in a closed chromatin domain, such as a nucleosome, indicated by shading. (B) Cells are lysed while nuclear integrity is preserved. Treatment with recombinant M.CviPI causes methylation of accessible GC sites. (C) Genomic DNA is extracted and treated with bisulfite. Unmethylated cytosine deaminates to sulfonyluracil, becoming uracil upon alkaline desulfonation. Methylated cytosine is unaffected. (D) In two rounds of amplification of the bottom molecule shown in (C), uracil is replaced with thymine, indicated in lower case. (E) Each sequence of an individual DNA molecule records its methylation pattern, that is, chromatin state, specifically, endogenous CG methylation in (A) and accessibility to GC M.CviPI in (B). The DNA sequences shown correspond to the two molecules of (C). (F) After deep sequencing, methylation information is extracted to make structure plots. (G) An assay that measures bulk accessibility at the two GC sites would not report the correlation between them seen when visualizing single molecules from a population as in (F).

techniques that evolved to genome-wide analysis, the future of MAPit undoubtedly lies with deep sequencing.

This chapter describes construction of an amplicon sequencing library for the Roche 454 system. The choice of sequencing platform is driven by the marked utility of read length in MAPit. A read with 20 GC sites has over a million possible methylation configurations, whereas a read with two GC sites has just four. Thus, the structural information per read scales dramatically with the number of sites sequenced per molecule. Furthermore, read length must exceed length of expected features, such as nucleosomes, to map both ends. At present, the 454 system is a good choice for mapping nucleosomes, with average read lengths of 400+ bp readily achievable. Because this platform produces hundreds of thousands of reads, we pool PCR of several loci and from several samples in equimolar amounts, using barcodes incorporated in the primers to track which sequence came from which sample.

Recently, it has become possible to sequence DNA methylation directly, without bisulfite conversion, using the Pacific Biosciences SMRT technology (Clark et al., 2012; Flusberg et al., 2010). Unfortunately, the low fidelity of SMRT recognition of cytosine methylation precludes single-molecule analysis at present. However, with further advances, this or another next-generation sequencing approach may soon revolutionize the way MAPit is done. In turn, a simpler and more direct approach to MAPit could have a broad impact on molecular biology.

2. MATERIALS

Methods and materials needed for the following common laboratory procedures are not detailed in this chapter: tissue culture, agarose gel electrophoresis, PCR both standard and quantitative real-time (qPCR), phenol: chloroform extraction, and ethanol precipitation.

2.1. Reagents

1. GC-sensitive restriction endonuclease such as *Hae*III and manufacturer's supplied buffer.
2. Hot-start Taq formulation, for example, HotStarTaq (Qiagen) or Pfu Turbo Cx (Stratagene), and manufacturer's supplied buffer.
3. M.CviPI, at least 20 U/µl, fused either to glutathione S-transferase (Zymo Research Corp.) or to maltose binding protein (New England Biolabs). Aliquot and store at -20 °C.

4. SAM (S-adenosylmethionine), 32 mM. Aliquot and store at -80 °C; do not reuse aliquots.
5. Trypan blue, 0.4% (w/v).

For homebrew bisulfite conversion (optional):
6. Hydroquinone
7. Sodium hydroxide
8. Sodium metabisulfite. Prepare 5 g aliquots in an oxygen-free environment. Store desiccated and in the dark

2.2. Solutions

Dissolve in deionized distilled water and store at room temperature indefinitely unless otherwise specified.
1. DNMT storage buffer. 15 mM Tris–HCl pH 7.4, 200 mM NaCl, 100 μM EDTA pH 8.0, 1 mM DTT, 200 μg/ml nuclease-free bovine serum albumin, 50% (v/v) glycerol. Store at -20 °C.
2. DTT (dithiothreitol), 1 M. Aliquot and store at -20 °C.
3. EDTA (ethylenediaminetetraacetic acid), 0.5 M pH 8.0.
4. PMSF (phenylmethanesulfonyl fluoride), 25 mM in ethanol. Aliquot and store at -20 °C.
5. PBS (phosphate-buffered saline). 137 mM NaCl, 5.37 mM Na$_2$HPO$_4$, 2.68 mM KCl, 1.76 mM KH$_2$PO$_4$, pH to 7.4.
6. Resuspension buffer. 20 mM HEPES (4-(2-hydroxyethyl)-1-piperazineethanesulfonic acid) pH 7.5, 70 mM NaCl, 0.25 mM EDTA pH 8.0, 0.5 mM EGTA (ethylene glycol tetraacetic acid) pH 8.0, 0.5% (v/v) glycerol. Store at 4 °C for up to a month. Add DTT to 10 mM and PMSF to 0.25 mM immediately before use and place on ice.
7. Lysis buffer. As resuspension buffer, plus 0.19% (v/v) Nonidet P-40 equivalent (octylphenoxypolyethoxyethanol). Store at 4 °C for up to a month. Add DTT to 10 mM and PMSF to 0.25 mM immediately before use and place on ice.
8. Stop buffer, 2×. 100 mM NaCl, 10 mM EDTA pH 8.0, 1% (w/v) SDS (sodium dodecyl sulfate). Store at room temperature (20–25 °C).
9. Proteinase K, 20 mg/ml in 5% (v/v) glycerol, 1 mM Tris pH 7.5, 250 μM CaCl$_2$. Do not vortex. Store aliquots at -20 °C.
10. Phenol:chloroform solution, phenol:chloroform:isoamyl alcohol (25:24:1), equilibrated with 1 M Tris–HCl pH 8.0.
11. TE, 0.1×. 1 mM Tris–HCl pH 8.0, 100 μM EDTA pH 8.0.

2.3. Supplies

1. Bisulfite conversion and/or clean-up kit, for example, EZ DNA Methylation (Zymo Research; catalog #D5001).
2. DNA purification kit, for example, QIAEX II Gel Extraction Kit (Qiagen; catalog #20021).
3. High-throughput DNA sequencer, for example, Roche 454 GS-FLX, and associated materials.

3. PROTOCOLS

3.1. MAPit in nuclei

This section describes MAPit with cultured fetal neural stem cells (NSCs) grown in suspension, taking enough cells for a two-point titration of M.CviPI concentration, and an untreated control. The protocol can be readily adapted to other cell types and growth conditions. For attached cells, trypsinization is preferred over scraping for cell detachment. If higher detergent concentration is needed for cell lysis to prepare nuclei, then it must be diluted to 0.07% (v/v) in the methylation reaction. Optimal M.CviPI concentration and reaction time may also vary with cell type. For more information on MAPit with reconstituted chromatin, see Darst et al. (2012), Miranda et al. (2010), or Pondugula, Gangaraju, and Bartholomew (2009). For *in vivo* MAPit using budding yeast, see Pardo, Hoose, Pondugula, and Kladde (2009).

3.1.1 Isolate nuclei

1. Pellet the cells from a single cell suspension in a sterile 50-ml conical tube by centrifugation at $1000 \times g$ for 5 min at 4 °C. From this point on, keep cells on ice.
2. Wash the cells twice with 5 ml ice-cold PBS by centrifugation at $1000 \times g$ for 5 min at 4 °C.
3. Resuspend the cells in 1 ml PBS and count, either on a hemocytometer by light microscopy, or with an automated counter. Aliquot 3.3×10^6 cells per sample to a 1.7-ml microcentrifuge tube.
4. Centrifuge at $1000 \times g$ for 5 min at 4 °C and resuspend pellet in 500 µl ice-cold resuspension buffer with freshly added DTT and PMSF. Centrifuge at $1000 \times g$ for 5 min at 4 °C and carefully remove supernatant.
5. Add 115 µl ice-cold lysis buffer with freshly added DTT and PMSF. Resuspend the cells by gentle tapping and incubate on ice.

6. After 8 min, quickly mix 2 μl cell suspension with 2 μl of 0.4% (v/v) trypan blue and examine by light microscopy. When lysis is complete, nuclei should be blue, round, and granular, with no attached debris. If intact cells are observed, continue lysis reaction and check again in 2 min.
7. Terminate the lysis by adding 171 μl resuspension buffer. Put the nuclei on ice.

3.1.2 Probe with M.CviPI

1. On ice, dilute DNMT to treat nuclei with 0, 30, and 100 U M.CviPI per 10^6 cells:
   a. 100 U/10 μl: 1 part 80 U/μl M.CviPI plus 7 parts resuspension buffer with freshly added DTT and PMSF;
   b. 30 U/10 μl: 24 parts 100 U/10 μl M.CviPI from step a, plus 7 parts DNMT storage buffer and 49 parts resuspension buffer;
   c. 0 U/10 μl: 1 part DNMT storage buffer plus 7 parts resuspension buffer.

   The dilution scheme ensures that M.CviPI concentration is the only variable.

2. Prewarm samples at 37 °C for 2 min. Prewarm stop buffer at 50 °C. Bring ice bucket with M.CviPI dilutions, timer, micropipette, and tips to 37 °C water bath.
3. Add SAM to 160 μM final concentration (1.5 μl of 32 mM SAM per 300 μl nuclei). Withdraw a 90-μl aliquot (about 10^6 cells) for each methylation reaction and transfer to a 1.7-ml microcentrifuge tube.
4. To ensure that all samples receive identical DNMT treatment, addition of DNMT to each sample should be staggered. Therefore, set timer to 15 min. After 30 s, add 10 μl DNMT dilution to the first sample. Mix by flicking the tube briefly and return to 37 °C. Repeat every 30 s until all samples have been treated.
5. 15 min after addition of DNMT (30 s after alarm sounds), add 100 μl of 2 × stop buffer to the first sample treated. Vortex for 5 s and keep at room temperature until all the samples have been processed. Treat the samples in the same order as before, once every 30 s.
6. Add proteinase K to 100 μg/ml to each sample (1 μl of 20 mg/ml per 200-μl sample). Mix by flicking the tube briefly and incubate 16–20 h at 50 °C.
7. Phenol:chloroform extract and ethanol-precipitate DNA, and then elute in 50 μl of 0.1 × TE at 4 °C for 12–16 h. Measure nucleic acid concentration by 260-nm absorbance.

3.2. MAPit validation

It is prudent to confirm that the methylation reaction was successful before investing in high-throughput library construction. Although there is no universal yardstick for measuring methylation efficiency, quantifying methylation of a single GC site provides a useful test for M.CviPI activity. The simplest such test is quantitative methyl-sensitive restriction endonuclease digest (qMSRE; Hashimoto, Kokubun, Itoi, & Roach, 2007). Percent GC methylation is proportional to percent protection from a restriction enzyme sensitive to GC methylation, which one can assay by real-time molecular amplification (qPCR), comparing digested to uncut. This protocol probes a HaeIII site at the human GAPDH promoter:

1. Digest 10 ng genomic DNA (from nuclei treated and untreated with M.CviPI) with HaeIII in 20 μl volume of 1× manufacturer's supplied buffer for 2 h at 37 °C. For mock digests, add 50% (v/v) glycerol solution in place of restriction enzyme.
2. Stop the reaction at 80 °C for 20 min.
3. Measure GAPDH promoter abundance by real-time PCR: three reactions per sample, each with 2 μl digestion or mock reaction, using primers 5'-TACTAGCGGTTTTACGGGCG-3' and 5'-TCGAACAGGAGGAGCAGAGAGCGA-3'.
4. Compute percent protection as: $2^{(\bar{C}_T \text{ mock} - \bar{C}_T \text{ digest})}$ where $\bar{C}_T$ is the average cycle at which signal exceeds the threshold for the indicated sample.

At least 95% of unmethylated DNA should be cut, whereas treatment with M.CviPI may give up to 80% protection. Percent protection depends both on M.CviPI activity and on the chromatin structure at the site.

3.3. Amplicon sequencing

In amplicon sequencing, libraries of selected loci are amplified by locus-specific primers with 5' extensions containing the universal sequencing primers used by the high-throughput sequencing platform of choice. For MAPit, the DNA must be bisulfite-treated first (Fig. 8.1, compare B to C). There are two challenges in amplification of deaminated DNA.

First, DNA strands fragment during bisulfite conversion, likely due to the temperature and acidity. It has been estimated that DNA recovered after bisulfite treatment averages 675 bp in length (Munson, Clark, Lamparska-Kupsik, & Smith, 2007). Thus, there is an inverse dependence of amplification efficiency with length. On the other hand, utility of MAPit sequence reads scales with length. Ideally, amplicon length matches the expected sequence length obtained, which is 400–500 bp with the current Roche 454 GS-FLX system.

The second challenge is reduced specificity of primers. Deaminated DNA is less complex, and once a primer has incorporated bases at a nonspecific site, it can no longer amplify the locus of interest. Primer design and optimization of amplification conditions are therefore critical for success. Below, we describe how to design primers, give an optional homebrew bisulfite conversion protocol, and discuss amplification optimization. After a PCR product is obtained, follow the manufacturer's protocol for deep sequencing.

3.3.1 Primer design

The 5′ end of each primer should contain one of a pair of adaptor sequences. With the Roche 454 platform, these adaptors are used to prime both emulsion PCR and the sequencing reaction. For the latest version of these sequences, consult the manufacturer of the deep sequencing platform of choice (e.g., Roche). Next, a short 5–10 bp barcode should be appended to sequence a locus from several different samples in one reaction. Roche has already designed and validated an array of barcodes, and other manufacturers may have as well.

The 3′ end of each primer must anneal to deaminated DNA or its complement. After bisulfite treatment, the two strands of a DNA helix are no longer complementary because C:G base pairings convert to U:G mismatches. During amplification, DNA polymerase pairs uracil with adenine (Fig. 8.1D). Thus, amplification products have different sequences depending on which strand originated them, and primer sequences to amplify each strand are different as well.

An easy way to design primers is to view the sequence of the locus within a text editor. To design primers to the top strand, replace all "CG" with "YG," all "GC" with "GY," then replace all "C" with lowercase "t." This is now the deaminated sequence of the top strand. Each lowercase "t" marks thymines that pair with adenines that had initially paired with uracils (converted cytosines) during the first extension. The "Y" marks cytosines that may or may not convert depending on methylation. Choose forward and reverse primer sequences that do not overlap a "Y," if possible, with melting temperatures in the range 52–58 °C. Naturally, the reverse primer has to be the reverse complement of a top strand sequence. If a "Y" cannot be avoided, replace it with an adenine in the forward primer (or replace its complement "R" with a thymine in the reverse primer). The mismatch minimizes biased amplification of methylated or unmethylated sequence.

To design primers to the bottom strand, go back to the original sequence and replace all "CG" with "CR," all "GC" with "RC," then replace all "G" with lowercase "a." In either case, the end result should be a forward primer in which thymine replaces cytosine, and a reverse primer in which adenine replaces guanine. Sometimes, one strand will become less complex than the other after deamination. For instance, the sequence 5′-AGGA-3′ is unchanged on the top deaminated stand, but the complement to the deaminated bottom strand becomes 5′-AaaA-3′. This could interfere with amplification or sequencing. Ideally, one would design and test primers to both strands in case of such problem sequences.

3.3.2 Bisulfite treatment

Bisulfite conversion can be performed with a number of commercially available kits, for example, EZ DNA Methylation (Zymo Research). Although these kits work well for most needs, for MAPit at CpG islands, we prefer a homebrew approach. We hypothesize that high GC density of deaminated DNA at these loci promotes reannealing, which interferes with bisulfite conversion. This homebrew method gives a better percent conversion than we have achieved with kits:

1. Degas 200 ml water by stirring 20 min either while boiling or under vacuum. Fill a 125-ml glass bottle to brim and cap. If boiled, let cool overnight before using.
2. Freshly prepare 3 N NaOH from pellets and 100 mM hydroquinone using the degassed water. Prepare 3× sample denaturation buffer: per sample, 6.5 µl degassed water, 3.0 µl of 3 N NaOH, and 0.5 µl of 500 mM EDTA pH 8.0.
3. Dissolve 0.2–4.0 µg total nucleic acid (DNA and RNA) in 20 µl of 0.1 × TE. Add 10 µl of 3 × sample denaturation buffer. Leave at room temperature.
4. Pipette 100 µl of 100 mM hydroquinone into a glass scintillation vial containing a small stir bar. Add quickly in order: 5 g sodium metabisulfite, 7 ml degassed water, and 1 ml of 3 N NaOH. Stir to dissolve, then adjust pH to 4.95–5.05 with 3 N NaOH (may take 100–200 µl). Place at 50 °C.
5. Denature the DNA samples at 98 °C in thermocycler. After 5 min, quickly add 200 µl metabisulfite solution to each sample and vortex. Keep the samples at 50 °C to prevent renaturation.
6. Incubate the samples at 50 °C in the dark for 6 h to deaminate cytosines to uracils.

7. To desulfonate and concentrate DNA, use a commercially available kit, for example, EZ Bisulfite Cleanup Kit (Zymo Research).
8. Elute the DNA in 25 μl of 0.1× TE prewarmed to 37 °C and store at −20 °C.

3.3.3 Amplification

Extensive denaturation is necessary with single-stranded, low complexity deaminated DNA. All thermocycling steps should be extended; the melt and anneal to 45 s, the extension to 3–4 min per kb, and the initial denaturation to 5 min. For this reason, a hot-start formulation of Taq DNA polymerase is preferred. Uracil in deaminated DNA stalls Pfu and other archaeal family B DNA polymerases (Lasken, Schuster, & Rashtchian, 1996). However, Pfu Turbo Cx (Stratagene) has been engineered to lack uracil recognition, providing another option.

For best amplification, the following parameters may need to be optimized: template DNA concentration, primer concentration, magnesium concentration, and annealing temperature. To begin, add 50 ng template DNA per reaction (i.e., 1–2% of the eluate) and primers to 250 nM each. Compare three magnesium concentrations (1.5, 2.5, and 3.5 mM) across an annealing temperature gradient from −3 to +3 °C of the calculated primer melting temperature, and electrophorese the products on a 1.5% (w/v) agarose gel. A smear or ladder of nonspecific product suggests that template DNA should be diluted. If no product is seen, increasing primer concentration up to 1 μM can sometimes help. In some cases, PCR additives such as trimethyl ammonium chloride (TMAC) are needed.

Once conditions have been optimized:
1. Amplify the library in at least three separate reactions to maximize sequence diversity.
2. Purify the product DNA from primers with a 1.5% (w/v) agarose gel.
3. Use Qiaex II (Qiagen) or other gel extraction kit to desalt and concentrate.
4. Run 20% on a second gel to quantify concentration.
5. Pool the samples for deep sequencing following platform manufacturer's protocol.

3.4. Sequence analysis

Having obtained several tens of thousands of DNA sequence reads, one must first find the locus and barcode of each sequence read and align the sequence against the wild-type reference for the locus. Once aligned, percent bisulfite conversion can be measured and sequences with insufficient deamination discarded. One may also choose to discard duplicate sequences.

To display the data, we recommend two methods. MethylViewer is best for inspecting a small number of molecules in depth. To display hundreds or thousands of molecules at once, we have developed a color-map scheme, employed by MethylTracker.

Further methods of analysis include graphing percent methylation of each site to compare population average to other measures of chromatin structure; counting footprints within various size ranges at each base pair position, to compare populations; and clustering by structure to identify qualitative differences in chromatin diversity.

3.4.1 Alignment

Common multipurpose alignment tools are not well suited for processing bisulfite-converted sequence, or for pyrosequencing as used by the Roche 454 platform. Single base pair insertions and deletions (indels) are common in pyrosequencing, especially at homopolymer sequences, but larger indels are very rare (and reads containing large insertions should be discarded). We have had success with both BLAST (http://blast.ncbi.nlm.nih.gov/Blast.cgi; Altschul, Gish, Miller, Myers, & Lipman, 1990) and MUSCLE (http://www.drive5.com/muscle/; Edgar, 2004) to align bisulfite-converted, 454-sequenced DNA, using some basic code in a language like Python to process sequences in batch.

A custom scoring matrix can be useful for aligning cytosine in the reference with thymine in the read. Although one should treat all cytosines in the read as thymines to prevent a methylation bias in mapping of sequences (Robinson, Statham, Speed, & Clark, 2010), some weak preference for cytosine–cytosine alignment is useful with the 454 sequencer. Otherwise, rare single-thymine insertions (GC to GTC) will make methylated GC sites appear unmethylated, that is, false negatives. We have not observed cytosine insertions on the deaminated strand, perhaps because there are few cytosine homopolymers. This suggests that there is little risk of false positives from 454 sequencer error.

3.4.2 Percent conversion

Incomplete deamination is the source of most false positives for methylation. Percent deamination of unmethylated cytosines can be estimated from percent conversion of cytosines not within potential methylation sites, in other words, HCH, where H is A, C, or T. Discard sequence reads with HCH to HTH conversion of less than 95%, or whatever cutoff desired.

3.4.3 Duplicate sequences

Optionally, duplicate sequences may be discarded. We have found that, given variable endogenous CG methylation, probe GC methylation, and HCH conversion, usually only 5–10% of sequences are exact duplicates. Since two or more identical sequences can result from duplication during the PCR amplification step, their number does not necessarily attest to the frequency of a chromatin structure in live cells. On the other hand, a combination of high bisulfite conversion efficiency with low methylation diversity increases the proportion of duplicates at some loci. In this case, many duplicates likely derive from identical chromatin structures present on separate molecules in the original population. If there is a high proportion of duplicate sequences and the two above-mentioned conditions do not apply, amplification efficiency was likely very poor and the data cannot be used.

3.4.4 PCR recombination

In theory, an incompletely extended PCR product could bind and prime replication of a different copy of the amplicon in a subsequent cycle. This would create a chimerical amplicon, with some sequence from one molecule and some from another. In some situations, chimerical amplicons would be detectable. For example, PCR recombination or crossover between molecules protected by a nucleosome and those that are not would make the footprint size variable. However, we have not detected chimerical amplicons, because we have observed variability in the positions of features, but not their sizes.

3.4.5 MethylViewer

The standard map of DNA methylation is the "lollipop" diagram. We have modified this scheme to depict GC and CG methylation together, using circles for CG and triangles for GC sites (as in Fig. 8.1F). By spacing these symbols according to the location of the sites, sizes of footprints are represented in the image.

The program MethylViewer automates the construction of these illustrations and is freely available at http://dna.leeds.ac.uk/methylviewer/ (Pardo, Carr, et al., 2011). MethylViewer takes sequences in either ABI or FASTA format, aligned or unaligned. Figure 8.2 shows the MethylViewer output. This experiment probed chromatin in cultured human NSCs. These cells were cultured under conditions that would preserve their phenotypic diversity, that is, NSCs and progenitor cells (Siebzehnrubl, Vedam-Mai, Azari,

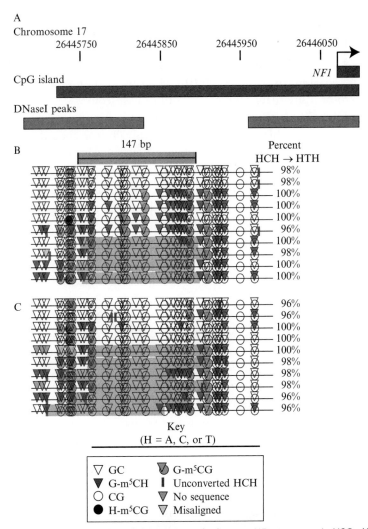

Figure 8.2 MethylViewer plot of chromatin at the human *NF1* promoter in NSCs. (A) The 303-bp amplicon was on the transcribed strand ~200 bp 5′ of the *NF1* transcription start site (bent arrow) and gene body (dark blue bar), overlapping both a CG island (annotated by UCSC Genome Browser; http://genome.ucsc.edu/; Dreszer et al., 2012; Fujita et al., 2011), and sites hypersensitive to DNase I in several cell lines (Sabo et al., 2006). Nucleotide coordinates on chromosome 17 are indicated at the top. (B) MethylViewer plot (to scale with A) of 10 molecules randomly chosen from over 1000 sequenced with forward primer. Of these, four featured a protected footprint on the nucleosome scale, marked by the light blue shading. Note that the top two molecules appear to be duplicates, having identical unconverted HCH site and methylation status of all sites. Nucleosome core particle length is indicated by bar with light blue shading. (C) Plot of 10 randomly chosen from over 6000 sequenced with the reverse primer. (See Color Insert.)

Reynolds, & Deleyrolle, 2011). Accordingly, we anticipated that these conditions would also conserve epigenetic diversity among the cell population. The promoter of the tumor suppressor gene *NF1* exhibited a diversity of chromatin structures including differential nucleosome positioning. Although MethylViewer is not suited for displaying large numbers of molecules, it is the best tool available for in-depth inspection of small numbers of MAPit reads. The MethylViewer image output displays all the relevant information for each molecule: CG methylation, GC methylation, methylation of GCG sites, percent HCH conversion, and location of unconverted HCH sites.

Note that GCG sites, which can be methylated both by endogenous mammalian DNMTs and by M.CviPI, are colored gray, to reflect their ambiguity. However, this is usually only a concern if significant levels of endogenous methylation are present at or within the vicinity of each GCG site, which can be assessed by a no M.CviPI control sample. MethylViewer can also display different types of methylation besides CG and GC. This is useful for instance for MAPit with M.HhaI, which methylates the 5' cytosine in GCGC. Up to four methylation types of any desired sequence can be displayed simultaneously.

3.4.6 MethylTracker

To display MethylViewer images for hundreds of molecules on one printed page is impractical. To extract qualitative data from hundreds or thousands of molecules simultaneously, the information has to be condensed. For this purpose, we map methylation to scale with contrasting colors painting blocks of successively methylated and unmethylated sites, and the borders between them (Fig. 8.3A–C; Delmas et al., 2011). Alternately, different shades of a single color can be used; in this case, our convention is that black marks unmethylated spans of sites (panels D–E). Although it is possible to overlay two different types of methylation (i.e., CG and GC), we generally plot only one at a time in this manner. We exclude GCG sites unless only one methylation type is present. Similar schemes have been used by others (e.g., Gal-Yam et al., 2006).

We have built an automated way to construct these plots, named MethylTracker (http://genome.ufl.edu/methyl/). The input must be a text file with aligned sequences in FASTA format. MethylTracker does not align the sequences itself. Currently, the interface allows the user to choose whether to plot CG or GC methylation, and whether to include GCG sites, and supports several ways of assigning patches. With the default settings, a patch is drawn between any two consecutive sites with the same methylation status. This

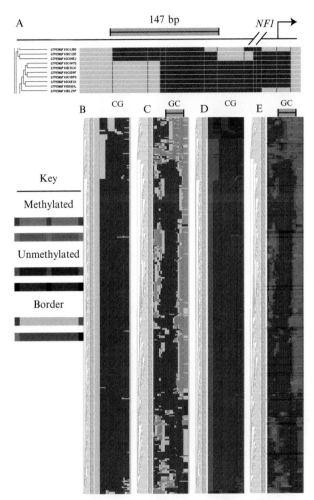

Figure 8.3 MethylTracker plot of chromatin at the human *NF1* promoter in NSCs. (A) Diagram of the same amplicon as in Fig. 8.2 and a close-up view of the MethylTracker plot of CG methylation with these settings: two sites to make a patch, one to break, and exclude GCG. Each row represents one sequence read. See key below: methylated and unmethylated sites are indicated by dark green and bright red vertical bars, respectively. Green patches span successive methylated sites and dark red patches span successive unmethylated sites. Yellow patches mark the borders between sites of different methylation status, as well as sequence not bordered on both sides by sites, that is, at each end. Blue bar indicates 147-bp scale. (B) The complete CG methylation plot for 500 randomly chosen molecules. (C) The GC methylation plot for the same molecules. Blue bar at the top indicates 147-bp nucleosome core particle-length scale. Note variable nucleosome positioning among the molecules inferred from ∼150 bp or longer spans of protection against GC methylation (dark red patches). As CG and GC methylation were clustered separately, the order of molecules differs from that in (B). (D) Molecules were

allows a three-color plot, in which borders between methylated and unmethylated zones are shaded to indicate ambiguity. One may instead ask that patches not break until they meet two consecutive sites of a different status, creating a two-color plot (because only one methylation status can be mapped at a time this way).

3.4.7 Data analysis

It is easy to compare quantitative data between experiments. For instance, one could plot the average percent methylation at each site at a locus in two cell populations, to see whether one has higher endogenous methylation or is more open to M.CviPI. This data is comparable to what one would obtain from a population assay, such as methyl-specific PCR or DNase I footprinting. Indeed, comparison of average data from MAPit to these techniques provides a way to double-check MAPit results.

To take advantage of the single-molecule MAPit information, different modes of analysis are needed. One method is to classify footprints based on length and count their occurrence at each position within the locus. Generally, one can infer that M.CviPI-protected footprints in the size range of 100–150 bp result from nucleosomes. Footprints smaller than 100 bp can be assigned to other DNA-binding factors, when occupying known sites, but could also result from incomplete methylation. M.CviPI-accessible patches of $\sim$100 bp or more in length represent nucleosome-free regions, as have been observed at promoters by deep sequencing of nucleosome core particle-length fragments (Jiang & Pugh, 2009). In this way, MAPit detects changes in nucleosome positioning or occupancy of subnucleosomal factors.

Another method is hierarchical clustering of methylation maps (Delmas et al., 2011). Suppose one population has a chromatin state that the other lacks. This may not be obvious from looking at the population average (especially if it is a rare state), or from looking at the methylation maps (if the vertical order of molecules is random). Hierarchical clustering brings together molecules

---

divided into three classes by CG methylation structure, tagged red, green, or blue, and clustered for CG methylation again. Dark shading now marks unmethylated spans, spans of bright color mark successively methylated sites, and intermediate shading represents the borders. (E) The GC methylation plot for the molecules barcoded by CG structure. Dispersed distribution of molecules from each of the three CG clusters in (D) over the range of chromatin structures clustered in (C) based on GC methylation shows that CG methylation does not correlate with nucleosome positioning at the locus. (See Color Insert.)

with similar chromatin structure. Thus, instead of thousands of molecules, the human eye sees a smaller number of structure clusters. These clusters can be viewed as chromatin states. For instance, the defining feature of one cluster might be a particular positioned nucleosome; it is the chromatin state in which that nucleosome is present. It may be possible to see a cluster that is present in one population but not in the other. If not, tagging the molecules by origin and mixing them, then clustering them, may reveal a cluster that has a statistically significant enrichment for tags from one population.

The MethylTracker program has a color barcoding system. To access this, the user must add the bar character " | " and a text string to the end of each sequence identifier line. MethylTracker tracks up to six of these text strings, assigning each a primary or secondary color. Each sequence with that barcode is then represented in the assigned color, as in Fig. 8.3D and E. In this example, three patterns of CG methylation were found in 500 molecules: a single site upstream (coded green), some sites downstream (red), or no methylation (blue). Molecules were barcoded by CG structure and reclustered. The GC methylation plot in Fig. 8.3E shows that red, green, and blue molecules are present in each structure cluster. Although this shows that these CG methylations do not affect nucleosome positioning, the opposite result would not support the reverse conclusion. A correlation between CG and GC methylation maps would indicate that chromatin structure variations are epigenetic, but would not necessarily define the epigenetic determinants.

ACKNOWLEDGMENTS

This work was supported by grants R01CA095525 and R01CA155390 from the National Institutes of Health, BC097648 from the Department of Defense, 1BD-03 and 2BT01 from the Florida Department of Health (N.H.N. and M. P. K.), and funds from the UF Genetics Institute (A. R.).

REFERENCES

Altschul, S. F., Gish, W., Miller, W., Myers, E. W., & Lipman, D. J. (1990). Basic local alignment search tool. *Journal of Molecular Biology, 215*, 403–410.

Andreu-Vieyra, C., Lai, J., Berman, B. P., Frenkel, B., Jia, L., Jones, P. A., et al. (2011). Dynamic nucleosome-depleted regions at androgen receptor enhancers in the absence of ligand in prostate cancer cells. *Molecular and Cellular Biology, 31*, 4648–4662.

Bouazoune, K., Miranda, T. B., Jones, P. A., & Kingston, R. E. (2009). Analysis of individual remodeled nucleosomes reveals decreased histone-DNA contacts created by hSWI/SNF. *Nucleic Acids Research, 37*, 5279–5294.

Clark, T. A., Murray, I. A., Morgan, R. D., Kislyuk, A. O., Spittle, K. E., Boitano, M., et al. (2012). Characterization of DNA methyltransferase specificities using single-molecule, real-time DNA sequencing. *Nucleic Acids Research, 40*, e29. http://dx.doi.org/10.1093/nar/gkr1146.

Darst, R. P., Pardo, C. E., Ai, L., Brown, K. D., & Kladde, M. P. (2010). Bisulfite sequencing of DNA. *Current Protocols in Molecular Biology*, chapter 7: Unit 7.9.1–17.
Darst, R. P., Pardo, C. E., Pondugula, S., Gangaraju, V. K., Nabilsi, N. H., Bartholomew, B., et al. (2012). Simultaneous single-molecule detection of endogenous C-5 DNA methylation and chromatin accessibility using MAPit. *Methods in Molecular Biology*, *833*, 125–141.
Dechassa, M. L., Sabri, A., Pondugula, S., Kassabov, S. R., Chatterjee, N., Kladde, M. P., et al. (2010). SWI/SNF has intrinsic nucleosome disassembly activity that is dependent on adjacent nucleosomes. *Molecular Cell*, *38*, 590–602.
Delmas, A. L., Riggs, B. M., Pardo, C. E., Dyer, L. M., Darst, R. P., Izumchenko, E. G., et al. (2011). WIF1 is a frequent target for epigenetic silencing in squamous cell carcinoma of the cervix. *Carcinogenesis*, *32*, 1625–1633.
Dreszer, T. R., Karolchik, D., Zweig, A. S., Hinrichs, A. S., Raney, B. J., Kuhn, R. M., et al. (2012). The UCSC Genome Browser database: Extensions and updates 2011. *Nucleic Acids Research*, *40*, D918–D923.
Edgar, R. C. (2004). MUSCLE: Multiple sequence alignment with high accuracy and high throughput. *Nucleic Acids Research*, *32*, 1792–1797.
Fatemi, M., Pao, M. M., Jeong, S., Gal-Yam, E. N., Egger, G., Weisenberger, D. J., et al. (2005). Footprinting of mammalian promoters: use of a CpG DNA methyltransferase revealing nucleosome positions at a single molecule level. *Nucleic Acids Research*, *33*, e176.
Flusberg, B. A., Webster, D. R., Lee, J. H., Travers, K. J., Olivares, E. C., Clark, T. A., et al. (2010). Direct detection of DNA methylation during single-molecule, real-time sequencing. *Nature Methods*, *7*, 461–465.
Fujita, P. A., Rhead, B., Zweig, A. S., Hinrichs, A. S., Karolchik, D., Cline, M. S., et al. (2011). The UCSC Genome Browser database: Update 2011. *Nucleic Acids Research*, *39*, D876–D882.
Gal-Yam, E. N., Jeong, S., Tanay, A., Egger, G., Lee, A. S., & Jones, P. A. (2006). Constitutive nucleosome depletion and ordered factor assembly at the *GRP78* promoter revealed by single molecule footprinting. *PLoS Genetics*, *2*, e160.
Han, H., Cortez, C. C., Yang, X., Nichols, P. W., Jones, P. A., & Liang, G. (2011). DNA methylation directly silences genes with non-CpG island promoters and establishes a nucleosome occupied promoter. *Human Molecular Genetics*, *20*, 4299–4310.
Hashimoto, K., Kokubun, S., Itoi, E., & Roach, H. I. (2007). Improved quantification of DNA methylation using methylation-sensitive restriction enzymes and real-time PCR. *Epigenetics*, *2*, 86–91.
Jessen, W. J., Hoose, S. A., Kilgore, J. A., & Kladde, M. P. (2006). Active *PHO5* chromatin encompasses variable numbers of nucleosomes at individual promoters. *Nature Structural & Molecular Biology*, *13*, 256–263.
Jiang, C., & Pugh, B. F. (2009). Nucleosome positioning and gene regulation: Advances through genomics. *Nature Reviews. Genetics*, *10*, 161–172.
Kilgore, J. A., Hoose, S. A., Gustafson, T. L., Porter, W., & Kladde, M. P. (2007). Single-molecule and population probing of chromatin structure using DNA methyltransferases. *Methods*, *41*, 320–332.
Kladde, M. P., Xu, M., & Simpson, R. T. (1996). Direct study of DNA-protein interactions in repressed and active chromatin in living cells. *The EMBO Journal*, *15*, 6290–6300.
Lasken, R. S., Schuster, D. M., & Rashtchian, A. (1996). Archaebacterial DNA polymerases tightly bind uracil-containing DNA. *The Journal of Biological Chemistry*, *271*, 17692–17696.
Lin, J. C., Jeong, S., Liang, G., Takai, D., Fatemi, M., Tsai, Y. C., et al. (2007). Role of nucleosomal occupancy in the epigenetic silencing of the *MLH1* CpG island. *Cancer Cell*, *12*, 432–444.

Lister, R., Pelizzola, M., Dowen, R. H., Hawkins, R. D., Hon, G., Tonti-Filippini, J., et al. (2009). Human DNA methylomes at base resolution show widespread epigenomic differences. *Nature*, 462, 315–322.

Miranda, T. B., Kelly, T. K., Bouazoune, K., & Jones, P. A. (2010). Methylation-sensitive single-molecule analysis of chromatin structure. *Current Protocols in Molecular Biology*, chapter 21: Unit 21.17.1–16.

Munson, K., Clark, J., Lamparska-Kupsik, K., & Smith, S. S. (2007). Recovery of bisulfite-converted genomic sequences in the methylation-sensitive QPCR. *Nucleic Acids Research*, 35, 2893–2903.

Pardo, C. E., Carr, I. M., Hoffman, C. J., Darst, R. P., Markham, A. F., Bonthron, D. T., et al. (2011). MethylViewer: Computational analysis and editing for bisulfite sequencing and methyltransferase accessibility protocol for individual templates (MAPit) projects. *Nucleic Acids Research*, 39, e5.

Pardo, C. E., Darst, R. P., Nabilsi, N. H., Delmas, A. L., & Kladde, M. P. (2011). Simultaneous single-molecule mapping of protein-DNA interactions and DNA methylation by MAPit. *Current Protocols in Molecular Biology*, chapter 21: Unit 21.22.1–18.

Pardo, C., Hoose, S. A., Pondugula, S., & Kladde, M. P. (2009). DNA methyltransferase probing of chromatin structure within populations and on single molecules. *Methods in Molecular Biology*, 523, 41–65.

Pondugula, S., Gangaraju, V. K., Bartholomew, B., & Kladde, M. P. (2009). *DNA methyltransferase-based single-molecule (MAPit) assay for mapping protein-DNA interactions in vitro*. Published online at http://www.epigenesys.eu/index.php/en/protcols/chromatin-biochemistry/

Pondugula, S., & Kladde, M. P. (2008). Single-molecule analysis of chromatin: Changing the view of genomes one molecule at a time. *Journal of Cellular Biochemistry*, 105, 330–337.

Robinson, M. D., Statham, A. L., Speed, T. P., & Clark, S. J. (2010). Protocol matters: Which methylome are you actually studying? *Epigenomics*, 2, 587–598.

Sabo, P. J., Kuehn, M. S., Thurman, R., Johnson, B. E., Johnson, E. M., Cao, H., et al. (2006). Genome-scale mapping of DNase I sensitivity *in vivo* using tiling DNA microarrays. *Nature Methods*, 3, 511–518.

Siebzehnrubl, F. A., Vedam-Mai, V., Azari, H., Reynolds, B. A., & Deleyrolle, L. P. (2011). Isolation and characterization of adult neural stem cells. *Methods in Molecular Biology*, 750, 61–77.

Wolff, E. M., Byun, H. M., Han, H. F., Sharma, S., Nichols, P. W., Siegmund, K. D., et al. (2010). Hypomethylation of a LINE-1 promoter activates an alternate transcript of the *MET* oncogene in bladders with cancer. *PLoS Genetics*, 6, e1000917.

Xu, M., Kladde, M. P., Van Etten, J. L., & Simpson, R. T. (1998). Cloning, characterization and expression of the gene coding for a cytosine-5-DNA methyltransferase recognizing GpC. *Nucleic Acids Research*, 26, 3961–3966.

Xu, M., Simpson, R. T., & Kladde, M. P. (1998). Gal4p-mediated chromatin remodeling depends on binding site position in nucleosomes but does not require DNA replication. *Molecular and Cellular Biology*, 18, 1201–1212.

CHAPTER NINE

Genome-Wide *In Vitro* Reconstitution of Yeast Chromatin with *In Vivo*-Like Nucleosome Positioning

Nils Krietenstein, Christian J. Wippo, Corinna Lieleg, Philipp Korber[1]

Adolf-Butenandt-Institut, Molecular Biology Unit, University of Munich, Munich, Germany
[1]Corresponding author: e-mail address: pkorber@lmu.de

Contents

1. Introduction	206
2. Preparation of Yeast WCE	207
3. Preparation of Histones	211
4. Expansion of Genomic Plasmid Library	214
4.1 Preparation of electrocompetent *E. coli* cells	216
4.2 Electroporation	216
4.3 Cultivation and plasmid preparation	217
5. Chromatin Reconstitution by Salt Gradient Dialysis	217
5.1 Pump and beakers	218
5.2 Dialysis mini chamber	218
5.3 Samples	220
5.4 Salt gradient dialysis	221
5.5 Titrating the assembly degree	221
6. Incubation of Salt Gradient Dialysis Chromatin with WCE or Purified Factors	227
7. Generation of Mononucleosomes by Limited MNase Digestion	228
8. Summary	229
Acknowledgments	230
References	230

Abstract

Recent genome-wide mapping of nucleosome positions revealed that well-positioned nucleosomes are pervasive across eukaryotic genomes, especially in important regulatory regions such as promoters or origins of replication. As nucleosomes impede access to DNA, their positioning is a primary mode of genome regulation. *In vivo* studies, especially in yeast, shed some light on factors involved in nucleosome positioning, but there is an urgent need for a complementary biochemical approach in order to confirm their direct roles, identify missing factors, and study their mechanisms.

Here we describe a method that allows the genome-wide *in vitro* reconstitution of nucleosomes with very *in vivo*-like positions by a combination of salt gradient dialysis reconstitution, yeast whole cell extracts, and ATP. This system provides a starting point and positive control for the biochemical dissection of nucleosome positioning mechanisms.

1. INTRODUCTION

Genome-wide nucleosome mapping in many species identified many nucleosomes with rather precise positions relative to the DNA sequence (Lantermann et al., 2010; Lee et al., 2007; Mavrich et al., 2008; Schones et al., 2008; Valouev et al., 2008, 2011; Yuan et al., 2005). This is especially true in regions important for genome regulation, for example, promoters or replication origins (Jiang & Pugh, 2009; Radman-Livaja & Rando, 2010; Segal & Widom, 2009). Indeed, genome regulation through chromatin often boils down to the question whether a certain DNA region is accessible to DNA binding factors or not (Bell, Tiwari, Thoma, & Schubeler, 2011). This in turn is mainly determined by the absence or presence of nucleosomes and by the regularity of their spacing, which may influence higher order compaction (Korber & Becker, 2010). In this sense, nucleosome positioning is a primary level of chromatin organization and genome regulation. It is therefore critical to understand the mechanisms that position nucleosomes along genomes. At some point, the study of molecular mechanisms usually requires the establishment of an *in vitro* reconstitution system that recapitulates the salient features of the *in vivo* situation but allows free control over factors and parameters. Even though there are several techniques available for the reconstitution and assembly of nucleosomes on DNA *in vitro* (Lusser & Kadonaga, 2004), *in vivo* nucleosome positions are usually not recapitulated. Overcoming this limitation, we established a yeast (*Saccharomyces cerevisiae*) extract-based *in vitro* system that is able to generate nucleosome positioning at several yeast loci such that it closely resembles the *in vivo* pattern (Ertel et al., 2010; Hertel, Längst, Hörz, & Korber, 2005; Korber & Hörz, 2004; Wippo & Korber, 2012; Wippo et al., 2009, 2011). Plasmids bearing these loci are reconstituted into chromatin by classical salt gradient dialysis, which leads to intrinsically DNA sequence-driven, but usually not *in vivo*-like nucleosome positions (Hertel et al., 2005; Wippo et al., 2011; Zhang et al., 2009). In a second step, such chromatin is incubated with yeast whole cell extract (WCE) in the presence of energy leading to remodeling of nucleosomes toward their *in vivo* positions. Very recently, we extended this system by using a plasmid-based genomic library and could reconstitute nucleosomes at *in vivo*-like positions throughout most of the yeast genome (Zhang et al., 2011). This method (Fig. 9.1) offers a unique tool to study the nucleosome positioning determinants in a cell-free system that is accessible to a whole range of biochemical techniques.

2. PREPARATION OF YEAST WCE

This protocol is based on previous work in the groups of Michael Schultz (Robinson & Schultz, 2003; Schultz, 1999; Schultz, Hockman, Harkness, Garinther, & Altheim, 1997), Jesper Svejstrup (personal communication), and our own (Ertel et al., 2010; Hertel et al., 2005; Korber & Hörz, 2004; Wippo et al., 2009, 2011), and was recently published (Wippo & Korber, 2012). We routinely use log-phase cultures of BY4741 for wild-type WCE preparation, but any yeast strain

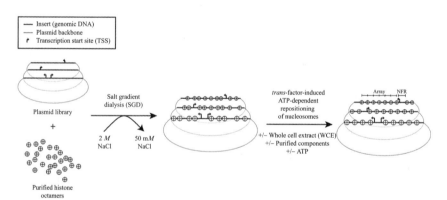

Figure 9.1 Flowchart for genome-wide *in vitro* chromatin reconstitution with *in vivo*-like nucleosome positioning. A genomic plasmid library (a schematic with three individual plasmids is shown as proxy) is reconstituted with purified histone octamers (crossed circles) into nucleosomes by salt gradient dialysis. The resulting nucleosome positions are mainly determined by DNA sequence preferences of the histone octamer (Kaplan et al., 2009; Zhang et al., 2009). In a second step, incubation with combinations of whole cell extract, purified components, and ATP may allow *trans*-factor-induced repositioning of nucleosomes toward their *in vivo* locations. This includes the formation of nucleosome-free regions (NFRs) at transcriptional start sites (broken arrows) and regular arrays over coding regions and can be analyzed by MNase-ChIP-seq (Zhang et al., 2011 and see Chapter 10).

and any biological state may be employed depending on the experimental question. Huge amounts of wild-type WCE can be prepared at very low cost by using household baker's yeast from a convenience store. The following protocol is written for 5–10 g of wet cell starting material (usually 2–5 l of mid to late log-phase culture) and can be easily adapted to other scales.

1. Wash cells in 200 ml ice cold ddH$_2$O and transfer into a 250-ml conical-bottom centrifuge tube. Pellet cells by centrifugation for 15 min at 4 °C and 6000 × g average (Heraeus Cryofuge 6000i, 4000 rpm).
2. Resuspend the cell pellet in 40 ml extraction buffer (200 mM HEPES–KOH, pH 7.5, 10 mM MgSO$_4$, 20% (v/v) glycerol, 1 mM EDTA, 390 mM (NH$_4$)$_2$SO$_4$, 1 mM freshly added DTT), collect cells by centrifugation for 10 min at 4 °C and 2047 × g average (Eppendorf 5810R, rotor A-4-62, 4000 rpm), and resuspend in 20 ml extraction buffer plus 1× protease inhibitors (Complete^{TM} (Roche Applied Science) or equivalent).
3. Pellet cells for 5 min at 4 °C and 2047 × g average (Eppendorf 5810R, rotor A-4-62, 4000 rpm) and determine wet cell weight again (usually less than before).
4. Scrape cell pellet with cold spatula into cold 5- or 10-ml plastic syringe with cut-off nozzle.
5. Cover a 400-ml plastic beaker with aluminum foil and poke a 50-ml conical tube through the foil such that it can stand upright in the beaker. Fill about 200 ml liquid nitrogen into the beaker and about 20 ml into the conical tube. Wear safety glasses and insulating gloves while handling liquid nitrogen.
6. Extrude cell pellet with syringe into the liquid nitrogen in the conical tube, such that it looks like "frozen spaghetti." Carefully pour off the liquid nitrogen without losing the frozen material, or let the nitrogen evaporate. The frozen spaghetti may be stored at −80 °C.
7. Fill porcelain mortar repeatedly with liquid nitrogen until the mortar is cold enough to keep the liquid nitrogen for a while. Have plenty of liquid nitrogen in a Dewar at hand for repletion during grinding. Add the frozen spaghetti into the liquid nitrogen in the mortar and carefully start to crush them into small pieces with a porcelain pestle. Grind the frozen cell material carefully (avoid spills), but forcefully, as this is the only cell lysis step, for 45 min. Always replenish liquid nitrogen shortly after it is all evaporated. This is somewhat hard work but can be interrupted at any moment by storing mortar with pestle and cell powder at −80 °C. After

about 20 min of grinding, add 0.4 ml extraction buffer with protease inhibitors per gram wet cell mass (as determined in step 3). The resultant ice particles are very crunchy and help with the lysis during the grinding. In the end, this will generate a very fine powder.

Alternatively and especially for larger amounts of starting material, we use an electric mortar (Retsch RM100). Fill the electric mortar with liquid nitrogen and close lid with pestle. After most of the liquid nitrogen has evaporated, open the lid and refill the mortar with liquid nitrogen. Immediately add the "frozen spaghetti", crunch them into smaller pieces with spoon or spatula, close lid, and start grinding at pestle setting of one (1). After grinding the material into a powder, add extraction buffer as above and increase pestle setting to ~5.5. Keep grinding at this setting for ~8–10 min (assuming 10 g material in step 3, shorter grinding for less material). Refill mortar with nitrogen through the small window at the top of the mortar each time shortly after its evaporation.

8. Let all liquid nitrogen evaporate and scrape powder into 100-ml beaker. Let warm quickly at room temperature under continuous stirring with a metal spatula until the powder turns into a thick paste. Place on ice immediately to avoid warming beyond 0 °C.

9. Scrape paste into precooled SW55Ti Ultra-Clear^{TM} tubes (Beckman or equivalent). If necessary, top off with cold mineral oil in order to fill the tube sufficiently and to avoid tube collapse during ultracentrifugation.

10. Spin for 2 h at 4 °C and 82,500 × g average (Beckman Coulter Optima LK-80k ultracentrifuge, SW55Ti rotor, 29,500 rpm) with brake on.

11. Carefully place the centrifuge tube in ice water bath without disturbing the phase separation. There are four different phases now: (a) the compact pellet of cell debris at the bottom, (b) a cloudy yellowish layer on top, which fades into a (c) clear supernatant, and finally a (d) whitish lipid-rich top phase at the meniscus.

12. Using a precooled 5-ml plastic syringe with needle (a rubber seal instead of all plastic plunger facilitates gentle suction), carefully remove the middle part of the clear supernatant (see c in step 11) by poking the needle through the lipid top layer or by puncturing the tube wall. Avoid as much of the yellowish cloudy layer (see b in step 11) as possible, but usually it is not possible to avoid all of it. It is possible to recentrifuge after this step to allow better phase separation, but this is usually not necessary. Transfer the withdrawn lysate into Microfuge® Polyallomer TLA55 (Beckman or equivalent) tubes on ice.

13. Determine the volume of the lysate, for example, by filling water into another tube up to the same level and determining the water volume with a pipette.
14. Grind ammonium sulfate into a fine powder and add in several portions (e.g., 50% + 25% + 25%) 337 mg of this powder per ml of lysate. A small funnel helps. After each addition, mix by twirling carefully with a plastic inoculation loop and place on rotating wheel at 4 °C until the added portion of ammonium sulfate has dissolved. Avoid foam generation. After all the ammonium sulfate has dissolved (check against the light if you still see tiny crystals sinking to the tube bottom), rotate tubes for an additional 30 min.
15. Spin the solution in TLA55 rotor for 20 min, 4 °C, at 26,000 rpm ($30,300 \times g$ average).
16. Carefully withdraw the supernatant with a cold 5-ml syringe with needle and discard it.
17. Redissolve the pellet in 0.2–0.5 ml of dialysis buffer (20 mM HEPES–KOH, pH 7.5, 80 mM KCl, 10 or 20% (v/v) glycerol, 1 mM EGTA and freshly added 5 mM DTT, 0.1 mM phenylmethylsulfonyl fluoride (PMSF), and 1 mM sodium metabisulfite) per gram starting material (according to step 3), depending on how well it dissolves and how concentrated the final extract shall be. Again, twirling with an inoculation loop helps.
18. Transfer the solution into a dialysis tube (MWCO 3.5 kDa) and dialyze twice for 1.5 h against 40- to 50-fold excess volume of dialysis buffer.
19. Remove dialyzed extract, flash freeze 50–1000 µl aliquots in liquid nitrogen, and store at −80 °C. The nucleosome positioning activity tolerates at least two freeze–thaw cycles and can be stored at −80 °C for at least 2 years.
20. Our yeast extracts usually contain 10–30 mg/ml protein as assayed by Bradford assay with BSA as standard. Of these, we usually take 15–5 µl per nucleosome positioning reaction. In contrast to the histone:DNA mass ratio (see Section 5), the amount of extract per nucleosome positioning reaction is much less critical, that is, variation by a factor of 2 or 0.5 usually has hardly any effect. Nonetheless, too much extract will lead to chromatin aggregation. We routinely adjust our extracts with dialysis buffer to 50 mg "protein" per ml according to Nanodrop (Thermo Scientific) reading at 280 nm and use 10 µl per 100 µl nucleosome positioning reaction. This usually corresponds to a protein concentration of ∼20 mg/ml as measured by Bradford assay.

The Nanodrop reading will be somewhat confounded by varying amounts of nucleic acids, especially RNA. Nonetheless, this procedure works just fine as a quick measure for how much extract to use per nucleosome positioning reaction. Avoid a too highly concentrated extract as it will be aggregation-prone.
21. The WCE may be fractionated in order to identify factors involved in nucleosome positioning at particular loci (Wippo et al., 2011).

3. PREPARATION OF HISTONES

Drosophila embryo histones can be easily prepared with high yield, high purity, and in form of histone octamers, that is, with the exact 1:1:1:1 molar ratio of all four histones. Nonetheless, these histones are heterologous to yeast and do contain some modifications (Bonaldi, Imhof, & Regula, 2004) and maybe histone variants. We also used recombinant *Drosophila*, *Xenopus*, and yeast histones for single-locus assemblies, but did not observe significant differences (Ertel et al., 2010 and unpublished observations). Like many others, we noted that recombinant yeast histones are more difficult to work with, that is, it is more difficult to achieve high assembly degrees and thus proper positioning of tricky loci such as the yeast *PHO5* promoter. Nonetheless, recombinant histones are modification-free and allow tailored mutations of single residues, shortened tails, etc.

The following hydroxylapatite-based protocol is derived from Faulhaber and Bernardi (1967) and Simon and Felsenfeld (1979). Unless indicated otherwise, all steps are done in the cold room at 4 °C or on ice and using cold solutions. Maybe keep aliquots of the various steps for gel documentation of the purification.

1. Harvest at least 50 g of 0–12 h *Drosophila* embryos (Kunert & Brehm, 2008; Shaffer, Wuller, & Elgin, 1994). For dechorionation, rinse the embryos from the apple juice agar plates with tap water into a three sieve embryo collection apparatus (Kunert & Brehm, 2008). Press the embryos carefully with a wooden spoon against the finest steel mesh sieve to remove excess liquid and transfer them into a beaker filled with 3% sodium hypochlorite in water. Stir vigorously with wooden spoon for 3 min, wash shortly with 0.7% (w/v) NaCl, 0.04% (v/v) Triton X-100, and once with tap water for further 5 min. Dechorionated embryos can be stored for years at −80 °C. The following protocol is designed for 50 g embryos. Up to 100 g,

embryos can be processed per 30 ml hydroxylapaptite if the protocol is scaled up accordingly.
2. Resuspend embryos in 40 ml of lysis buffer (15 mM HEPES–KOH, pH 7.5, 10 mM KCl, 5 mM MgCl$_2$, 0.1 mM EDTA, 0.5 mM EGTA, 17.5% (w/v) sucrose, 1 mM DTT, 0.2 mM PMSF, 1 mM sodium metabisulfite) and let thaw in cold room if they were frozen.
3. Homogenize six times with Yamato LSC LH-21 homogenizator at 1000 rpm and filter through Miracloth (Calbiochem-Novabiochem Corporation, La Jolla, CA).
4. Transfer into 30-ml Corex glass centrifuge tubes (No8445, Corex®, USA) and spin down for 15 min at 8000 rpm in a GSA rotor (Sorvall) or equivalent (6573 × g average). This yields three phases: dark and very solid bottom, soft and light brown jelly on top (the nuclei for further use), and a turbid supernatant with a lipid skim layer at the meniscus.
5. Take off the supernatant with a 10-ml glass pipette as completely as possible without removing any of the nuclei.
6. Fill a fresh 10-ml pipette with suc-buffer (15 mM HEPES–KOH, pH 7.5, 10 mM KCl, 5 mM MgCl$_2$, 0.05 mM EDTA, 0.25 mM EGTA, 1.2% (w/v) sucrose, 1 mM DTT, 0.1 mM PMSF) and wash out the nuclei by pipetting up and down. Use 50 ml suc-buffer in total. The dark bottom phase is usually not disturbed but sits there like a rock. Transfer these resuspended nuclei into fresh, cold Corex tubes.
7. Spin again as in step 4.
8. Wash again in 50 ml suc-buffer as in step 6. Avoid carrying over remnants of the dark bottom pellet. Spin again as in steps 4 and 7.
9. Resuspend nuclei in suc-buffer, ad 30 ml final volume, and dounce 20 times with a glass dounce homogenizer (Dounce Tissue Grinder, Wheaton/Fisher Scientific GmbH) fitted with a B pestle.
10. Add 90 µl of 1 M CaCl$_2$ and warm up for 5 min at 26 °C in a waterbath.
11. Add 125 µl of 0.59 U/µl MNase and protease inhibitors (Complete™ (Roche Applied Science) or equivalent) in addition to the PMSF from the suc-buffer. Incubate 10 min at 26 °C. This step will digest most of the chromatin into mononucleosomes. Besides the degree of dechorionation (step 1) and the clear separation of the three phases (step 4), the MNase digestion degree is most critical for the final yield of histone octamers. Therefore, one should determine the right amount of MNase and/or time for this digest using small nuclei aliquots such that as many mononucleosomes as possible are obtained without further degradation to subnucleosomal particles. The digestion degree can be

checked by running a deproteinized aliquot on a 1.5% agarose gel with ethidium bromide staining and the 2log ladder (NEB) as standard. Use Orange G instead of bromophenol blue as marker dye as the former migrates ahead of the mononucleosomal band and the latter migrates within the MNase ladder, roughly at the position of the dinucleosomal fragment band. If time or material is limiting, the above conditions should work just fine if the MNase stock is made in the following way: Add 850 μl of 10 mM HEPES-KOH, pH 7.6, 1.5 mM MgCl$_2$, 0.5 mM EGTA, 10% (v/v) glycerol, 50 mM KCl, 1 mM DTT, 0.2 mM PMSF to 1 vial of Micrococcal Nuclease with 500 U (Sigma-Aldrich, N5386), dissolve the MNase powder by light twirling and flash freeze in 50 μl aliquots, and store at $-20\ °C$.

12. Stop MNase digest by addition of 600 μl of 0.5 M EDTA.
13. Spin immediately as in steps 4, 7, and 8.
14. Resuspend pellet in 6 ml TE (10 mM Tris-HCl, 1 mM EDTA) pH 7.6, 1 mM DTT, 0.2 mM PMSF (added freshly) for hypotonic lysis of nuclei and rotate for 30–45 min.
15. In the meantime, wash 30 ml hydroxylapatite resin (Biorex or equivalent) on a rotating wheel five times for 5 min in a 50-ml conical tube filled up with 0.63 M KCl, 0.1 M potassium phosphate, pH 7.2, 1 mM DTT. Each time spin down lightly the hydroxylapatite for 1 min at maximal $1000 \times g$ (Hereaus Megafuge 2.0), avoid compacting the resin too much by too harsh centrifugation. Pour the washed material into a 30-ml chromatography column.
16. Spin lysed nuclei for 30 min at $15{,}322 \times g$ (Sorvall RC 6 plus, SS-34 rotor, 14,000 rpm). Keep supernatant and estimate volume with 10-ml glass pipette.
17. Adjust to 0.63 M KCl by adding the appropriate volume of 2 M KCl, 0.1 M potassium phosphate, pH 7.2 and centrifuge for 15 min at $15{,}322 \times g$ average (SS-34 rotor or equivalent).
18. Sterile filter the supernatant with a 0.45-μm and a 0.22-μm filter before the supernatant is applied onto the equilibrated hydroxylapatite column (step 15).
19. Wash the column at maximally 1 ml/min with 0.63 M KCl, 0.1 M potassium phosphate, pH 7.2 for at least two column volumes or until steady UV baseline.
20. Elute with 2 M KCl, 0.1 M potassium phosphate, pH 7.2 at 0.5 ml/min and collect 3-ml fractions in 4-ml tubes. The fractions can be kept in the cold room over night.

21. Test 15 μl each of peak and surrounding fractions on 18% SDS-PAGE. If available, load some reference histone preparation for comparison on the gel.
22. Pool fractions that contain the highest histone concentrations (purity is usually not an issue). Concentrate by ultrafiltration (10 kDa MWCO) to a volume of about 0.15–0.5 ml.
23. Add equal volume of 87% (v/v) glycerol and adjust to 1 × protease inhibitors (Complete™ (Roche Applied Science) or equivalent) and 5 mM DTT. Histones are very sticky proteins. Use siliconized tubes.
24. Store at $-20\,°C$.
25. Estimate histone octamer concentration on SDS-PAGE by comparison of an appropriate dilution series with dilutions of BSA standard and, if available, against a histone octamer preparation that was previously titrated for chromatin assembly (Fig. 9.2). Make sure that the uppermost histone H3 band is present at the same stoichiometry, that is, H3 is not degraded yielding a fifth band migrating below the H2A/H2B bands. You may also get a rough estimate by photometry (ε_{280} (*Drosophila* histone octamers) $= 0.42\ cm^{-1}\ ml/mg$).

4. EXPANSION OF GENOMIC PLASMID LIBRARY

Any DNA may be reconstituted into chromatin by salt gradient dialysis. However, we and others (Patterton & von Holt, 1993; Pfaffle & Jackson, 1990) noted that substantially higher assembly degrees in terms of nucleosomes per DNA can be obtained with circular supercoiled templates, probably as formation of a nucleosome corresponds to about one negative supercoil (Germond, Hirt, Oudet, Gross-Bellark, & Chambon, 1975; Shure & Vinograd, 1976). Further, we found that for some loci, for example, the *PHO5* (Hertel et al., 2005) or *PHO84* (Wippo et al., 2009) promoter regions, a high assembly degree is necessary in order to generate the *in vivo*-like nucleosome positioning by incubation with the WCE. Finally, average spacing of nucleosomes on DNA *in vivo* varies roughly between 165 and 200 bp (van Holde, 1988) corresponding to a histone octamer:DNA mass ratio of 0.8:1 to 1:1, which is physiological "fully assembled" chromatin. Similar to others (Kaplan et al., 2009), we have not been able to obtain such high assembly degrees without aggregation by using linear DNA templates like sheared genomic DNA. Nonetheless, genomic plasmid libraries work just fine. In addition to the circular and supercoiled nature of the template, the

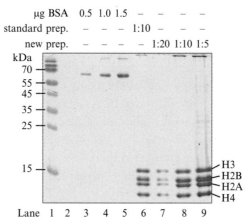

Figure 9.2 Initial estimation of histone concentration by SDS-PAGE. 10 μl of appropriate dilutions of a new *Drosophila* embryo histone octamer preparation ("new prep.," lanes 7-9) were loaded onto a 17.5% SDS-PAGE gel alongside with increasing amounts of BSA (lanes 3-5) and an appropriate dilution of a previously tested histone octamer preparation ("standard prep.," lane 6). Protein bands were visualized by Coomassie blue staining. Size marker (lane 1) was peqGOLD Protein Marker IV (Peqlab). This comparison should allow estimating which volume of the new histone preparation corresponds to about 10 μg of histone octamers necessary for a full assembly of 10 μg plasmid DNA (see Section 5.3.). In this example, each of the four histone bands in lane 8 has approximately the same intensity as the BSA band in lane 5. So 1 μl of the new histone preparation contains about 6 μg of histones, 1.7 μl corresponds to 10 μg. As the effective histone concentration is usually a bit lower than estimated from the gel, typically 10-20% more than calculated, for example, 2 μl of the new histone preparation (10 μl of a 1:5 dilution, see Fig. 9.4), will be the starting point for the fine titration of the assembly degree (see Sections 5.3 and 5.5). Further, with such a gel, the integrity and purity of the purified histones can be checked. All four histones should be present at the same stoichiometry, for example, there should be no fifth band below the H2A band corresponding to clipped histone H3.

prokaryotic DNA in the plasmid backbone may be important, too. Eukaryotic DNA has on average higher affinity for histone octamers than prokaryotic DNA (Zhang et al., 2009) so that the prokaryotic vector may serve as intramolecular buffer once the eukaryotic portion is fully assembled (Huynh, Robinson, & Rhodes, 2005). A similar strategy was used by Struhl and colleagues as they titrated prokaryotic competitor DNA into a salt gradient assembly reaction with sheared genomic DNA and reported to achieve a 1:1 histone octamer:DNA mass ratio (Zhang et al., 2009).

Any plasmid library can be used. For the sake of efficiency in high-throughput sequencing, inserts should be rather large (10–30 kb) and the percentage of empty vectors small. You may generate your own custom yeast plasmid library according to published protocols (*Current Protocols in Molecular Biology*) or obtain one of the many available ones (e.g., Jones et al., 2008; Rose, Novick, Thomas, Botstein, & Fink, 1987). In contrast to classical genetics work, large amounts of plasmid DNA are needed for *in vitro* reconstitution, whereas full genome coverage is usually less important. The following protocol is therefore geared toward high DNA yield and not necessarily toward high genome coverage. The protocol should be conducted under sterile conditions.

4.1. Preparation of electrocompetent *E. coli* cells
(adapted from Adams, Schmoldt, & Kolmar, 2004)
1. Inoculate 200 ml dYT medium (1.6% (w/v) tryptone, 1% (w/v) yeast extract, 85.56 mM NaCl) with 200 μl of an overnight culture of a typical cloning strain (e.g., DH5α, XL1-Blue (Stratagene), TOP10 (Invitrogen)) and grow cells at 37 °C to a density of 1.6×10^8 cells/ml.
2. Split the culture into four 50-ml tubes and incubate for 30 min on ice.
3. Pellet the cells by centrifugation for 10 min at 4 °C and $2047 \times g$ average (Eppendorf 5810R, rotor A-4-62, 4000 rpm), resuspend in 50 ml ice cold ddH_2O, and incubate for 20 min on ice.
4. Repeat step 3 with 25 and 10 ml ice cold ddH_2O.
5. Pellet cells by centrifugation as in step 3, decant supernatant, and incubate on ice until the remaining ddH_2O is collected at the bottom. Resuspend cells in the remaining ddH_2O (about 200 μl) and transfer into precooled electroporation cuvettes.

4.2. Electroporation
1. Add a maximum of 5 μl of plasmid library (sufficient material to generate enough clones, see determination of clone number in step 4.3.2) that has been desalted, for example, by desalting column with ddH_2O as eluent or dialysis against ddH_2O.
2. Apply a pulse, for example, 2500 V, 25 μF, 200 Ω with "Gene Pulser"(BIO-RAD).
3. Add 800 μl of room-temperature dYT medium immediately and transfer the cell suspension into a sterile 15-ml tube.
4. Incubate cells for 1 h at 37 °C under shaking or rotating.

4.3. Cultivation and plasmid preparation

1. Prepare a serial dilution (1:10, 1:100, 1:1000) in sterile ddH$_2$O using 20 μl of each transformation batch and plate 100 μl of each dilution on separate selective LB agar plates. Plate the remaining cell suspension (2 × ca. 450 μl) on two selective LB plates (145 mm diameter, 20 mm height). Incubate all plates over night at 37 °C.
2. Count single colonies on the serial dilution plates to estimate the number of independent clones per microliter transformation batch, that is, the maximal number of independent plasmids in your expanded library. This number should be 10 to 100 times higher than the diversity of the original library. If the number of clones is not sufficient, prepare more transformations, rather than using more plasmid solution per transformation. Transformations can be pooled afterward to achieve sufficient numbers.
3. Wash off the bacteria from the large preparative LB agar plates with 2–3 ml LB medium and an L-shaped glass rod. Pool bacteria suspension from all plates into a sterile 50-ml tube. Use 4 × 1 ml of this cell suspension to prepare a glycerol stock for future library preparations. *Note*: Library diversity decreases mostly because of each transformation step. Therefore, library diversity is most likely maintained by using glycerol stocks to expand your library for new plasmid preparations. However, for long-term storage (>2 years), the library should be stored as plasmid preparation.
4. For plasmid preparation, use 200 μl of the remaining washed-off cell suspension to inoculate 4 × 200 ml LB medium with appropriate selection and incubate over night at 37 °C while shaking.
5. Prepare plasmid library using a Qiagen Maxiprep kit or equivalent. We routinely filter (filter paper) the supernatant after the neutralization and centrifugation step before it is applied onto the column in order to avoid clogging of the column. In order to fit into the standard pipetting scheme of salt gradient dialysis, adjust the DNA concentration to no less than 0.3 μg/μl (see Section 5).

5. CHROMATIN RECONSTITUTION BY SALT GRADIENT DIALYSIS

The histone octamer interacts with the DNA mainly through electrostatic interactions (Luger, Mader, Richmond, Sargent, & Richmond, 1997). Just mixing histones and DNA at physiological ionic strength will lead to aggregation due to the high concentration of charges both on the histones as well as on the DNA. The classical protocol in order to avoid aggregation and to reconstitute nucleosomes *in vitro* is salt gradient dialysis. Histones and DNA are first mixed at high-salt concentration, for example, 2 M NaCl, that prevents the electrostatic interactions. The salt is then very slowly and gradually dialyzed away such that conditions are generated along the way (probably around 1–0.75 M salt) where the histone–DNA interactions are strong enough to allow nucleosome formation but weak enough to prevent aggregation. This protocol has been developed a long time ago and is published several times (Germond et al., 1975; Luger, Rechsteiner, & Richmond, 1999; Oudet, Gross-Bellard, & Chambon, 1975; Rhodes & Laskey, 1989; Stein, 1989). There is also an extensive body of work regarding biophysical aspects of this procedure, such as the questions of equilibrium and DNA sequence preferences for nucleosome positioning under such conditions (Schnitzler, 2008; Widom, 2001). We present here our version of this classical protocol, which was also published in Wippo, C. J., & Korber, P. (2012). In vitro reconstitution of in vivo-like nucleosome positioning on yeast DNA. *Methods in Molecular Biology*, *833*, 271–287, with kind permission of Springer Science+Business Media (Fig. 9.3).

5.1. Pump and beakers

1. Set up the salt gradient dialysis in a hood as high concentrations of β-mercaptoethanol are used.
2. Fill one 3-l beaker with 3 l of 1 × low-salt buffer (10 mM Tris–HCl, pH 7.6, 50 mM NaCl, 1 mM EDTA, 0.05% (w/v) IGEPAL CA630; prepare as 20 × stock) and another 3-l beaker with 300 ml of 1 × high-salt buffer (as low-salt buffer but with 2 M NaCl). Add 300 μl β-mercaptoethanol to the beaker with low-salt buffer and mix well.
3. Place the beaker with high-salt buffer on a magnetic stirrer and add a large (e.g., 4 cm) stir bar.
4. Set up a peristaltic pump and place into each of the 3-l beakers one end of the tube. Fix the tube at each 3-l beaker with tape such that the tube cannot slide off. Make sure that the tube end in the beaker with the low-salt buffer is situated at the bottom of the beaker such that all buffer can be pumped out.

5.2. Dialysis mini chamber

1. Cut off the end of a siliconized 1.5-ml tube, just above the 0.5-ml mark.
2. Using pointed scissors, puncture the thin center part of the tube lid that is circumscribed by the elevated edge that fits into the tube upon closing

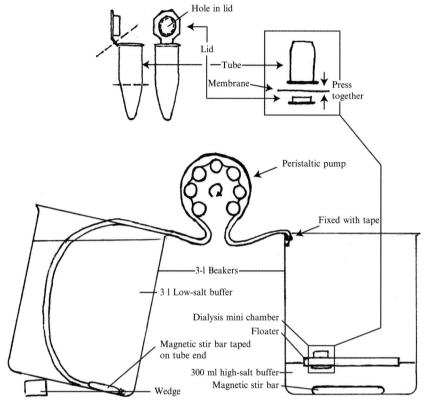

Figure 9.3 Schematics of salt gradient dialysis set up. Dialysis mini chamber: A siliconized 1.5-ml reaction tube is cut at the stippled lines, and a hole is punctured into the lid along the stippled circle. The dialysis membrane is wedged in between the perforated lid and the truncated tube. Pump and beakers: Prepare one 3-l beaker filled with 3 l low-salt buffer and a second with 300 ml high-salt buffer. The tube of a peristaltic pump is fixed in the beakers as drawn in order to allow complete transfer of low-salt buffer into high-salt buffer. Salt gradient dialysis: The dialysis mini chamber floats in a floater on the high-salt buffer such that the perforated lid faces downward and the membrane is in contact with the buffer. Usually there are air bubbles that have to be removed. The buffer in the high-salt beaker is slowly mixed with a magnetic stirrer (not shown), while the 3 l low-salt buffer is pumped over a course of about 15 h into the high-salt buffer. Both beakers are covered, for example, with cling film and a glass plate (not shown).

the lid, and scrape out the plastic up to the elevated edge. Basically, you generate a lid with a central hole of about 8 mm in diameter. Make sure not to generate sharp edges that could puncture or rip the dialysis membrane later on.

3. Cut off the such perforated lid from the previously truncated tube.
4. Cut off about 1.5–2 cm of dialysis tubing (MWCO 3.5 kDa) and place in a small beaker filled with ddH$_2$O for about 10 min. Cut the tubing open at one side so that the dialysis membrane can be folded open as a single layer.
5. Place the perforated lid top down onto a sheet of cling film, which serves as a convenient and clean surface to prepare the dialysis mini chamber. Place the dialysis membrane centered on top of the lid. Press the truncated siliconized tube with its top over the dialysis membrane onto the lid such that the membrane becomes wedged in between lid and tube like a drumhead. To prevent leakage, it is important that the membrane is tightly sealed between lid and tube and that the membrane surface is tense and smooth. If several dialysis mini chambers are prepared at the same time, make sure that the membranes do not dry out at any point. You can make a small puddle of ddH$_2$O onto the cling film and store there the dialysis mini chambers lid-down, which will keep the membranes wet. Do not allow any water into the dialysis chamber as this will dilute your sample. Cut away most of the excess membrane sticking outward from the tube.
6. Use a floater to let the dialysis mini chamber float on top of the high-salt buffer in the 3-l beaker with lid and membrane facing downward and the truncated tube facing upward. Air bubbles right underneath the membrane have to be removed. It is convenient to suck away the bubbles with a drawn out Pasteur pipette that has been bent twice into U-shape over a Bunsen burner flame.

5.3. Samples

Combine 10 μg plasmid library DNA (determined photometrically, e.g., by Nanodrop (Thermo Scientific)), 20 μg BSA (fraction V (Roth)), variable amounts of histone octamers (often diluted appropriately with 0.1 M KCl for more accurate pipetting), 50 μl of 2 × high-salt buffer without IGEPAL CA630, 5 μl of 1% (w/v) IGEPAL CA630, and ddH$_2$O to make up 100 μl. Mix thoroughly by pipetting but avoid foam generation. Use about 10–12 μg histone octamers as estimated in Section 4, step 25 (see also Fig. 9.2) as starting point and titrate up and down in order to optimize the assembly degree (see Section 5.5). Both a too low and a too high assembly degree can be problematic, the latter mainly because of chromatin aggregation.

5.4. Salt gradient dialysis

1. Pipet samples through the open end of the floating dialysis mini chamber onto the membrane. Be careful not to damage the membrane with the pipette tip.
2. Adjust magnetic stirrer underneath the high-salt buffer beaker such that slow mixing is achieved without compromising easy floating of the dialysis mini chambers.
3. Add 300 µl β-mercaptoethanol to the high-salt buffer beaker and cover the beaker with cling film. Make sure, for example, by using tape or placing a glass plate on top, that the beaker is properly sealed. As this is in the hood and runs overnight, the sample volume decreases substantially due to evaporation if the beaker is not properly covered.
4. Set speed of peristaltic pump such that all of the 3 l low-salt buffer will be pumped into the high-salt buffer over the course of at least 15 h. A trial run with water and without samples is advisable.
5. After complete transfer of the low-salt buffer, transfer the floater with the dialysis mini chambers to a jug with 1 l fresh low-salt buffer plus 300 µl β-mercaptoethanol. Remove air bubbles again from underneath the membranes.
6. Dialyze for 1–2 h with slow stirring to ensure complete buffer exchange.
7. Transfer the samples with a pipette from the dialysis mini chambers into fresh siliconized 1.5-ml tubes and determine the volume with the pipette. The volume sometimes increases to 120–130 µl. The salt gradient dialyzed chromatin samples can be stored at 4 °C for several weeks up to a few months, but must not be frozen.

5.5. Titrating the assembly degree

For each new preparation of histones, the amount needed for a certain assembly degree has to be titrated. It is not sufficient, for example, to determine the concentrations of DNA and histones by whichever means and then mix them at a calculated histone octamer:DNA mass ratio of 1:1 in order to obtain "fully assembled" chromatin. The actual degree of assembly after salt gradient dialysis has to be determined. Once this is done for a given histone preparation, the assembly degree is usually well reproducible and need not be checked again for each reconstitution. Technically, the maximal assembly degree *in vitro* is limited by increasing precipitation of chromatin at increasing histone:DNA ratios. Fortunately, with our method, this limit is very close to physiological "fully assembled" chromatin (see Fig. 9.4).

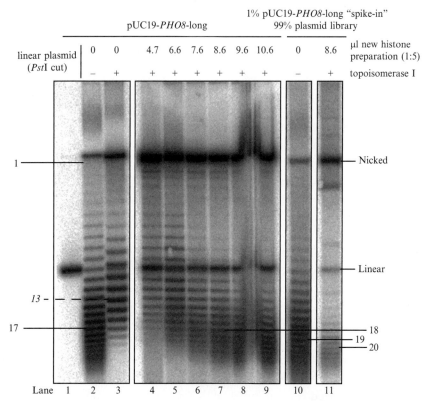

Figure 9.4 Plasmid pUC19-*PHO8*-long (6168 bp, Wippo et al., 2011) was linearized with *Pst*I (2.3 ng DNA loaded in lane 1), left untreated (5.6 ng DNA loaded in lane 2), relaxed with topoisomerase I (5.6 ng DNA loaded in lane 3), or assembled into chromatin by salt gradient dialysis with the indicated amounts of 1:5 diluted histones (same new histone preparation as in Fig. 9.2) per 10 µg plasmid, treated with topoisomerase I and deproteinized (56 ng DNA loaded in lanes 4-9). A mix of plasmid library with pUC19-*PHO8*-long at a mass ratio of 99:1 was either left untreated (0.5 µg DNA loaded in lane 10) or assembled into chromatin by salt gradient dialysis with 8.6 µl of 1:5 diluted histones per 10 µg DNA mix, treated with topoisomerase I and deproteinized (0.75 µg DNA loaded in lane 11). Samples were electrophoresed in a 1.3% Tris-glycine agarose gel with 9 µM chloroquine. DNA was visualized by Southern blotting, hybridization with a *PHO8*-specific probe, and exposure to a PhosphorImager screen (Fuji FLA 3000). All samples were electrophoresed alongside in the same gel, but lanes were rearranged in the figure using Photoshop CS2. Numbered lines denote the center of gravity of the supercoil distribution as estimated from a one-dimensional lane profile of the PhosphorImager scan using the AIDA v4.15.025 software (Raytest). If very low amounts of DNA are loaded, it may become advisable to add carrier DNA, e.g., 1 µg/lane sonicated salmon sperm DNA in TE. The sum of the absolute value of negative supercoils (throughout the figure indicated by solid lines and numbered with normal font) of the untreated plasmids in the presence of chloroquine (17 supercoils in lane 2) plus the positive supercoil number

The assembly degree of each reconstitution can be quantitatively measured by counting the plasmid supercoils that remain after topoisomerase I treatment and subsequent deproteinization of the chromatin. Each nucleosome constrains one negative supercoil (Germond et al., 1975; Shure & Vinograd, 1976), which can be resolved in chloroquine-containing agarose gels. Unfortunately, this is difficult for plasmid libraries due to the very large and heterogeneous template sizes. Therefore, we do the initial histone titration with a small plasmid as proxy. We usually keep the DNA concentration constant (10 μg) and titrate the histone octamer concentration. As internal control for the actual assembly degree of the plasmid library, we spike the proxy plasmid into the library and detect its superhelicity after Southern blotting and specific hybridization (see Fig. 9.4).

A simpler but less quantitative method to find the maximally possible assembly degree employs limited MNase digestion of the salt gradient dialysis chromatin. This will generate increasingly extensive "MNase ladders" (successive bands of mono-, di-, tri-, etc., nucleosomal fragments in the gel lane) with increasing assembly degree. If overassembled,

---

(indicated by stippled line and numbered with italic font) of the relaxed plasmid in the presence of chloroquine (13 supercoils in lane 3) reflects the number of negative supercoils introduced by *E. coli* (30 supercoils). Titration of histones in the salt gradient dialysis reconstitution of plasmid pUC19-*PHO8*-long led to increasingly more nucleosome-constrained negative supercoils (lanes 4 to 9). Assembly with 8.6 μl of the 1:5 diluted new histone preparation (lane 7) gave about the same number of negative supercoils as in the untreated plasmid (lane 2), and even a bit higher assembly degrees were possible (lanes 8 and 10). However, chromatin began to precipitate with the addition of 11 or more microliters of 1:5 diluted histones leading to the appearance of empty lanes (not shown). In order to stay away from the precipitation limit and as too tightly packed nucleosomes are more refractive for efficient remodeling, we chose 8.6 μl of the 1:5 histone dilution (= 1.7 μl undiluted) in this experiment for assembly of a plasmid library with spiked-in pUC19-*PHO8*-long. For unknown reasons, the center of gravity of the untreated spike-in plasmid in the context of the library was shifted by two negative supercoils (19 instead of 17 in lane 10 vs. 2, respectively). The assembled spike-in (lane 11) yielded a maximum at 20 negative supercoils, which corresponds to the 18 negative supercoils in lane 7 taking the two-supercoils-difference between lanes 10 and 2 into account. This very similar assembly degree for both independent salt dialysis reconstitutions (lanes 7 vs. 11) demonstrates that a histone:DNA ratio determined for one proxy plasmid can be applied in further reconstitution reactions yielding similar results. Note that chromatinized plasmids, especially with increasing size, are prone to DNA strand breaks, probably by shear forces during pipetting. Therefore, both the nicked as well as the linear form become more abundant for chromatinized samples.

uncomplexed histones will lead to aggregation of chromatin, which is detected as decreasing signal strength in the gel lane and sometimes material stuck in the gel well. So the maximally possible assembly degree corresponds to the point where MNase ladders are as extensive as possible without substantial aggregation.

Sometimes there is a difference between the maximal assembly degree that can be obtained without chromatin precipitation in salt gradient dialysis alone or after incubation with a WCE, that is, highly assembled salt gradient dialysis chromatin may be just soluble but precipitate if WCE is added. Therefore, the MNase ladder assay, for example, should be done with chromatin under the same conditions as the actual application later on. Maybe a lower than maximal assembly degree has to be chosen, depending on the downstream application.

5.5.1 Supercoil assay of nucleosome assembly degree after salt gradient dialysis

Each nucleosome constrains about one negative supercoil against relaxation by topoisomerase I. Incubation of chromatinized circular plasmids with topoisomerase I under equilibrium conditions will therefore result in plasmids that bare as many negative supercoils as originally present nucleosomes. These supercoils may be resolved and counted by gel electrophoresis, especially in the presence of chloroquine. Chloroquine is a DNA-intercalating molecule, similar to ethidium bromide but without the same fluorescent properties, that will introduce positive supercoils. Titrating the chloroquine concentration allows to counteract as many of the nucleosome-derived negative supercoils by the chloroquine-derived positive supercoils such that the mean supercoil distribution is shifted into a range that is well resolved in agarose gels. The number of chloroquine-introduced supercoils can be derived from electrophoresing a topoisomerase I-relaxed naked circular plasmid alongside in the same gel, that is, at the same chloroquine concentration. This number plus the number of remaining negative supercoils in the topoisomerase I-treated chromatinized and then deproteinized plasmid gives the total number of originally present negative supercoils and thus nucleosomes. Interestingly, the number of negative supercoils as introduced by the prokaryotic DNA gyrase in *E. coli* grown at 37 °C, from which the plasmids are usually purified, is roughly the same as the nucleosome number on a plasmid fully assembled into nucleosomes (Krajewski & Becker, 1999). The untreated plasmid prepared from *E. coli* can therefore serve as a guideline for full assembly degree.

1. Reconstitute a plasmid that contains a major portion of yeast DNA, but also an empty cloning vector works, by salt gradient dialysis with varying histone octamer:DNA mass ratios (see Section 5.3).
2. Sample preparation for chloroquine gel electrophoresis:
   2.1. Incubate 15 µl of a reconstitution reaction, corresponding to 1.5 µg of chromatinized plasmid DNA, with 40 U topoisomerase I (NEB or equivalent) in 150 µl adjusted to 1 × topoisomerase I buffer and incubate over night at 37 °C.
   2.2. Linearize 1.5 µg of the same but unassembled plasmid by digestion with an appropriate restriction enzyme.
   2.3. Relax 1.5 µg of the same but unassembled plasmid with 40 U topoisomerase I in 150 µl adjusted to 1 × topoisomerase buffer over night at 37 °C.
   2.4. Dilute 1.5 µg of the same but unassembled plasmid with TE buffer into 30 µl final volume (untreated plasmid control without topoisomerase I).
3. Proceed with the first three sample types as followed: Add EDTA to a final concentration of 10 mM to stop the topoisomerase I.
4. For deproteinization, prewarm samples to 55 °C and add 4.5 µl of 20% (w/v) SDS, followed by addition of 15 µl proteinase K (20 mg/ml in ddH$_2$O (Roche Applied Science)) and 1.5 µl glycogen (20 mg/ml in ddH$_2$O (PCR-grade; Roche Applied Science)). Incubate over night at 55 °C.
5. Add 149 µl ddH$_2$O and 17.5 µl of 5 M NaCl.
6. Precipitate with 2.5 volumes of room-temperature 95–100% ethanol, wash with 1 volume of room-temperature 70% ethanol, and resuspend in 30 µl TE buffer. This is enough for two gel lanes. Samples may be stored at -20 °C.
7. Prepare a 1.3% agarose gel in Tris–glycine buffer (28.8 g/l glycine, 6 g/l Tris–base). If called for, add chloroquine (Sigma-Aldrich) just before pouring the gel. For plasmid sizes between 2 and 6 kb, chloroquine concentrations in the range of 1.5–3 µM are fine. As chloroquine is sensitive to light, keep chloroquine-containing gels in the dark. Consult material safety and data sheet for chloroquine.
8. As chloroquine gel electrophoresis takes a long time, it is important for even DNA migration to keep the gel running buffer well mixed throughout electrophoresis. For example, slowly stir the buffer in both reservoirs of the gel chamber during electrophoresis using magnetic stirrers.

9. As running buffer use 1× Tris–glycine buffer with the same chloroquine concentration as used for gel preparation.
10. Load the DNA samples from step 6 (amounts per lane as in the legend to Fig. 9.4) in a volume of 15 μl (made up with TE pH 7.6) plus 5 μl Orange G loading dye (40% (w/v) sucrose, 10 mM Tris–HCl pH 8.0, 0.25% (w/v) Orange G (Sigma-Aldrich)).
11. Electrophorese for 3 days at 1.1 V/cm in the dark, for example, by covering the gel apparatus with a large cardboard box.
12. The gel may be analyzed by Southern blotting (Sambrook, Fritsch, & Maniatis, 1989) with specific hybridization against the spike-in plasmid.

5.5.2 MNase ladder assay of assembly degree

Omit steps 1 and 2 if plain salt gradient dialysis chromatin shall be assayed.
1. Incubate with WCE from wild-type cells and energy (see Section 6, omit steps 6 and 7).
2. Remove ATP by addition of 4 μl of 50 U/ml apyrase (NEB M0393L) and incubation for 30 min at 30 °C.
3. Add 1/50 volume of 75 mM CaCl$_2$.
4. Add MNase (see Section 7.2) and incubate for 5 min at 30 °C. Stop the reaction by adding EDTA to a final concentration of 10 mM.
5. For deproteinization, heat samples to 55 °C, add 10 μl proteinase K (20 mg/ml in ddH$_2$O (Roche Applied Science)), 1.5 μl glycogen (20 mg/ml in ddH$_2$O (PCR-grade; Roche Applied Science)), and 3 μl of 20% (w/v) SDS, and incubate at 55 °C over night. Keep the samples at 55 °C to avoid precipitation of potassium SDS.
6. Increase the volume to 250 μl with ddH$_2$O and add NaClO$_4$ to a final concentration of 1 M.
7. Add 1 volume of phenol (equilibrated with Tris–HCl pH 8.0 (Sigma-Aldrich BioUltra Nr. 77607)), vortex thoroughly, and add another volume of 24:1 (v/v) chloroform:isoamyl alcohol mix. Vortex thoroughly again and centrifuge for 5 min at room temperature and full speed in a table-top centrifuge. Collect the upper aqueous phase containing the DNA.
8. Concentrate the sample by ethanol precipitation (see Section 5.5.1, step 6) and resuspend in 100 μl TE buffer.
9. Remove RNA by addition of 1 μl RNase A (10 mg/ml in 5 mM Tris–HCl, pH 7.5 (Roche Applied Science)) and incubation for 3 h at 37 °C.

10. Precipitate the DNA by addition of first 1/20 volume of 5 M NaCl and then 0.7 volumes of room-temperature isopropanol. Wash with 1 volume of room-temperature 70% ethanol and resuspend pellet in 30 µl TE buffer.
11. Prepare an 1.3% TAE-agarose gel with 0.33 µg/ml ethidium bromide (electrophoresis in presence of ethidium bromide sharpens the DNA bands).
12. Load 15 µl sample supplemented with 5 µl Orange G loading buffer (see Section 5.5.1, step 10) per lane.
13. Electrophorese for 2 h at 2.7 V/cm.
14. Visualize by UV light or Southern blotting (Sambrook et al., 1989).

6. INCUBATION OF SALT GRADIENT DIALYSIS CHROMATIN WITH WCE OR PURIFIED FACTORS

1. Block siliconized 1.5-ml tubes by pipetting 1 ml block solution (2 mg/ml BSA, 0.1% (w/v) IGEPAL CA630, 20 mM HEPES–KOH, pH 7.5) into and out of the tubes. The block solution can be reused many times. Collect remaining solution in the tubes by short centrifugation in table-top centrifuge, remove the last droplet with yellow tip and let the tubes air dry. Such blocked tubes can be prepared in large quantities beforehand and stored indefinitely.
2. Dissolve lyophilized creatine kinase (CK) powder (Roche Applied Science) in 0.1 M imidazole–HCl, pH 6.6 at 20 mg/ml, and flash freeze in liquid nitrogen as 20-µl aliquots. Store at $-80\,^{\circ}$C. Prepare a fresh 1:20 dilution of a freshly thawed 20-µl CK aliquot by adding 380 µl of 0.1 M imidazole–HCl, pH 6.6. Prepare CK dilution freshly before use and always use a fresh aliquot. Do not refreeze. Mix by pipetting and keep on ice.
3. Spin down the salt gradient dialysis chromatin and thawed WCE aliquot for 3 min at full speed in a cooled table-top centrifuge. The WCE may show a visible pellet, which is not a problem. Avoid disturbance of the pellet when taking out aliquots. However, if the salt gradient dialysis chromatin shows more than a tiny pellet, it is aggregated and should not be used.
4. Combine 25 µl of 4× reconstitution mix (80 mM HEPES–KOH, pH 7.5, 320 mM KCl, 12 mM MgCl$_2$, 2 mM EGTA, 10 mM DTT, 48% (v/v) glycerol, 12 mM ATP, 40 mM creatine phosphate, store at $-20\,^{\circ}$C. For conditions without energy, omit ATP, MgCl$_2$, and creatine phosphate.), 4 µl of 0.25 M ammonium sulfate, 2 µl of 1:20 CK dilution, salt gradient dialysis chromatin corresponding to 0.5–1 µg of

preassembled DNA, 10 μl of WCE (if the protein content is about 20 mg/ml, see Section 2, step 19), and ddH$_2$O to make up a volume of 100 μl. The reaction can be scaled down to 50 μl for assembly of 0.5 μg DNA. Start with ddH$_2$O, 4 × reconstitution mix, and ammonium sulfate, all three of which can be combined to a master mix if several reactions are done in parallel. If called for, any purified component, for example, the transcription factor Pho4 or a remodeling enzyme (Ertel et al., 2010; Wippo et al., 2011), may be added. Addition of protease inhibitors is usually not necessary. We compared several single-locus reconstitution reactions with and without inhibitors and never saw a difference in DNaseI indirect end-labeling patterns. Nonetheless, depending on the application and readout, addition of protease inhibitors may become advisable.

5. Incubate between 1 h and overnight at 30 °C. A 2-h incubation is our standard.
6. If cross-linking is required, add formaldehyde to a final concentration of 0.05% (v/v) and incubate for 15 min at 30 °C. Quench the reaction by addition of glycine to a final concentration of 125 mM and incubate for 5 min at 30 °C.
7. We routinely add 4 μl of 50 U/ml apyrase (NEB M0393L) and incubate for 30 min at 30 °C in order to deplete ATP and therefore stop all remodeling in addition to the cross-linking. The concentration of ATP may be determined at any point during this protocol using a luciferase-based assay (e.g., Enliten, Promega, FF2021, in connection with a Berthold Lumat luminometer). *Attention*: This assay is very sensitive and therefore easily saturated. Measure serial 10-fold dilutions (up to 10^{-6}) in ddH$_2$O in order to find the actual linear working range of the assay. The high dilutions will also slow down ATPases from further depleting ATP if you are interested in the ATP concentration at a particular point in your procedure.

7. GENERATION OF MONONUCLEOSOMES BY LIMITED MNase DIGESTION

The resulting chromatin may be analyzed in many ways. We describe in the following the preparation of mononucleosomes by limited MNase digestion that are suitable for the analysis of nucleosome positions by anti-H3 immunoprecipitation and high-throughput sequencing (see Chapter 10).

Table 9.1 MNase concentration guidelines for typical assembly reactions

Sample	Volume MNase solution (μl)	Concentration MNase solution (U/μl)
No WCE, no energy	12.5	0.0059 (1:100 dilution of stock)
No WCE, with energy	9	0.12 (1:5 dilution of stock)
With WCE, no energy	3.5	0.59 (undiluted stock)
With WCE, with energy	12	0.59 (undiluted stock)

1. Add 1/50 volume of 75 mM CaCl$_2$.
2. Add MNase. The optimal concentration of MNase has to be found by titration. The values in Table 9.1 are guidelines for highly assembled chromatin. Lower assembly degrees require less MNase, although this effect is rather small in the presence of WCE. WCE generally calls for higher MNase concentrations, presumably due to many proteins binding unspecifically to chromatin or the MNase, or due to competing RNA. The 0.59 U/μl MNase stock solution is prepared as described in Section 3, step 11, and diluted in 10 mM HEPES–KOH, pH 7.5, 0.1 mg/ml BSA in 5- or 10-fold steps to facilitate pipetting of small amounts. "No energy" refers to a reconstitution without ATP, MgCl$_2$, and creatine phosphate (see Section 6, step 4). The presence of this energy mix will require more MNase. We assume this is either because ATP binds Ca^{2+} (without apyrase treatment) or because the apyrase digestion generates inorganic phosphate, which in turn complexes Ca^{2+}. Therefore, the MNase concentration has to be higher for samples with energy and/or WCE.
3. Incubate for 5 min at 30 °C.
4. Stop MNase digest by adding EDTA to a final concentration of 10 mM.
5. Add protease inhibitors (Complete™ (Roche Applied Science) or equivalent) and proceed with or without antihistone H3 immunoprecipitation, DNA purification, and library generation for high-throughput sequencing (see Chapter 10).

8. SUMMARY

Even though the here described method cannot generate a perfectly faithful replica of the *in vivo* nucleosome organization so far, it is clear that incubation of salt gradient dialysis assembled nucleosomes with yeast extract and energy makes their positioning dramatically more *in vivo*-like than salt gradient dialysis alone. This—in connection with *in vivo* evidence, for example (Badis et al., 2008; Gkikopoulos et al., 2011; Hartley & Madhani, 2009; Parnell, Huff, & Cairns, 2008)—underscores that factors beyond the histone octamer and the DNA sequence are necessary to determine nucleosome positions *in vivo*. Our method provides a positive reference point for the biochemical isolation of involved factors, for testing combinations of purified candidate factors, and for elucidating their mechanisms.

ACKNOWLEDGMENTS

Work in our laboratory is funded by the German Research Community (DFG, grant within the SFB/Transregio 5 and individual grant KO 2945/1-1). We thank Dr. Natascha Steffen for technical advice on the preparation of *Drosophila* embryo histones.

REFERENCES

Adams, T. M., Schmoldt, H.-U., & Kolmar, H. (2004). FACS screening of combinatorial peptide and protein libraries displayed on the surface of Escherichia coli cells. In S. Brakmann & A. Schiwenhorst (Eds.), *Evolutionary Methods in Biotechnology*. Weinheim: Wiley-VCH.

Badis, G., Chan, E. T., van Bakel, H., Pena-Castillo, L., Tillo, D., Tsui, K., et al. (2008). A library of yeast transcription factor motifs reveals a widespread function for Rsc3 in targeting nucleosome exclusion at promoters. *Molecular Cell*, *32*, 878–887.

Bell, O., Tiwari, V. K., Thoma, N. H., & Schubeler, D. (2011). Determinants and dynamics of genome accessibility. *Nature Reviews Genetics*, *12*, 554–564.

Bonaldi, T., Imhof, A., & Regula, J. T. (2004). A combination of different mass spectroscopic techniques for the analysis of dynamic changes of histone modifications. *Proteomics*, *4*, 1382–1396.

Ertel, F., Dirac-Svejstrup, A. B., Hertel, C. B., Blaschke, D., Svejstrup, J. Q., & Korber, P. (2010). In vitro reconstitution of PHO5 promoter chromatin remodeling points to a role for activator-nucleosome competition in vivo. *Molecular and Cellular Biology*, *30*, 4060–4076.

Faulhaber, I., & Bernardi, G. (1967). Chromatography of calf-thymus nucleoprotein on hydroxyapatite columns. *Biochimica et Biophysica Acta*, *140*, 561–564.

Germond, J. E., Hirt, B., Oudet, P., Gross-Bellark, M., & Chambon, P. (1975). Folding of the DNA double helix in chromatin-like structures from simian virus 40. *Proceedings of the National Academy of Sciences of the United States of America*, *72*, 1843–1847.

Gkikopoulos, T., Schofield, P., Singh, V., Pinskaya, M., Mellor, J., Smolle, M., et al. (2011). A role for Snf2-related nucleosome-spacing enzymes in genome-wide nucleosome organization. *Science*, *333*, 1758–1760.

Hartley, P. D., & Madhani, H. D. (2009). Mechanisms that specify promoter nucleosome location and identity. *Cell*, *137*, 445–458.

Hertel, C. B., Längst, G., Hörz, W., & Korber, P. (2005). Nucleosome stability at the yeast PHO5 and PHO8 promoters correlates with differential cofactor requirements for chromatin opening. *Molecular and Cellular Biology*, *25*, 10755–10767.

Huynh, V. A., Robinson, P. J., & Rhodes, D. (2005). A method for the in vitro reconstitution of a defined "30 nm" chromatin fibre containing stoichiometric amounts of the linker histone. *Journal of Molecular Biology*, *345*, 957–968.

Jiang, C., & Pugh, B. F. (2009). Nucleosome positioning and gene regulation: Advances through genomics. *Nature Reviews Genetics*, *10*, 161–172.

Jones, G. M., Stalker, J., Humphray, S., West, A., Cox, T., Rogers, J., et al. (2008). A systematic library for comprehensive overexpression screens in *Saccharomyces cerevisiae*. *Nature Methods, 5*, 239–241.

Kaplan, N., Moore, I. K., Fondufe-Mittendorf, Y., Gossett, A. J., Tillo, D., Field, Y., et al. (2009). The DNA-encoded nucleosome organization of a eukaryotic genome. *Nature, 458*, 362–366.

Korber, P., & Becker, P. B. (2010). Nucleosome dynamics and epigenetic stability. *Essays in Biochemistry, 48*, 63–74.

Korber, P., & Hörz, W. (2004). In vitro assembly of the characteristic chromatin organization at the yeast PHO5 promoter by a replication-independent extract system. *The Journal of Biological Chemistry, 279*, 35113–35120.

Krajewski, W. A., & Becker, P. B. (1999). Reconstitution and analysis of hyperacetylated chromatin. *Methods in Molecular Biology, 119*, 207–217.

Kunert, N., & Brehm, A. (2008). Mass production of Drosophila embryos and chromatographic purification of native protein complexes. *Methods in Molecular Biology, 420*, 359–371.

Lantermann, A. B., Straub, T., Stralfors, A., Yuan, G. C., Ekwall, K., & Korber, P. (2010). Schizosaccharomyces pombe genome-wide nucleosome mapping reveals positioning mechanisms distinct from those of Saccharomyces cerevisiae. *Nature Structural and Molecular Biology, 17*, 251–257.

Lee, W., Tillo, D., Bray, N., Morse, R. H., Davis, R. W., Hughes, T. R., et al. (2007). A high-resolution atlas of nucleosome occupancy in yeast. *Nature Genetics, 39*, 1235–1244.

Luger, K., Mader, A. W., Richmond, R. K., Sargent, D. F., & Richmond, T. J. (1997). Crystal structure of the nucleosome core particle at 2.8 A resolution. *Nature, 389*, 251–260.

Luger, K., Rechsteiner, T. J., & Richmond, T. J. (1999). Expression and purification of recombinant histones and nucleosome reconstitution. *Methods in Molecular Biology, 119*, 1–16.

Lusser, A., & Kadonaga, J. T. (2004). Strategies for the reconstitution of chromatin. *Nature Methods, 1*, 19–26.

Mavrich, T. N., Jiang, C., Ioshikhes, I. P., Li, X., Venters, B. J., Zanton, S. J., et al. (2008). Nucleosome organization in the Drosophila genome. *Nature, 453*, 358–362.

Oudet, P., Gross-Bellard, M., & Chambon, P. (1975). Electron microscopic and biochemical evidence that chromatin structure is a repeating unit. *Cell, 4*, 281–300.

Parnell, T. J., Huff, J. T., & Cairns, B. R. (2008). RSC regulates nucleosome positioning at Pol II genes and density at Pol III genes. *The EMBO Journal, 27*, 100–110.

Patterton, H. G., & von Holt, C. (1993). Negative supercoiling and nucleosome cores. I. The effect of negative supercoiling on the efficiency of nucleosome core formation in vitro. *Journal of Molecular Biology, 229*, 623–636.

Pfaffle, P., & Jackson, V. (1990). Studies on rates of nucleosome formation with DNA under stress. *The Journal of Biological Chemistry, 265*, 16821–16829.

Radman-Livaja, M., & Rando, O. J. (2010). Nucleosome positioning: How is it established, and why does it matter? *Developmental Biology, 339*, 258–266.

Rhodes, D., & Laskey, R. A. (1989). Assembly of nucleosomes and chromatin in vitro. *Methods in Enzymology, 170*, 575–585.

Robinson, K. M., & Schultz, M. C. (2003). Replication-independent assembly of nucleosome arrays in a novel yeast chromatin reconstitution system involves antisilencing factor Asf1p and chromodomain protein Chd1p. *Molecular and Cellular Biology, 23*, 7937–7946.

Rose, M. D., Novick, P., Thomas, J. H., Botstein, D., & Fink, G. R. (1987). A Saccharomyces cerevisiae genomic plasmid bank based on a centromere-containing shuttle vector. *Gene, 60*, 237–243.

Sambrook, J., Fritsch, E. F., & Maniatis, T. (1989). *Molecular cloning: A laboratory manual* (2nd ed.). Cold Spring Harbor, NY: Cold Spring Harbor Laboratory Press.
Schnitzler, G. R. (2008). Control of nucleosome positions by DNA sequence and remodeling machines. *Cell Biochemistry and Biophysics, 51*, 67–80.
Schones, D. E., Cui, K., Cuddapah, S., Roh, T. Y., Barski, A., Wang, Z., et al. (2008). Dynamic regulation of nucleosome positioning in the human genome. *Cell, 132*, 887–898.
Schultz, M. C. (1999). Chromatin assembly in yeast cell-free extracts. *Methods, 17*, 161–172.
Schultz, M. C., Hockman, D. J., Harkness, T. A., Garinther, W. I., & Altheim, B. A. (1997). Chromatin assembly in a yeast whole-cell extract. *Proceedings of the National Academy of Sciences of the United States of America, 94*, 9034–9039.
Segal, E., & Widom, J. (2009). What controls nucleosome positions? *Trends in Genetics, 25*, 335–343.
Shaffer, C. D., Wuller, J. M., & Elgin, S. C. (1994). Raising large quantities of Drosophila for biochemical experiments. *Methods in Cell Biology, 44*, 99–108.
Shure, M., & Vinograd, J. (1976). The number of superhelical turns in native virion SV40 DNA and minicol DNA determined by the band counting method. *Cell, 8*, 215–226.
Simon, R. H., & Felsenfeld, G. (1979). A new procedure for purifying histone pairs H2A + H2B and H3 + H4 from chromatin using hydroxylapatite. *Nucleic Acids Research, 6*, 689–696.
Stein, A. (1989). Reconstitution of chromatin from purified components. *Methods in Enzymology, 170*, 585–603.
Valouev, A., Ichikawa, J., Tonthat, T., Stuart, J., Ranade, S., Peckham, H., et al. (2008). A high-resolution, nucleosome position map of C. elegans reveals a lack of universal sequence-dictated positioning. *Genome Research, 18*, 1051–1063.
Valouev, A., Johnson, S. M., Boyd, S. D., Smith, C. L., Fire, A. Z., & Sidow, A. (2011). Determinants of nucleosome organization in primary human cells. *Nature, 474*, 516–520.
van Holde, K. E. (1988). *Chromatin*. New York: Springer.
Widom, J. (2001). Role of DNA sequence in nucleosome stability and dynamics. *Quarterly Reviews of Biophysics, 34*, 269–324.
Wippo, C. J., Israel, L., Watanabe, S., Hochheimer, A., Peterson, C. L., & Korber, P. (2011). The RSC chromatin remodelling enzyme has a unique role in directing the accurate positioning of nucleosomes. *The EMBO Journal, 30*, 1277–1288.
Wippo, C. J., & Korber, P. (2012). In vitro reconstitution of in vivo-like nucleosome positioning on yeast DNA. *Methods in Molecular Biology, 833*, 271–287. With kind permission of Springer Science + Business Media.
Wippo, C. J., Krstulovic, B. S., Ertel, F., Musladin, S., Blaschke, D., Sturzl, S., et al. (2009). Differential cofactor requirements for histone eviction from two nucleosomes at the yeast PHO84 promoter are determined by intrinsic nucleosome stability. *Molecular and Cellular Biology, 29*, 2960–2981.
Yuan, G. C., Liu, Y. J., Dion, M. F., Slack, M. D., Wu, L. F., Altschuler, S. J., et al. (2005). Genome-scale identification of nucleosome positions in S. cerevisiae. *Science, 309*, 626–630.
Zhang, Y., Moqtaderi, Z., Rattner, B. P., Euskirchen, G., Snyder, M., Kadonaga, J. T., et al. (2009). Intrinsic histone-DNA interactions are not the major determinant of nucleosome positions in vivo. *Nature Structural and Molecular Biology, 16*, 847–852.
Zhang, Z., Wippo, C. J., Wal, M., Ward, E., Korber, P., & Pugh, B. F. (2011). A packing mechanism for nucleosome organization reconstituted across a eukaryotic genome. *Science, 332*, 977–980.

CHAPTER TEN

Genome-Wide Mapping of Nucleosome Positions in Yeast Using High-Resolution MNase ChIP-Seq

Megha Wal, B. Franklin Pugh[1]
Center for Eukaryotic Gene Regulation, Department of Biochemistry and Molecular Biology,
The Pennsylvania State University, University Park, Pennsylvania, USA
[1]Corresponding author: e-mail address: bfp2@psu.edu

Contents

1. Introduction	233
2. Methodology	235
2.1 Harvesting	235
2.2 Cell lysis, MNase digestion	236
2.3 ChIP and library preparation	240
2.4 Sequencing library preparation	245
Acknowledgments	249
References	249

Abstract

Eukaryotic DNA is packaged into chromatin where nucleosomes form the basic building unit. Knowing the precise positions of nucleosomes is important because they determine the accessibility of underlying regulatory DNA sequences. Here we describe a detailed method to map on a genomic scale the locations of nucleosomes with very high resolution. Micrococcal nuclease (MNase) digestion followed by chromatin immunoprecipitation and facilitated library construction for deep sequencing provides a simple and accurate map of nucleosome positions.

1. INTRODUCTION

Nucleosomes form the building blocks of chromatin, consisting of ~147 base-pair (bp) DNA that is tightly wrapped around a histone octamer core (Richmond & Davey, 2003). This wrapping requires substantial bending of DNA. At some nucleosome positions, DNA sequences such as the AA/TT dinucleotides may be enriched every 10 bp and along with regions of higher GC content near the nucleosome dyad (midpoint) collaborate to favor nucleosome formation (Albert et al., 2007; Satchwell, Drew, & Travers, 1986; Segal & Widom, 2009; Tillo & Hughes, 2009). In contrast, Poly dA:dT tracts tend to be intrinsically rigid and hence disfavor nucleosome formation (Anderson & Widom, 2001; Bao, White, & Luger, 2006; Iyer & Struhl, 1995).

In addition to the role of DNA sequence, adenosine triphosphate (ATP)-dependent chromatin remodeling enzymes use the energy of ATP hydrolysis to alter histone–DNA contacts and control nucleosome positions. *In vivo*, the action of these enzymes may lead to diverse outcomes, ranging from shifting nucleosome positions to partial and/or complete eviction of nucleosomes and histone replacement (Hartley & Madhani, 2009; Langst & Becker, 2001; Whitehouse, Rando, Delrow, & Tsukiyama, 2007). Recent work has suggested specific roles of distinct chromatin remodelers. Isw2 functions adjacent to promoters, where it repositions nucleosomes at the interface between genic and intergenic regions (Whitehouse et al., 2007). The RSC complex is responsible for maintaining nucleosome-free promoter regions and its flanking nucleosomes (Hartley & Madhani, 2009). Isw1, Isw2, and Chd1 remodelers might work together in maintaining genome-wide nucleosome organization (Gkikopoulos et al., 2011). *In vitro* studies have also indicated the role of ATP-dependent activities in reconstituting proper nucleosome positioning across the genome (Zhang et al., 2011).

Nucleosome locations affect every cellular process requiring access to DNA, from influencing evolution to regulating gene expression, development, aging, and human health (Zhang & Pugh, 2011). Thus, determining precise nucleosome positions is of importance and an area of active research. Changes in nucleosome positions, even by a few base pairs, can have significant outcomes.

In this chapter, we describe the experimental details of mapping nucleosome positions across the yeast genome with high precision. In brief, standard chromatin immunoprecipitation (ChIP) methods are initially employed, whereby cell cultures are first treated with formaldehyde, which cross-links proteins to DNA (Fragoso & Hager, 1997). Cells are then harvested, and the chromatin was isolated and fragmented by micrococcal nuclease (MNase) digestion. Histone H3 is immunopurified under semidenaturing conditions to ensure that isolated nucleosomal DNA is cross-linked to histone H3 *in vivo*, as cross-linking can be quite inefficient. Importantly, if such a requirement is not imposed, then there remains a possibility that at least some nucleosomes, particularly mutants, may become repositioned during the chromatin isolation and MNase digestion. A key novel feature of the protocol described here is the construction of the nucleosomal sequence library while it is present within the immunoprecipitate. This facilitates subsequent sample clean up.

2. METHODOLOGY

2.1. Harvesting

We use $\sim 10^7$–10^8 yeast cells as the starting material for mapping nucleosomes. Yeast cells are grown in YPD growth medium and utilize dextrose as the carbon source. We autoclave dextrose for 20 min and have not encountered any issues with crystallization of the media.

Reagents
  Prepare ahead of time.
  1× YPD
    10 g Yeast extract
    20 g Bacto-Peptone
    20 g Dextrose
    In 1 l water.
    Autoclave at 121 °C for 20 min.
  ST buffer
    10 mM Tris–Cl, pH 7.5
    100 mM NaCl
    Filter 0.22 μm, store at 4 °C.
  Aliquot stocks
    Complete, Mini, EDTA-free Protease Inhibitor Cocktail (CPI) tablets—Roche, catalog #04693159001
    2.5 M Glycine
    37% Formaldehyde
  Other
    Liquid nitrogen

Day 1
1. Make a primary inoculum by inoculating 5–10 ml of YPD with cells of interest.
2. Grow in shaker at 25 °C, 250–300 rpm until cultures have grown sufficiently to start large cultures at desired $OD_{600\ nm}$. This will take usually at least 6 h or overnight. The doubling time of wild-type yeast when grown at 25 °C is around 2 h. Alternatively, you can also grow yeast at 30 °C with a doubling time of 90 min.

Day 2

3. Check and record the $OD_{600\ nm}$ of starter cultures.
4. Inoculate 500 ml YPD with enough of starter culture to bring the culture to the desired starting $OD_{600\ nm}$.
   Calculation: (desired large culture OD)/(starter culture OD) × volume of large culture = ml's to inoculate.
5. Grow in shaker at 25 °C, 250–300 rpm.

Day 3

6. Monitor growth carefully until $OD_{600\ nm}$ reaches ~0.8. Do not go higher than OD of 1.0.
7. Add 37% formaldehyde to a final concentration of ~1.0%.
8. Return flasks to a room-temperature shaker for 15 min to cross-link.
9. Quench cross-linked culture by adding 2.5 M glycine to a final concentration of 125 mM. Incubate for 5 min at room temperature with constant shaking.
10. Transfer the contents of the flask to GS3 tubes or equivalent (maximum of 500 ml per tube).
11. Centrifuge in a Sorvall RC6+ centrifuge at 4 °C for 3 min at 4000 rpm or $\sim 2800 \times g$.
12. Resuspend one cell pellet with 1 ml ice-cold ST buffer to which fresh protease inhibitor cocktail is added.
13. Resuspend the second pellet with this cell lysate. Continue to repeat until all pellets are combined and resuspended. All the above steps are done on ice to keep the cells cold.
14. Divide the resuspended cells into microcentrifuge tubes (USA Scientific, catalog #1420-8700) to get the equivalent of ~100 ml of cells per tube.
15. Spin the tubes in bench-top microcentrifuge at 4 °C for 2 min at $\sim 9300 \times g$.
16. Resuspend each cell pellet in 0.5 ml of ST buffer.
17. Spin the tubes at 4 °C for 2 min at $\sim 9300 \times g$. Aspirate off the supernatant and freeze the cells in liquid nitrogen.

2.2. Cell lysis, MNase digestion

We use mechanical force (by agitating ceramic zirconia beads at high speed) to break yeast cell walls, when isolating chromatin. Alternatively, one can use glass beads coupled to a vortexer, or enzymatic treatment (zymolase or other enzymes) for cell wall digestion. Efficiency of cell disruption can be monitored by observing the cells under a microscope. The isolated chromatin is subsequently fragmented using MNase treatment. MNase is a unique endo–exonuclease, which results in cleavage at nucleosome linker regions. We include an additional step of brief sonication to completely solubilize the fragmented chromatin from the pellet into the supernatant. This sonication does not further fragment the nucleosomal DNA. A portion of the digestion chromatin is then extracted and visualized by agarose gel electrophoresis to obtain an MNase ladder corresponding to multiples of the nucleosomes core with the linker DNA.

Reagents
  Prepare ahead of time.
  FA lysis buffer
    50 mM HEPES/KOH, pH 8.0
    150 mM NaCl
    2.0 mM EDTA
    1.0% (v/v) Triton X-100
    0.1% (w/v) Sodium deoxycholate
    Filter 0.22 μm, store at 4 °C.
  NP-S buffer
    0.5 mM Spermidine
    0.075% (v/v) IGEPAL
    50 mM NaCl
    10 mM Tris–Cl, pH 7.5
    5 mM MgCl$_2$
    1 mM CaCl$_2$
    Filter 0.22 μm, store at 4 °C
  2 × Proteinase K buffer
    40 mM of 1 M Tris–Cl, pH 7.5
    40 mM EDTA
    2% (w/v) SDS
    Filter 0.22 μm, store at 4 °C
  TE buffer
    10 mM Tris–Cl, pH 8
    1 mM EDTA
    Filter 0.22 μm, store at 4 °C
  Enzyme stocks
    • 40 U/μl MNase—Worthington Biochemical.

MNase is supplied as a lyophilized powder, which is resuspended in NP-S buffer containing 30% glycerol to a final concentration of 40 U/µl and stored at −20 °C.
- Proteinase K—Roche
- DNase-free RNase—Roche

Aliquot stocks
- 20% (w/v) SDS
- 14.3 M β-Mercaptoethanol (BME)
- 0.1 M Phenylmethylsulfonyl fluoride (PMSF)
- 0.5 M EDTA
- Phenol:chloroform:iso-amyl alcohol (PCIA).

PCIA aliquots should be kept at 4 °C, stored in aluminum foil wrapped tubes as it is light sensitive. PCIA should be handled with care as it can cause skin burns.
- 20 mg/ml Glycogen
- Isopropanol, stored at room temperature
- 70% (v/v) Ethanol, stored at room temperature
- 6× Xylene cyanol DNA-loading dye.
   30% (v/v) Glycerol.
   0.02% (w/v) Xylene cyanol
- CPI stock

Other
- Zirconia/silica beads—BioSpec Products, catalog #11079105z
- Agarose—OmniPur, EMD Chemicals, catalog #2125

2.2.1 Day1: Bead-beating lysis

1. Briefly (around 2 min) thaw 100-ml cell pellet aliquots of cross-linked cells. Vortex and spin the cells before use.
2. Resuspend the cell pellet in 1.0 ml of FA lysis buffer to which fresh protease inhibitor cocktail and PMSF (to a final concentration of 0.2 mM) are added.
3. Add 1.0 ml of zirconia beads.
4. To lyse the cells, use the Mini-Beadbeater-96 machine (BioSpec Products, catalog #1001) for three cycles of 3 min each with a gap of 5 min in between the cycles. (Keep the Eppendorf holder at −20 °C for 10 min before lysis and on an iced block for the 5-min intervals.) Alternatively, one can use glass beads and vortex at maximum speed for 2 h at 4 °C.
5. Pierce the top and bottom of the tube with a hot 22-gauge needle and spin the extracts into a collection tube using a bench-top clinical centrifuge for 5 s at ∼3500 × g. Mix and transfer into 1.5-ml tube.

6. Spin the tubes in bench-top centrifuge at 4 °C for 10 min at $\sim 16000 \times g$. Discard the supernatant.

2.2.2 MNase digestion of lysed cells

7. Resuspend the chromatin pellet in 600 μl NP-S buffer by pipetting (or vortex 30 s at maximum speed). Spin down the chromatin at 4 °C for 5 min at $16,000 \times g$ and discard the supernatant.
8. Add BME to NP-S buffer to a final concentration of 1 mM. BME is not stable in solution, so it must be added to the buffer on the day of experiment.
9. Resuspend the chromatin pellet in 300 μl NP-S buffer to which BME is added. Mix thoroughly.
10. Remove 5 ml cell equivalent of the sample as "input" to MNase digestion for visualizing on the gel.
11. For each sample, titrate the amount of MNase added to determine the optimum MNase concentration where $\sim 80\%$ of the material is in mononucleosomal form.
12. Incubate for 20 min at 37 °C in thermomixer with shaking. Alternatively, one can use a heat-block or thermocycler set at a particular temperature and mix by pipetting every 5 min.
13. Halt digestion by transferring samples to 4 °C and adding 0.5 M EDTA to a final concentration of 10 mM. Incubate the samples on ice for 10 min.
14. Spin at 4 °C for 10 min at $16,000 \times g$. Collect the supernatant in a fresh tube and label as supe1.
15. To completely solubilize chromatin from the pellet, wash the pellet with 300 μl of NP-S buffer containing 0.2% SDS and then sonicate for four cycles (30-s pulse) at medium strength.
16. Spin at 4 °C for 15 min at $16,000 \times g$ and add this second supernatant to supe1.
17. Remove 5 ml cell equivalent of the pellet and supernatant sample for visualization on gel.
18. Store the remaining supernatant and the pellet sample at -80 °C. The pellet sample can be discarded after DNA extraction once you establish that it does not have any significant amount of insoluble DNA.

2.2.3 Reversing cross-links for sample checks

Exposure to high temperature reverses the protein–DNA cross-links. Proteinase K treatment digests all DNA-bound proteins.

19. Take 5 ml cell equivalent of the input, pellet and supernatant sample.
20. Make up the volume of all samples to 200 μl by adding NP-S buffer.
21. Add equal volume of 2× proteinase K buffer.
22. Add 3 μl of 20 mg/ml proteinase K enzyme to each sample.
23. Incubate the samples at 65 °C overnight for proteinase K treatment.

2.2.4 Day 2: DNA extraction for sample checks
24. Extract the samples with 450 μl of PCIA. Vortex the samples for 20 s.
25. Spin in a bench-top microcentrifuge at 16,000 × g for 6 min at room temperature.
26. Carefully remove the upper aqueous layer and transfer to fresh Eppendorf tube.
27. Repeat "PCIA" extraction once more.
28. Add 1 μl of 20 mg/ml glycogen.
29. Precipitate DNA from aqueous layer with 0.6 volume of isopropanol (room temperature). Mix well and then incubate for 30 min at room temperature.
30. Pellet by microcentrifugation for 30 min, 16,000 × g at room temperature. Check for the presence of pellet and then decant the supernatant into the sink.
31. Wash the pellet with 500 μl of 70% ethanol (room temperature). Mix by pipetting. Spin for 10 min, 16000 × g at room temperature.
32. Repeat 70% ethanol wash once more.
33. Decant the supernatant. Dry the pellet for ~15 min (or until dry) in a speed vac (medium temperature setting).
34. Add 30 μl TE buffer containing 50 μg/ml RNase to each sample.
35. Incubate the samples at 37 °C for 2 h (minimum of 30 min).
36. Dissolve the samples in 6× xylene cyanol loading dye and run on 2% agarose gel at 135 V for 35 min.
37. Visualizing DNA from the supernatant sample provides a determination of whether optimum MNase digestion had occurred—see Fig. 10.1A. Any DNA present in the pellet is an indication of unsolubilized DNA—see Fig. 10.1B.
38. Proceed with the best titration sample for immunoprecipitation.

2.3. ChIP and library preparation

We enrich for nucleosomal DNA by immunoprecipitation (ChIP) with anti-H3 antibody and select using magnetic Protein A beads. We use 0.05% SDS during IP to remove uncross-linked DNA fragments. Enriched

Genome-Wide Mapping of Nucleosome Positions in Yeast 241

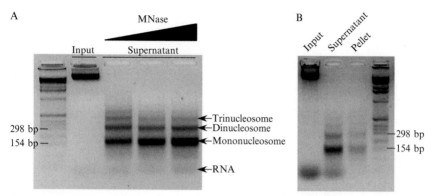

Figure 10.1 MNase digestion results. The input lane indicates undigested chromatin prior to MNase treatment. (A) An MNase ladder is visible in all three MNase-treated supernatant lanes. As indicated in the gel, with increasing MNase concentration, there is an increase in mononucleosomal DNA and decrease in higher molecular weight bands corresponding to "polynucleosomes." Left-most lane indicates a 1-kb ladder. (B) The pellet lane indicates a small amount of unsolubilized DNA trapped in the pellet prior to sonication.

ChIP DNA is then processed for sequencing library preparation while on the magnetic beads. Here we describe library preparation when using Illumina as the sequencing platform, but this procedure can be modified according to the sequencing platform used.

Reagents

Prepare ahead of time.

FA lysis buffer, see above

FA high-salt wash buffer
50 mM HEPES–KOH, pH 7.5
1 M NaCl
1.0% (v/v) Triton X-100
1.0% (w/v) Sodium deoxycholate
2 mM EDTA, pH 8.0
Filter 0.22 μm. Store at 4 °C.

FA wash buffer 3
10 mM Tris–Cl, pH 8.0
25 mM LiCl
1.0% (v/v) IGEPAL
1.0% (w/v) Sodium deoxycholate
2 mM EDTA, pH 8.0
Filter 0.22 μm. Store at 4 °C.

ChIP buffers
- FA lysis buffer + 0.2% (w/v) SDS
- 10 mM Tris–HCl, pH 7.5

Antibody
- Anti-histone H3 antibody—Abcam, catalog #ab1791

Enzymes
- T4 DNA polymerase (NEB)
- T4 DNA ligase (NEB)
- Phi29 DNA polymerase (NEB)
- Klenow fragment exo- (NEB)
- Taq DNA polymerase (NEB)

Enzyme buffers
- 10× NEBuffer 2 (NEB)
- 10× T4 DNA ligase buffer (NEB)
- 10× Phi29 polymerase buffer (NEB)
- 10× Taq polymerase buffer (NEB)

Aliquot stocks
- 20% (w/v) SDS
- 3 mM dNTPs
- 25 mM dNTPs
- 3 mM dATP
- 1× BSA (1 mg/ml)
- CPI stock

Adaptors and primers
- Sequencing adaptors (15 μM)
- PCR primers (20 μM)
- Index (20 μM)

Other
- Magna ChIP Protein A Magnetic Beads (Millipore)
- DynaMag—15 magnet—Invitrogen, catalog #123-01D
- DynaMag—2 magnet—Invitrogen, catalog #123-21D
- Agarose, see above
- Qiagen QIAquick Gel Extraction Kit

2.3.1 Day 1: Antibody–histone attachment

1. Thaw the MNase-digested supernatant from −80 °C storage. Make sure samples go on ice as soon as thawing is complete.
2. Adjust the sample with NP-S buffer to have a minimal volume of 500 μl. Then add 20% SDS for a final concentration of 0.05%. Add 1.25 μl of

20% SDS to a 500-µl sample without SDS. Dilute the sample with NP-S buffer if it contains an SDS concentration higher than 0.05%.
3. Thaw one aliquot of anti-H3 antibody for each sample. Each antibody aliquot should contain 10 µg.
4. Add a 10-µg aliquot of antibody to each sample and incubate the samples on inverting rototorque, 4 °C overnight (12–16 h). Check after ∼10 min to make sure samples are not leaking.

2.3.2 Day 2: ChIP
Notes
- Do not let the beads remain "dry" for longer than about 30 s. If necessary, aspirate a subset of tubes, then add buffer.
- Only spin the samples down before and during enzymatic reactions.
- Follow these instructions for all washes (not just the ChIP washes).
- Wash buffers should be kept on ice, do not set buffers out at room temperature.

1. Briefly resuspend Protein A Magna beads, then immediately add 50 µl Magna beads to each sample in a 15-ml Falcon tube.
2. Mix the sample by inverting the tube.
3. Incubate the reactions for 1.5 h at 4 °C on rototorque (slowly rotating).
4. After incubation, spin down the samples briefly.
5. Place in a DynaMag—15 magnet (which can hold 15-ml Falcon tubes) for 1 min.
6. Using P1000 pipette, carefully transfer the supernatant (flow-through) to new tubes labeled "FT."
7. Store flow-through at −80 °C until the resulting data set is validated, then it may be discarded.

2.3.3 ChIP wash series
1. Wash the beads with FA lysis buffer + 0.025% SDS.
   a. Add 0.8 ml FA lysis buffer + 0.025% SDS + CPI (4 °C) to each sample.
   b. Mix by pipetting up and down and transfer to a new 1.5-ml tube (including beads).
   c. Place in a DynaMag—2 magnet (magnetic rack) for ∼1 min.
   d. Aspirate off the supernatant (Wash 1).
   e. Add 1.4 ml FA lysis buffer + 0.025% SDS + CPI.
   f. Mix gently by repeated inversions.
   g. Spin down the sample briefly, place on magnetic rack for ∼1 min.

h. Aspirate off the supernatant (Wash 2).
  i. Add 1.4 ml FA lysis buffer + 0.025% SDS + CPI.
  j. Mix gently by repeated inversions, rotate on rototorque for 15 min.
  k. Spin down the sample briefly, place on magnetic rack for ~1 min.
  l. Aspirate off the supernatant (Wash 3).
2. Wash the beads with FA lysis buffer
  a. Add 1.4 ml FA lysis buffer + CPI.
  b. Mix gently by repeated inversions.
  c. Spin down the sample briefly, place on magnetic rack for ~1 min
  d. Aspirate off the supernatant (Wash 4).
3. Wash the beads with FA high-salt buffer.
  a. Add 1.4 ml FA high-salt buffer + CPI.
  b. Mix gently by repeated inversions.
  c. Spin down the sample briefly, place on magnetic rack for ~1 min.
  d. Aspirate off the supernatant (Wash 5).
  e. Add 1.4 ml FA high-salt buffer + CPI.
  f. Mix gently by repeated inversions, rotate on rototorque for 15 min.
  g. Spin down the sample briefly, place on magnetic rack for ~1 min.
  h. Aspirate off the supernatant (Wash 6).
4. Wash the beads with FA wash buffer 3
  a. Add 1.4 ml FA wash buffer 3 + CPI.
  b. Mix gently by repeated inversions.
  c. Place on magnetic rack for ~1 min.
  d. Aspirate off the supernatant (Wash 7).
  e. Add 1.4 ml FA wash buffer 3 + CPI.
  f. Mix gently by repeated inversions.
  g. Spin down the sample briefly, place on magnetic rack for ~1 min.
  h. Aspirate off the supernatant (Wash 8).
5. Wash the beads with 10 mM Tris–HCl, pH 8.0
  a. Add 1 ml of 10 mM Tris–HCl, pH 8.0 + CPI.
  b. Transfer the sample to fresh 1.7-ml LoBind tube.
  c. Place on magnetic rack for ~1 min.
  d. Carefully remove the supernatant with P1000 pipette (Wash 9).
6. Wash the beads with double-distilled water + CPI.
  a. Add 0.5 ml double-distilled water + CPI.
  b. Mix gently by repeated inversions, place on magnetic rack for ~1 min.
  c. Carefully remove the supernatant with P1000 pipette.

2.4. Sequencing library preparation

2.4.1 Kinase reaction

This step is essential as MNase leaves a 3′ phosphate, whereas a 5′ phosphate is needed. T4 polynucleotide kinase (PNK) removes 3′ phosphates and adds 5′ phosphates in the presence of ATP.
1. Preheat thermomixer to 37 °C.
2. Prepare Master Mix with 1× T4 DNA ligase buffer and 1.5 U T4 PNK to a total volume of 18 μl per reaction tube.
   Ligase buffer contains DTT, which can form a white precipitate when cold. Ensure all DTT is resuspended before use.
3. Add 18 μl Kinase Mix to 2 μl bead-bound DNA for total of 20 μl. Pipette up/down ∼10 times gently.
4. Incubate the samples on thermomixer at 37 °C for 30 min.
5. Wash the beads with FA high-salt buffer.
   a. Add 0.5 ml FA high-salt buffer.
   b. Mix gently by repeated inversions, place on magnetic rack for ∼1 min.
   c. Aspirate off the supernatant.
6. Wash the beads with 10 mM Tris–HCl, pH 8.0.
   a. Add 1 ml of 10 mM Tris–HCl, pH 8.0.
   b. Mix gently by repeated inversions, place on magnetic rack for ∼1 min.
   c. Carefully remove the supernatant with P1000 pipette.

2.4.2 A-tailing

This step adds a single adenosine nucleotide to the 3′ end of the DNA, which will increase the efficiency of ligation to the Illumina adapters (which have a T overhang). Other sequencing platforms may not use this strategy, in which case the A-tailing step can be skipped.
1. Prepare Master Mix with 1× NEBuffer 2, 100 μM dATP, and 10 U Klenow fragment, exo- to a total volume of 28 μl per reaction tube.
2. Add 28 μl A-Tail Mix to 2 μl bead-bound DNA for total of 30 μl. Pipette up/down ∼10 times gently.
3. Incubate the samples on shaking thermocycler at 37 °C for 30 min.
4. Wash the beads with FA high-salt buffer.
   a. Add 0.5 ml FA high-salt buffer,
   b. Mix gently by repeated inversions, place on magnetic rack for ∼1 min.

    c. Aspirate off the supernatant.
5. Wash the beads with 10 mM Tris–HCl, pH 8.0.
    a. Add 1 ml of 10 mM Tris–HCl, pH 8.0.
    b. Transfer the sample to fresh 1.7-ml LoBind tube.
    c. Mix gently by repeated inversions, place on magnetic rack for ~1 min.
    d. Carefully remove the supernatant with P1000 pipette.

2.4.3 Adaptor ligation

Here, it is necessary to ligate the sequencing adaptor that is appropriate for the sequencing instrument. If planning to multiplex, then each sample gets assigned a unique identity by ligating an individual index sequence for each sample.
1. Preheat thermomixer to 15 °C.
2. Prepare Master Mix with 15 pmol of each adaptor, 30 pmol index (if multiplexing), 1× T4 DNA ligase buffer, and 1800 U T4 ligase to a total volume of 48 μl.

    We use nonphosphorylated adapters to minimize primer dimer. This necessitates a later phi29 fill-in step.
3. Add 48 μl ligation mix to bead-bound DNA for a total of 50 μl. Pipette up/down ~10 times gently.
4. Incubate the samples on thermomixer at 15 °C overnight.

Day 3: Continuation of Section 2.4.3
5. Wash the beads with FA high-salt buffer.
    a. Add 0.5 ml FA high-salt buffer.
    b. Mix gently by repeated inversions, place on magnetic rack for ~1 min.
    c. Aspirate off the supernatant.
6. Wash the beads with 10 mM Tris–HCl, pH 8.0.
    a. Add 1 ml of 10 mM Tris–HCl, pH 8.0.
    b. Transfer the sample to a fresh 1.7-ml LoBind tube.
    c. Mix gently by repeated inversions, place on magnetic rack for ~1 min.
    d. Carefully remove the supernatant with P1000 pipette.

2.4.4 Phi29 fill in

Phi29 DNA polymerase activity will start at a nick and continue to the end of the DNA. It has a strand displacement activity.

1. Preheat thermomixer to 30 °C.
2. Prepare Master Mix with 200 μg/ml BSA, 1× phi29 Buffer, 200 μM dNTPs, and 10 U phi29 polymerase to a total volume of 48 μl.
3. Add 48 μl Fill in Mix to 2 μl bead-bound DNA for total of 50 μl. Pipette up/down ~10 times gently.
4. Incubate the samples on thermomixer at 30 °C for 20 min.
5. Wash the beads with FA high-salt buffer.
   a. Add 0.5 ml FA high-salt buffer.
   b. Mix gently by repeated inversions, place on magnetic rack for ~1 min.
   c. Aspirate off the supernatant.
6. Wash the beads with 10 mM Tris–HCl, pH 8.0.
   a. Add 1 ml of 10 mM Tris–HCl, pH 8.0.
   b. Mix gently by repeated inversions, place on magnetic rack for ~1 min.
   c. Carefully remove the supernatant with P1000 pipette.
7. Resuspend the 2-μl bead-bound DNA in 48 μl double-distilled water + CPI.

2.4.5 Ligation-mediated PCR

1. Prepare Master Mix with 1× Taq buffer, 0.25 mM each dNTP, and 0.3 μM PCR primers to a total volume of 49 μl.
2. Add 49 μl PCR Mix to 50 μl bead-bound DNA for a total of 99 μl. Pipette up/down ~10 times gently.
3. Run the samples in thermocycler with the following program: "Hot start" the samples by adding 1 μl of Taq DNA polymerase to the reaction mix at the "forever" 95 °C step and then continue. Alternatively, a latent heat-activated DNA polymerase may be used.

Time	Temperature (°C)	Cycles
Forever	72	1
20 min	72	1
Forever	95	1
5 min	95	1
15 s	95	18
15 s	52	
1 min	72	
5 min	72	1
Forever	4	Hold

It is not recommended to go above 18 cycles of PCR amplification, rather < 18 cycles is preferable if sufficient sample is present.

2.4.6 Gel purification

It is important to size select for appropriate size DNA and remove small molecular weight DNAs that preferentially get amplified.
1. Thoroughly clean and rinse an appropriate size gel box.
2. Pour 2% agarose gel with thick combs.
3. Mix 1/6th volume of 6× sequencing-grade xylene cyanol dye with each sample.
4. Load 7 μl of 1-kb ladder. Load the samples (region B) on agarose gel between 1-kb ladders (regions A and C) as shown in Fig. 10.2.
5. Excise Lanes A and C (dashed black lines) from the gel. Leave "region B" of the gel at your bench.
6. Visualize Lanes A and C on a short-wavelength UV transilluminator and mark the desired DNA fragment size on the DNA ladders encompassing the 200 and 400 bp markers (blue boxes) by cutting the gel in that area.
7. Reassemble the gel at your bench; using Lanes A and C as markers, excise the sections of agarose-containing DNA fragments of the desired size (red box).
8. Place each gel excision into a 1.7-ml LoBind tube

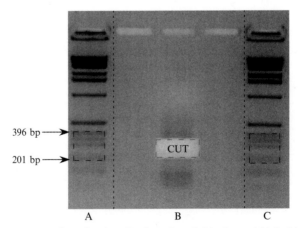

Figure 10.2 Agarose gel purification. Regions A and C indicate 1-kb ladder. Region B: red box indicates the excised gel portion. (See Color Insert.)

9. Record the weight of each excised piece. Write this weight directly onto the LoBind tube.
10. Take short-wavelength UV image of excised gel.

2.4.7 Qiagen cleanup
Notes
- All buffers and centrifuge spins should be at room temperature.
- Do not heat the gel to dissolve the gel slice as it may result in denaturation of the library. Instead, dissolve the gel slice by vortexing at room temperature until completely dissolved.

Use QIAquick Gel Extraction Kit to purify the DNA from the gel. Quantify DNA using bioanalyzer and/or qPCR and then send for deep sequencing. It is recommended that you make clear to the sequencing facility that your library has already been constructed, so that they do not attempt to construct a library with your sample. The sequencing facility should be able to map your DNA sequences to the reference genome that you specify. Those sequencing reads can then be clustered into peak calls using publicly available peak calling algorithms, and nucleosome positions and occupancy levels defined, as discussed elsewhere (Zhang & Pugh, 2011).

ACKNOWLEDGMENTS
We thank previous lab member Elissa Ward for initial standardization of the protocol. We also thank Matt Rossi and Vinesh Vinayachandran for their suggestions.

REFERENCES
Albert, I., Mavrich, T. N., Tomsho, L. P., Qi, J., Zanton, S. J., Schuster, S. C., et al. (2007). Translational and rotational settings of H2A.Z nucleosomes across the *Saccharomyces cerevisiae* genome. *Nature, 446*, 572–576.
Anderson, J. D., & Widom, J. (2001). Poly(dA-dT) promoter elements increase the equilibrium accessibility of nucleosomal DNA target sites. *Molecular and Cellular Biology, 21*, 3830–3839.
Bao, Y., White, C. L., & Luger, K. (2006). Nucleosome core particles containing a poly(dA.dT) sequence element exhibit a locally distorted DNA structure. *Journal of Molecular Biology, 361*, 617–624.
Fragoso, G., & Hager, G. L. (1997). Analysis of in vivo nucleosome positions by determination of nucleosome-linker boundaries in crosslinked chromatin. *Methods, 11*, 246–252.
Gkikopoulos, T., Schofield, P., Singh, V., Pinskaya, M., Mellor, J., Smolle, M., et al. (2011). A role for Snf2-related nucleosome-spacing enzymes in genome-wide nucleosome organization. *Science, 333*, 1758–1760.
Hartley, P. D., & Madhani, H. D. (2009). Mechanisms that specify promoter nucleosome location and identity. *Cell, 137*, 445–458.
Iyer, V., & Struhl, K. (1995). Poly(dA:dT), a ubiquitous promoter element that stimulates transcription via its intrinsic DNA structure. *The EMBO Journal, 14*, 2570–2579.

Langst, G., & Becker, P. B. (2001). Nucleosome mobilization and positioning by ISWI-containing chromatin-remodeling factors. *Journal of Cell Science, 114*, 2561–2568.

Richmond, T. J., & Davey, C. A. (2003). The structure of DNA in the nucleosome core. *Nature, 423*, 145–150.

Satchwell, S. C., Drew, H. R., & Travers, A. A. (1986). Sequence periodicities in chicken nucleosome core DNA. *Journal of Molecular Biology, 191*, 659–675.

Segal, E., & Widom, J. (2009). What controls nucleosome positions? *Trends in Genetics, 25*, 335–343.

Tillo, D., & Hughes, T. R. (2009). G+C content dominates intrinsic nucleosome occupancy. *BMC Bioinformatics, 10*, 442.

Whitehouse, I., Rando, O. J., Delrow, J., & Tsukiyama, T. (2007). Chromatin remodelling at promoters suppresses antisense transcription. *Nature, 450*, 1031–1035.

Zhang, Z., & Pugh, B. F. (2011). High-resolution genome-wide mapping of the primary structure of chromatin. *Cell, 144*, 175–186.

Zhang, Z., Wippo, C. J., Wal, M., Ward, E., Korber, P., & Pugh, B. F. (2011). A packing mechanism for nucleosome organization reconstituted across a eukaryotic genome. *Science, 332*, 977–980.

CHAPTER ELEVEN

Preparation of *Drosophila* Tissue Culture Cells from Different Stages of the Cell Cycle for Chromatin Immunoprecipitation Using Centrifugal Counterflow Elutriation and Fluorescence-Activated Cell Sorting

Nicole E. Follmer, Nicole J. Francis[1]

Department of Molecular and Cellular Biology, Harvard University, Cambridge, Massachusetts, USA
[1]Corresponding author: e-mail address: francis@mcb.harvard.edu

Contents

1.	Introduction	252
2.	Centrifugal Elutriation of *Drosophila* Cells to Obtain G1, S, and G2 Populations	253
	2.1 Materials, buffers, and reagents	254
	2.2 Cell culture conditions	254
	2.3 Elutriation	254
	2.4 Expected results	261
3.	FACS Sorting of *Drosophila* Cells to Obtain Mitotic Cell Populations	261
	3.1 Materials, buffers, and reagents	261
	3.2 G2/M arrest with colchicine treatment	263
	3.3 Staining procedure	264
	3.4 Pre- and postsorting procedures	267
	3.5 Analyze pre- and postsorted populations on a FACS analyzer	267
	3.6 Preparation of control cells	267
	3.7 Expected results	268
4.	Summary	268
Acknowledgments		268
References		268

Abstract

Many nuclear proteins alter their localization during the cell cycle. This includes proteins which regulate and execute cell cycle events and proteins involved in transcription and DNA repair. The core components of chromatin, the histone proteins, also change their modification state through the cell cycle. Chromatin immunoprecipitation (ChIP) makes it possible to localize chromatin-associated proteins to specific sequences in the genome and has revolutionized studies of transcription. Fewer studies have used ChIP to analyze protein localization or modification at specific stages in the cell cycle. This is in part because these studies require isolation of pure populations of cells at each stage of the cell cycle, which is challenging for many cell types. However, the ability to carry out ChIP from cells at specific stages in the cell cycle in some systems has revealed cell cycle regulation of chromatin localization, and cell cycle stage-specific functions and modification of chromatin proteins, providing incentive to pursue these experiments.

This chapter presents protocols for isolating *Drosophila* S2 cells from all phases of the cell cycle using centrifugal elutriation and fluorescent-activated cell sorting. These cells are suitable for ChIP analysis.

1. INTRODUCTION

Chromatin immunoprecipitation (ChIP) is a powerful tool for localizing chromatin-bound proteins. Localization of proteins to particular sites on chromatin can be queried by qPCR, or microarray or high-throughput sequencing can be used to reveal global localization of a protein on the genome. Localization of many proteins to chromatin varies through the cell cycle. This includes proteins involved in DNA replication and mitotic chromosome condensation, but likely also includes many proteins that regulate transcription. ChIP on pure populations of cells from specific stages of the cell cycle is a powerful way to study cell cycle-dependent chromatin protein dynamics (Blobel et al., 2009 and studies reviewed in Young & Kirchmaier, 2012). In some cells, such as yeast, cell cycle-staged populations can be achieved with simple, well-characterized treatments (Fox, 2004; Manukyan, Abraham, Dungrawala, & Schneider, 2011). In other cell types, such as *Drosophila* tissue culture cells, cell cycle-synchronizing drugs work inefficiently (Nicole E. Follmer, Nicole J. Francis unpublished results). This chapter describes protocols to generate such pure populations of cells from all stages of the cell cycle in quantities sufficient for ChIP studies.

Centrifugal counterflow elutriation is a method for separating heterogeneous cell populations by sedimentation rate, a property that depends in part on the diameter and density of the cell (Lindahl, 1948). As cell size correlates well with DNA content and thus with the cell cycle, this method can separate an asynchronously growing population of cells according to cell cycle stage. Cells are placed in a specialized rotor in which centrifugal force is balanced by counterflow of buffer or media. Stepwise increase in flow rate or decrease in centrifugal force allows for separation of the heterogeneous cell population into successive fractions of uniform size, starting with the smallest, most slowly sedimenting cells and ending with the largest, most rapidly sedimenting cells (Sanderson & Bird, 1977). This method allows for the separation of cells into G1, S, and G2 populations without the use of pharmacological or other perturbations of the cells. While centrifugal counterflow elutriation can be performed on living cells, the protocol presented here is for the separation of formaldehyde fixed cells, which are suitable for ChIP.

Mitotic cells do not fractionate as a single population via centrifugal elutriation (Hengstschlager, Pusch, Soucek, Hengstschlager-Ottnad, & Bernaschek, 1997; Kauffman, Noga, Kelly, & Donnenberg, 1990), and thus other methods must be employed to obtain pure populations of mitotic cells in large quantities. Only a small percentage of cells in an asynchronously growing culture are in mitosis (4–5% for *Drosophila* S2 cells) so that large-scale separation by any means is challenging. Mitotic shake-off, a method used for many adherent cell lines to obtain purely mitotic populations (Terasima & Tolmach, 1963), is not feasible in *Drosophila* S2 cells or other cell types that grow in suspension or semi-adherent. Mitotic-arrest-inducing drugs can be used, but the cells respond with incomplete mitotic arrest. A solution arises with a combination of strategies: arrested mitotic cells can be separated from the nonmitotic cells in a drug-treated culture with the use of a mitotic marker in conjunction with flow cytometry.

This chapter describes protocols for isolating *Drosophila* S2 cells from each stage of the cell cycle that are suitable for ChIP. G1, S, and G2 cells are isolated by centrifugal elutriation, while mitotic cells are isolated by fluorescent-activated cell sorting (FACS). This protocol gives conditions specific to *Drosophila* S2 cells with attention to critical parameters. This should allow the protocol to be adapted to other cell types.

2. CENTRIFUGAL ELUTRIATION OF *DROSOPHILA* CELLS TO OBTAIN G1, S, AND G2 POPULATIONS

The protocol below describes a method for isolating G1, S, and G2 populations of fixed *Drosophila* S2 cells using centrifugal elutriation. Protocols are given for preparing cells and the elutriation system, conducting the elutriation, and processing the fractions. This protocol does not describe the protocol for installing, cleaning, and maintaining the elutriation system. For this information, see Coulter (2007–2009) and Banfalvi (2011). This protocol is loosely based on Banfalvi et al. (2007).

2.1. Materials, buffers, and reagents

2.1.1 Materials
Drosophila S2 cells
Centrifugal elutriation system: rotor with a single large chamber and a counterbalance, centrifuge with strobe assembly and specialized lid with viewing port, bubble trap, sample reservoir, silicone tubing, and variable-speed pump
4-L Buffer reservoir
Hemocytometer
Flow cytometer for analysis
Cell cycle analysis software

2.1.2 Disposables
225-mL Conical bottom plastic collection tubes
15- and 50-mL Conical tubes
1.5-mL Microfuge tubes
10-mL Pipettes
0.2-μm Bottle top filters
10-mL Syringe 18 gauge needle

2.1.3 Buffers and reagents
12 L of Elutriation buffer at 4 °C: 1 × PBS, 5% horse serum
Formaldehyde 2.5 M Glycine, pH 7.9
1 × PBS
95% Ethanol
Propidium iodide (0.5 mg/mL)
RNase A (50 mg/mL)

2.2. Cell culture conditions
Drosophila S2 cells are grown in shaking flasks at 27 °C at a density between 2×10^6 and 10×10^6 cells/mL in protein-free insect cell media containing 34 mM L-glutamine.

2.3. Elutriation

2.3.1 Elutriation system set up
The protocol described below has been optimized using one large chamber (40 mL) with a counterbalance in a JE-5.0 rotor in an Avanti J-26XP centrifuge (Beckman Coulter). Our system is set up as in Fig. 11.1A. A 4-L

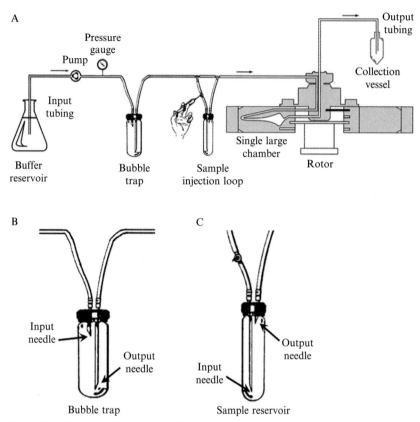

Figure 11.1 The elutriation system. (A) Overview of the configuration of the elutriation system used to derive the protocol described here. Adapted from Figure 2–6 (Coulter, 2007–2009). (B) Configuration of input and output needles in the bubble trap. (C) Configuration of input and output needles in the sample reservoir. Images courtesy of Beckman Coulter, Inc.

Erlenmeyer flask serves as a buffer reservoir. A variable-speed pump, a pressure gauge, and a bubble trap are placed between the buffer reservoir and a sample injection loop, which consists of a three-way valve to divert flow through the injection loop or to allow flow to bypass the loop, a three-way valve for syringe attachment for sample injection, and a sample reservoir. The tubing then passes into the rotor, through a single large chamber and out to a collection vessel. It should be noted that the input needle of the bubble trap should be inserted so the tip is near the top of the trap while the tip of the output needle should be near the bottom of the trap, as in Fig. 11.1B. The sample reservoir assembly has the reverse configuration: the tip of the input needle is near the bottom of the reservoir while the tip of the output needle is at the very top.

2.3.2 Preparing the elutriator for a run

1. Check the tubing running through the pump to make sure it is not worn. If it is worn, run an unworn length of tubing through the pump.
2. Set the centrifuge to 4 °C. Set the timer to HOLD. Place the input tubing on the tip of a 10-mL pipette with the cotton plug removed. Place the 10-mL pipette in the 4-L buffer reservoir containing cold elutriation buffer.
3. Purge air from the system: with the output tubing placed in the buffer reservoir, set the pump to 200 mL/min. Invert bubble trap to allow it to fill with buffer. Open valve to injection loop to allow it to fill with buffer. Flick tubes to release any air bubbles stuck along the tubing or in any of the valves. With the pump running, set the centrifuge to 1000 rpm ($880 \times g$). Bubbles should be purged from the elutriation chamber. Increase the speed of the centrifuge to 3500 rpm ($2300 \times g$) to test for an imbalance. If an imbalance error occurs, stop the centrifuge and reset the quick-release assembly.
4. Before each run, check pump calibration. With the system filled with buffer and the centrifuge spinning at 3500 rpm ($2300 \times g$), set pump speed to 100 mL/min. Place the output tubing into a 250-mL graduated cylinder and collect buffer for 1 min. The amount of buffer collected in 1 min should be 100 mL. If the volume collected is not 100 mL, recalibrate the pump, if the pump model allows, or simply adjust flow rate settings on the pump as needed to give the actual flow rate desired.
5. Set pump speed to 80 mL/min. Set centrifuge speed to 3500 rpm ($2300 \times g$) (see Note 2.1).

Note:

2.1 While counterbalance between flow rate and centrifugal force determines whether or not particles will be retained in the rotor, higher flow rates and faster centrifuge speeds afford greater resolution of separation (see Fig. 11.2).

2.3.3 Loading the cells and equilibration

1. Prepare 225-mL conical tubes on ice in which to collect fractions.
2. Fix $1–1.5 \times 10^9$ cells by adding formaldehyde directly to the tissue culture media to a final concentration of 1% for 10 min, rocking or shaking at room temperature. Quench formaldehyde by adding 2.5 M glycine, pH 7.9 to a final concentration of 130 mM. Centrifuge for 5 min at 4 °C at $1920 \times g$. Resuspend in 1 × PBS to wash. Remove 1.5×10^6 cells for FACS analysis and centrifuge the rest of the cells for 5 min at 4 °C at $1920 \times g$.

Isolation of Drosophila S2 Cells by Cell Cycle Stage

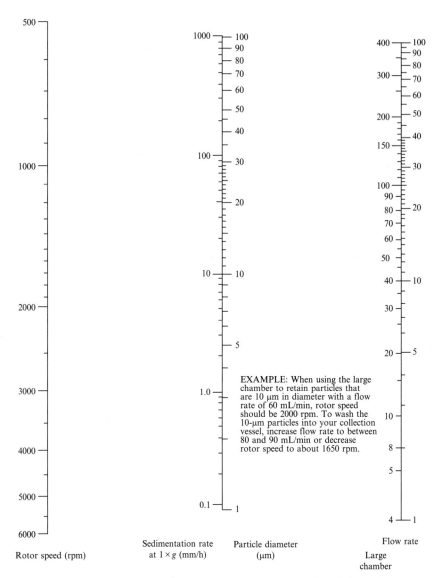

Figure 11.2 Rotor speed and flow rate nomogram. Reproduced from Figure 4-1 (Coulter, 2007–2009). Rotor speed and flow rate nomogram allows for the determination of the smallest, lightest particles retained at a given flow rate and rotor speed. A line connecting a given flow rate and rotor speed will intersect the center axis, indicating the lowest sedimentation rate and smallest particle diameter retained under those conditions.

3. Resuspend the cells in elutriation buffer to a final volume of 10 mL. Pass the cells slowly through an 18-gauge needle attached to a 10-mL syringe 10 times to disperse cell clumps. Do not introduce air bubbles into the cells. Finally, fill the syringe with the cells and remove the needle.
4. Close the injection loop three-way valve so that the buffer bypasses the injection loop. Close the sample injection three-way valve to the input (Fig. 11.3A). Back pressure should push some buffer out of the side Luer fitting, purging air from the valve. Attach the syringe containing the cells to the side Luer fitting of the sample injection three-way valve and slowly inject the cells into the sample reservoir (Fig. 11.3B). Cells will be collected at the bottom of the sample reservoir. Close the valve to the syringe. Slowly invert the sample reservoir to allow the cells to move as a bolus to the top of the sample reservoir where the output is (Fig. 11.3C). Place the output tubing into the first collection tube and open the injection loop three-way valve to the injection loop to allow the cells to move out of the sample reservoir into the rotor (Fig. 11.3D). After all cells have passed out of the injection loop, close the injection loop three-way valve so that buffer bypasses the injection loop. Examine the cells in the elutriation chamber. A clear front of cells should be visible near the top of the chamber (see Note 2.2).
5. Over the first two 200-mL fractions, incrementally increase pump speed to 96 mL/min-increase pump speed to 84 mL/min after 100 mL has been collected, to 88 mL/min after 200 mL has been collected, to 92 mL/min after 300 mL has been collected, and up to 96 mL/min after 400 mL has been collected. The first two fractions should only contain cell debris and a few ($< 1 \times 10^6$) very small cells (see Note 2.3).
6. Let cells equilibrate at 96 mL/min at 3500 rpm ($2300 \times g$) for 2 h. To insure that cells are not escaping the changer during the equilibration, collect and monitor the first several 200-mL fractions during the equilibration. Centrifuge the fractions at 4 °C at $1920 \times g$ for 5 min, gently pour off the supernatant, and resuspend the pellet in 1 mL of cold 1 × PBS. Examine a sample of each resuspended pellet on a hemocytometer under a light microscope to verify the absence of cells. If no cells are present, the rest of the fractions collected during the 2 h equilibration can be passed through a 0.2-μm filter and reused (see Note 2.4).

Notes:

2.2 The appearance of swirling of the cells, known as the Coriolis jetting effect, is undesirable. This causes mixing of cells in the chamber, which decreases the resolution of separation. We have found that increasing

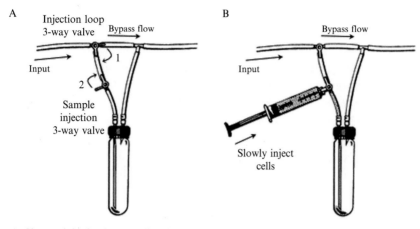

1. Close to injection loop to allow bypass.

2. Close to input to open to side Luer fitting and sample reservoir.

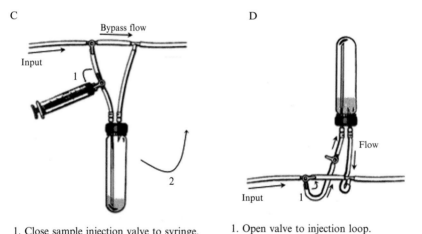

1. Close sample injection valve to syringe.
2. Slowly invert sample reservoir.

1. Open valve to injection loop.

Figure 11.3 Loading cells into the elutriation system. (A) The order and orientation of three-way valve switching required to load cells into the sample reservoir. (B) An illustration of sample injection. (C) After sample injection, the sample injection three-way valve is closed to the syringe and the sample reservoir is slowly inverted. (D) Opening the injection loop three-way valve to the injection loop allows cells to leave the sample reservoir and enter the rotor. All adapted from Figure 3-1 (Coulter, 2007–2009).

the number of cells loaded into the chamber greatly decreases the Coriolis jetting effect.

2.3 If cells are escaping the chamber during the equilibration, one of two things may be happening: (1) the chamber is overloaded or (2) the centrifugal and counterflow forces are not balanced. This means that either the flow rate is too high or the centrifuge speed is too low. The flow rate should be as high as possible to avoid pelleting during equilibration (see Note 2.4) while still retaining the smallest cells in the chamber.

2.4 The cells should not pellet during equilibration. If they do, the pump speed is too low or the centrifuge speed is too high.

2.3.4 Collecting and processing fractions

Collect and process all fractions on ice. All centrifugation is carried out at 4 °C.

1. To begin collecting the cells after the equilibration, decrease the rotor speed by 50 rpm ($0.5 \times g$) and collect two 200-mL fractions. Continue decreasing rotor speed by 50 rpm ($0.5 \times g$) increments, collecting two 200-mL fractions at each step, until the rotor speed is 2200 rpm. There will be 44 postequilibration fractions.

2. Collect the cells from each fraction by centrifugation for 5 min at $1920 \times g$ in a swinging bucket rotor. Gently decant the supernatant, taking care not to disrupt the cell pellet. Resuspend the cells in 2–15 mL of ice cold 1× PBS (see Note 2.5). Record the volume for each fraction and count the cells using a hemocytometer to determine the total number of cells in each fraction. Remove 1.5×10^6 cells per fraction for FACS analysis.

3. Pellet the cells in each fraction by centrifugation for 5 min at $1920 \times g$ in a swinging bucket rotor. Remove the supernatant and flash freeze the pellets in liquid nitrogen and store at $-80\,°C$.

Note:

2.5 Fractions can be processed as the elutriation run is in progress. Ideally, the elutriation centrifuge is situated near the low-speed centrifuge used to pellet the fractions so that one person can both process the fractions and monitor the elutriation. Supernatants from fractions after centrifugation can be filtered through a 0.2-μm filter and reused.

2.3.5 FACS analysis of fractions

1. Pellet all samples saved for FACS analysis (including the input) by centrifugation for 5 min at $1920 \times g$ in a swinging bucket rotor at 4 °C. Resuspend in 100 μL of 1× PBS and add 900 μL ice cold 95% ethanol. Incubate at $-20\,°C$ for ≥ 30 min to fix the cells.

2. Pellet fixed cells by centrifugation in a swinging bucket rotor (see Note 2.6) for 5 min at 560 × g. Resuspend in 1 × PBS with 5 μg/mL propidium iodide and 50 μg RNase A. Incubate for 2 h at room temperature (see Note 2.7). Analyze immediately or store at 4 °C up to 3–4 days.
3. Analyze on a flow cytometer. Perform cell cycle analysis on the FACS profiles using appropriate software.

Notes:

2.6 It is important to spin the ethanol-fixed cells in a swinging bucket rotor to avoid damaging them against the side of the microfuge tube.

2.7 It is important to allow the RNase treatment to go to completion. Incomplete RNA degradation results in a contribution of RNA to the FACS profile of propidium iodide-stained cells, which may render the cell cycle analysis uninterpretable. In this case, samples can be saved and RNase treatment repeated.

2.4. Expected results

Representative results for separation of asynchronous populations composed of ∼15% G1, ∼50% S, ∼30% G2, and ∼5% M are given in Fig. 11.4. Using this protocol, the purity of fractions ranges from 75% to 85% G1, 100% S, and 85–92% G2. Roughly 20–30% of the total number of cells in each phase are recovered in these highly enriched fractions.

3. FACS SORTING OF *DROSOPHILA* CELLS TO OBTAIN MITOTIC CELL POPULATIONS

The protocol below describes a method for isolating mitotic cell populations using FACS sorting of fixed *Drosophila* S2 cells that are immunostained for an intracellular mitotic marker, phosphorylated histone H3 (H3Ser10p). The protocol allows for the isolation of >95% pure mitotic cells in quantities sufficient for ChIP studies. Protocols are given for mitotic arrest with colchicine, evaluation of arrest, immunostaining for flow cytometry, preparing cells for sorting, and recovering cells from the sorting procedure. This protocol does not describe the protocol for running the flow cytometer.

3.1. Materials, buffers, and reagents

3.1.1 Materials

Drosophila S2 cells
15-mL Round-bottom glass tubes
Hemocytometer

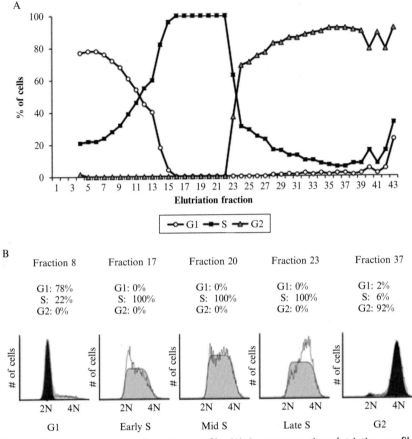

Figure 11.4 Representative elutriation profile. (A) A representative elutriation profile given over all post-equilibration fractions collected. The decrease in homogeneity seen in fractions 35 and up is most likely due to cell aggregates that have higher sedimentation rates than single cells. (B) Representative FACS profiles of individual fractions from the various cell cycle stages.

Flow cytometer for analysis
Flow cytometer for sorting

3.1.2 Disposables

15- and 50-mL Conical tubes
1.5-mL Microfuge tubes
40-μm Nylon cell strainers

3.1.3 Buffers and reagents

Colchicine
20% Sucrose, 1× PBS
Formaldehyde 2.5 M Glycine, pH 7.9 95% Ethanol
10% Formaldehyde, 90% methanol
α-Histone H3 antibody, FITC conjugated
α-Histone H3Ser10p antibody, FITC conjugated
1× PBS
1× PBS, 0.1% Triton X-100, 1% BSA
1× PBS, 0.03% Triton X-100
Protease inhibitors. Stocks: 10 mg/mL TLCK in 1 mM HCl, 0.2 M PMSF in isopropanol, protease inhibitor cocktail of 5 mg/mL aprotinin, 8 mg/mL benzamidine, 5 mg/mL leupeptin, 1 mg/mL pepstatin, 5 mg/mL phenanthroline in methanol. Working concentrations: 50 μg/mL TLCK, 0.4 mM PMSF, 1:500 dilution of protease inhibitor cocktail.
Horse serum

3.2. G2/M arrest with colchicine treatment

Add colchicine to cells to a final concentration of 350 ng/mL (88 nM). The cells should be at a concentration of $\geq 5 \times 10^6$ cells/mL. 15 h later, take out a sample for FACS analysis of DNA content and to count condensed chromosomes (see Note 3.1). Fix the cells with 1% formaldehyde in media for 10 min, rocking or shaking at room temperature. Quench the formaldehyde by adding 2.5 M glycine pH 7.9 to a final concentration of 130 mM. Centrifuge the cells at 730 × g at 4 °C for 5 min. Resuspend in ice cold 1 × PBS to a concentration of 5–15 × 10^7 cells/mL. Slowly layer cells over a 9-volume of 20% sucrose cushion to remove cell debris. Centrifuge at 480 × g for 5 min at 4 °C. Resuspend the pellet in PBS and repeat with a second cushion. Resuspend the cells in ice cold 1 × PBS + protease inhibitors to 2 × 10^7 cells/mL. Examine the cells under a microscope to ensure cell debris has been removed. If much cell debris still remains, pass through a third cushion and recheck for loss of debris (see Note 3.2). Proceed with the staining procedure.

3.2.1 FACS analysis of DNA content

1. Pellet the sample of cells saved for FACS analysis by centrifugation for 5 min at 560 × g in a swinging bucket rotor. Resuspend 100 μL of 1 × PBS and add 900 μL of ice cold 95% ethanol. Incubate at −20 °C for ≥30 min.

2. Pellet fixed cells by centrifugation in a swinging bucket rotor for 5 min at 560 × g. Resuspend in 1× PBS with 5 µg/mL propidium iodide and 50 µg RNase A. Incubate for 2 h at room temperature. Analyze immediately or store at 4 °C up to 3–4 days.
3. Analyze on a flow cytometer. Perform cell cycle analysis on the FACS profiles using appropriate software. The DNA content is typically ~100% 4 N, indicating a 100% G2/M population (Fig. 11.5B).

3.2.2 Determination of mitotic index by counting condensed chromosomes

1. Pellet the sample of cells saved for chromosome counting by centrifugation for 5 min at 560 × g in a swinging bucket rotor. Resuspend cells in 100 µL of 1× PBS and fix by adding 900 µL of 10% formaldehyde/90% methanol. Incubate at room temperature for 8 min.
2. Pellet fixed cells by centrifugation in a swinging bucket rotor for 5 min at 560 × g. Resuspend in 1× PBS to a concentration of ~2.5×10^7 cells/mL. Mount 3 µL of cells on a slide with 5 µL of 80% glycerol and 0.5 µL of 0.5 µg/mL Hoechst stain. Cover with a coverslip.
3. Count the cells under a microscope, using the Hoechst stain to distinguish nuclei and condensed chromosomes. Tally the number of cells with condensed chromosomes. Calculate the mitotic index as number of cells with condensed chromosomes/total cells (see Fig. 11.5C and D).

Notes:

3.1 Drug concentration, time of treatment, and cell density may have to be titrated to obtain optimal arrest. In our lab, these conditions reproducibly yield 60–66% mitotic cells (Fig. 11.5D and E).

3.2 The sucrose cushion conditions presented are very stringent, that is, a large percentage of the cells are lost at each step. Increase centrifuge speed if too many cells are being lost or decrease centrifuge speed if cell debris remains after passing cells through the cushions.

3.3. Staining procedure

Cells should be kept on ice at all times. All centrifugation is performed at 4 °C and all buffers should be at 4 °C and supplemented with protease inhibitors immediately before use.

1. Take fixed and purified colchicine-treated cells at a concentration of 2×10^7 cells/mL and add an equal volume of 0.03% Triton X-100 in 1× PBS for a final concentration of 0.016% Triton X-100 (see Note 3.3). Incubate on ice 15 min. It is important that the starting

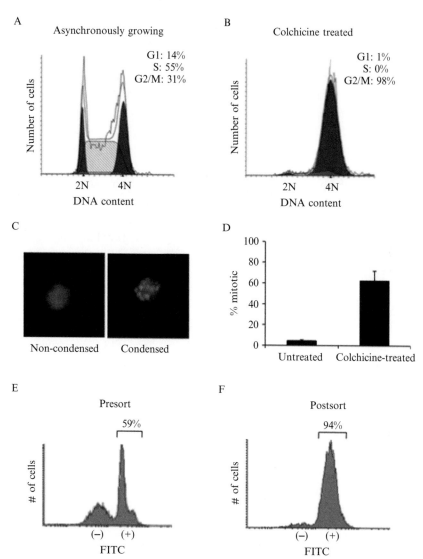

Figure 11.5 Colchicine treatment and FACS profiles. (A) Representative cell cycle profile of asynchronously growing S2 cells stained with propidium iodide for DNA content. (B) Representative cell cycle profile of colchicine-treated S2 cells stained with propidium iodide for DNA content. (C) Representative images of colchicine-treated cells stained with Hoechst with non-condensed chromosomes (left panel) or with condensed chromosomes (right panel). (D) Quantification of percent mitotic for untreated and colchicine-treated S2 cells based on counting cells with condensed chromosomes. (E) Representative FACS profile of colchicine-treated, H3Ser10p-FITC-stained cells before FACS sorting. (F) Representative FACS profile of H3Ser10p-FITC-stained cells after FACS sorting.

concentration of the cells is accurate, as the cells will be sticky after permeabilization making counting them impossible.
2. Centrifuge the cells at $730 \times g$ for 5 min. Wash the cells with 1% BSA, 0.1% Triton X-100 in 1× PBS. Resuspend the cells at a concentration of 1×10^7 cells/mL in 1× PBS with FITC-conjugated α-H3Ser10p antibody at a concentration of 2.7–3 μg/mL (see Notes 3.4 and 3.5).
3. Incubate for 30 min on ice in the dark. Centrifuge the cells at $730 \times g$ for 5 min and wash with 1% BSA, 0.1% Triton X-100 in 1× PBS. Resuspend in a volume of horse serum equivalent to one-tenth of the volume required to give 1.7×10^7 cells/mL. Add nine-tenths volume of 1× PBS to give a final concentration of 1.7×10^7 cells/mL (see Note 3.6).
4. Incubate the cells overnight at 4 °C in the dark. This step greatly improves staining, giving greater separation of FITC-positive and FITC-negative cells, allowing for a more efficient sort.

Notes:
3.3 Higher concentrations of detergent may give better staining leading to greater separation between FITC-positive and FITC-negative cells, which should allow for a more efficient sort. However, higher concentrations of detergent can also lead to more damaged cells which causes cell clumping, resulting a higher abort rate during the sort. It can also lead to greater adherence of cells to surfaces. Both of these effects decrease the overall recovery of cells.
3.4 Titration of antibody may be necessary to achieve maximal separation of FITC-positive and FITC-negative cells.
3.5 Nonconjugated antibodies can be used in conjunction with an FITC-conjugated secondary antibody. After the cells are incubated in primary antibody, wash with 1% BSA, 0.1% Triton X-100 in 1× PBS and resuspend in 1× PBS with 1:500 dilution of FITC-conjugated secondary antibody (0.75 mg/mL) to a final concentration of 1×10^7 cells/mL. Incubate on ice for 30 min in the dark. Wash again with 1% BSA, 0.1% Triton X-100 in 1× PBS and resuspend in horse serum and 1× PBS as in step 3. Continue with the protocol as described.
3.6 This is the concentration at which the cells will be sorted. The concentration may need to be varied to achieve the most efficient sorting—maximal events/second with the lowest abort rate.

3.4. Pre- and postsorting procedures

The cells will be very sticky so it is extremely important to coat everything that will come into contact with the cells with horse serum.

Make all efforts to keep the cells in the dark and on ice when not running through the cytometer.

1. Pass the cells through a 40-μm nylon cell strainer. Load into input tubes in 0.5–1 mL aliquots.
2. Sort and collect FITC-positive cells. We sort on a MoFlo Legacy Cell Sorter (Beckman Coulter). Gate conservatively to obtain the purest population of FITC-positive cells. If possible, check the cells coming off the cytometer on an analyzer to make sure the gating is stringent enough. If the cells are <95% FITC-positive, regate more conservatively.
3. Collect the sorted cells in 14-mL round-bottom glass tubes coated with horse serum, placed on ice.
4. Pool all the collected cells. Count the postsorted cells using a hemocytometer. The count should be similar to the counts given by the cytometer for the number of particles selected. If the counted number is much less than the number given by the cytometer, cells may be sticking to collection tubes, centrifuge tubes, or pipette tips. Be sure to coat all surfaces with horse serum prior to bringing them into contact with the cells.
5. Aliquot the cells in 10×10^6 cell aliquots, centrifuge at 4 °C for 5 min at $1920 \times g$ in a spinning bucket rotor. Remove the supernatant and flash freeze the pellets in liquid nitrogen and store at -80 °C.

3.5. Analyze pre- and postsorted populations on a FACS analyzer

Pre- and postsorted cells can be analyzed directly on a FACS analyzer. Representative results are shown in Fig. 11.4D and E.

3.6. Preparation of control cells

It is important to have a control population that is processed in the same way as the mitotic cells. Asynchronously growing cells can be stained with α-histone H3 antibody and sorted to serve as a control. Asynchronously growing cells are fixed and stained similarly to the colchicine-treated cells with the following modifications:
– Sucrose cushions are spun at $1920 \times g$ since the asynchronously growing cells are smaller in size.

- Staining is done with FITC-conjugated α-histone H3 antibody at a concentration of 3 μg/μL.
- After permeabilization and washing (steps 1 and 2 of Section 3.3), separate out 10% of the total cells and incubate in 1× PBS without antibody. Process in parallel with the α-H3-stained cells for the rest of the staining procedure. Before sorting, mix unstained cells with the α-H3-stained cells to a final ratio of 1–2% unstained. The unstained cells guide the gating of FITC-positive cells.

3.7. Expected results

As stated above, >95% pure mitotic cells should be attainable using this protocol. Representative recovery of sorted cells (# of cells recovered/total cells sorted) ranges from 11% to 45%, with an average recovery of 23%. Actual cells recovered as a percentage of flow cytometer counts are >85%, averaging 98%. Our average sorting rate is 29×10^6 cells/h, while our average recovery rate is 7×10^6 cells/h.

Additional Notes:

ChIP performed with sorted cells should be performed with biotinylated antibodies and streptavidin-coated beads rather than protein A or G beads to avoid pulling down the antibodies used for sorting.

4. SUMMARY

Here, we describe a protocol for isolation of *Drosophila* S2 cells from every stage of the cell cycle suitable for ChIP.

ACKNOWLEDGMENTS

This work was supported by NIGMS (GM078456-01, GM078456-04S1).

REFERENCES

Banfalvi, G. (2011). Synchronization of mammalian cells and nuclei by centrifugal elutriation. *Methods in Molecular Biology, 761*, 25–45.

Banfalvi, G., Trencsenyi, G., Ujvarosi, K., Nagy, G., Ombodi, T., Bedei, M., et al. (2007). Supranucleosomal organization of chromatin fibers in nuclei of Drosophila S2 cells. *DNA and Cell Biology, 26*, 55–62.

Blobel, G. A., Kadauke, S., Wang, E., Lau, A. W., Zuber, J., Chou, M. M., et al. (2009). A reconfigured pattern of MLL occupancy within mitotic chromatin promotes rapid transcriptional reactivation following mitotic exit. *Molecular Cell, 36*, 970–983.

Coulter, B. (2007–2009). The JE-5.0 Elutriation System for use with Avanti J-20XP Series, J-26-XP Series, and Jj6-MI Centrifuges. *Instruction manual*. JE5-IM-13. Fullerton, CA: Beckman Instruments Inc.

Fox, M. H. (2004). Methods for synchronizing mammalian cells. *Methods in Molecular Biology, 241,* 11–16.
Hengstschlager, M., Pusch, O., Soucek, T., Hengstschlager-Ottnad, E., & Bernaschek, G. (1997). Quality control of centrifugal elutriation for studies of cell cycle regulations. *Biotechniques, 23*(232–4), 236–237.
Kauffman, M. G., Noga, S. J., Kelly, T. J., & Donnenberg, A. D. (1990). Isolation of cell cycle fractions by counterflow centrifugal elutriation. *Analytical Biochemistry, 191,* 41–46.
Lindahl, P. E. (1948). Principle of a counter-streaming centrifuge for the separation of particles of different sizes. *Nature, 161,* 648.
Manukyan, A., Abraham, L., Dungrawala, H., & Schneider, B. L. (2011). Synchronization of yeast. *Methods in Molecular Biology, 761,* 173–200.
Sanderson, R. J., & Bird, K. E. (1977). Cell separations by counterflow centrifugation. *Methods in Cell Biology, 15,* 1–14.
Terasima, T., & Tolmach, L. J. (1963). Growth and nucleic acid synthesis in synchronously dividing populations of HeLa cells. *Experimental Cell Research, 30,* 344–362.
Young, T. J., & Kirchmaier, A. L. (2012). Cell cycle regulation of silent chromatin formation. *Biochimica et Biophysica Acta, 1819,* 303–312.

CHAPTER TWELVE

Genome-Wide Polyadenylation Site Mapping

Vicent Pelechano[1], Stefan Wilkening[1], Aino Inkeri Järvelin, Manu M. Tekkedil, Lars M. Steinmetz[2]

Genome Biology Unit, European Molecular Biology Laboratory, Heidelberg, Germany
[1]Equal contribution to this project.
[2]Corresponding author: e-mail address: lars.steinmetz@embl.de

Contents

1. Introduction	272
2. General Recommendations	275
3. Sample Preparation	276
3.1 Extraction of total RNA	276
3.2 Removal of DNA contamination	280
3.3 Fragmentation of total RNA	281
3.4 Purification of the fragmented RNA	281
4. Library Construction	282
4.1 Reverse transcription	282
4.2 Purification of the cDNA	282
4.3 Second-strand synthesis	283
4.4 Purification of the double-stranded cDNA	283
4.5 Capture of the 3′ terminal cDNA fragments	284
4.6 End repair	285
4.7 Addition of dA overhang	285
4.8 Ligation of the sequencing adapter	286
4.9 Enrichment PCR	286
4.10 PCR purification	287
4.11 Size selection	288
5. Bioinformatic Analysis of Polyadenylation Site Sequencing Reads	289
5.1 Demultiplexing and recovering polyadenylated reads	289
5.2 Alignment of trimmed reads to a reference genome	289
5.3 Optional: Filtering for high-confidence reads	289
5.4 Assignment of reads to a reference transcriptome and differential expression analysis	290
5.5 Further steps for downstream biological analyses	291
6. Quality Control	291
6.1 Library size	291
6.2 Library cloning and Sanger sequencing	292

Methods in Enzymology, Volume 513
ISSN 0076-6879
http://dx.doi.org/10.1016/B978-0-12-391938-0.00012-4

© 2012 Elsevier Inc.
All rights reserved.

6.3 Use of *in vitro* transcripts as internal controls	292
7. Preparation of Double-Stranded Linkers	293
8. Summary	294
Acknowledgments	294
References	294

Abstract

Alternative polyadenylation site usage gives rise to variation in 3′ ends of transcripts in diverse organisms ranging from yeast to human. Accurate mapping of polyadenylation sites of transcripts is of major biological importance, since the length of the 3′UTR can have a strong influence on transcript stability, localization, and translation. However, reads generated using total mRNA sequencing mostly lack the very 3′ end of transcripts. Here, we present a method that allows simultaneous analysis of alternative 3′ ends and transcriptome dynamics at high throughput. By using transcripts produced *in vitro*, the high precision of end mapping during the protocol can be controlled. This method is illustrated here for budding yeast. However, this method can be applied to any natural or artificially polyadenylated RNA.

1. INTRODUCTION

Transcription is a complex and highly regulated process in which isoform variation and chromatin organization play a pivotal role (Di Giammartino, Nishida, & Manley, 2011; Li, Carey, & Workman, 2007; Luco, Allo, Schor, Kornblihtt, & Misteli, 2011; Venters & Pugh, 2009). Recent genome-wide investigations have revealed an unexpected diversity of transcripts in eukaryotic RNA populations (Carninci, 2010; Jacquier, 2009). This highlights the importance of developing new techniques for the study of transcriptional features that cannot be identified using tiling arrays (David et al., 2006; Xu et al., 2009) or conventional RNA-seq methods (Nagalakshmi et al., 2008; Wang, Gerstein, & Snyder, 2009). One of the transcriptional features requiring specialized methods is the identification of polyadenylation site (PAS) usage during transcription termination.

The exact mapping of the PAS provides relevant biological information as it influences the length of the 3′UTR, where multiple signals that regulate transcript stability, localization, and translation efficiency are located (reviewed in Andreassi & Riccio, 2009; Di Giammartino et al., 2011;

Lutz & Moreira, 2011). Transcription termination site selection has been an active area of research for many years (Proudfoot, 2011). However, only recent genome-wide studies have revealed the dimensions of alternative transcript end usage during several biological processes, spanning from development (Ji, Lee, Pan, Jiang, & Tian, 2009; Mangone et al., 2010) to environmental response (Yoon & Brem, 2010) and cancerous transformation (Mayr & Bartel, 2009) in various organisms. Moreover, there appears to be cross talk between PAS usage and chromatin structure, as transcript 3'UTRs are depleted of nucleosomes in both human (Spies, Nielsen, Padgett, & Burge, 2009) and yeast (Mavrich et al., 2008). This depletion, however, is much lower than the one associated with the transcription start sites, and it has been previously overestimated due to technical reasons (Fan et al., 2010).

Standard mRNA sequencing methods are quite ineffective for PAS mapping. Assuming a random fragmentation, the fragments obtained from the ends of cDNA molecules are often shorter than those obtained from internal parts and are subsequently more likely to be lost during library preparation. Additionally, both sequencing and mapping favor fragments without the poly(A) tail. Therefore, different sequencing methods have recently been developed to sequence the 3' terminal fraction of mRNAs (Beck et al., 2010; Fox-Walsh, Davis-Turak, Zhou, Li, & Fu, 2011; Ozsolak et al., 2010; Shepard et al., 2011; Yoon & Brem, 2010; Jan, Friedman, Ruby, & Bartel, 2011; Derti et al., 2012). Most of these methods, however, do not perform well at mapping the precise PAS. This is mainly due to the presence of the poly (A) tail that is generally used to prime the first-strand cDNA synthesis. When reading from the 3' end through the poly(A) tail, the presence of this homopolymeric sequence causes a drastic decrease in sequencing quality and thus a drop in the method's performance. The simplest solution to this problem is to begin sequencing from the 5' end of a fragment until the poly(A) site is reached (Beck et al., 2010; Fox-Walsh et al., 2011). However, in this case, only a fraction of the reads will contain the exact PAS. An alternative possibility is to begin sequencing from the poly(A) tail into the 3'UTR by shortening the poly(A) tail (Jan, Friedman, Ruby, & Bartel, 2011) or priming the poly(A) tail before the sequencing reaction (Ozsolak et al., 2010; Shepard et al., 2011; Wilkening et al., in preparation).

Here, we present a straightforward method designed to simultaneously obtain a genome-wide map of PASs and quantification of gene expression

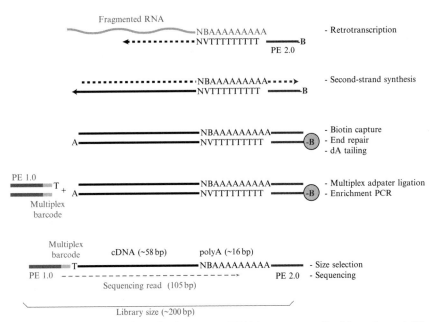

Figure 12.1 Protocol outline. The fragmented RNA is retrotranscribed from the poly(A) tail using a sequencing-compatible oligo(dT) primer. The second strand is generated, and the 3′ terminal fragments are immobilized onto magnetic beads. A multiplexed adapter compatible with Illumina sequencing is added to the 5′ end and an enrichment PCR is performed. After a stringent size selection, the library should have the displayed structure allowing the identification of the PAS.

levels (Fig. 12.1). This protocol uses total RNA as input and can be performed in 2 days. It has been designed to enable simultaneous handling of multiple samples and can be easily scaled up using multipipettes and 96-well plates. Although it has been designed for the Illumina sequencing platform, it can be adapted to any other technology. The first step includes chemical fragmentation of the total RNA, followed by production of cDNA primed from the poly(A) tail with a sequencing-compatible oligo(dT) primer. In the second step, the cDNA is converted to double-stranded cDNA, and the 3′ terminal fragments are immobilized on magnetic beads. Then a multiplexed adapter compatible with Illumina sequencing is ligated to the 5′ end, and an enrichment PCR is performed. Finally, a stringent size selection of the library assures that sequencing from the 5′ end toward the 3′ end will reach the poly(A) tail and hence be informative as to both the cDNA identity and the exact position of the PAS.

2. GENERAL RECOMMENDATIONS

- We use 10 μg total RNA as starting amount for this protocol, but we have tested different amounts as low as 500 ng total RNA without significant decrease in quality.
- This protocol has been designed to process multiple samples simultaneously. It is advisable to use 0.2-mL PCR strips or 96-well plates for library preparation in combination with multichannel pipettes.
- A thermocycler with a heated lid is used for all incubations.
- For the cleanups with magnetic beads in 0.2-mL PCR strips, we use handmade magnetic stands, which consist of neodymium magnets (Webcraft GmbH, Gottmadingen, Germany) mounted on trimmed 96-well plates (Fig. 12.2).
- Throughout the protocol, different volume-to-sample ratios of Agencourt AMPure XP beads (Beckman Coulter Genomics) are used.

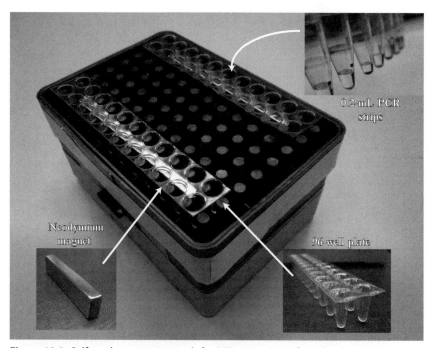

Figure 12.2 Self-made magnetic stands for PCR strips. Bars of neodymium magnets and trimmed 96-well plates are assembled on the bottom of a pipette tip box. The trimmed 96-well plates serve as a support for fixing the 0.2-mL PCR tube strips during the magnetic separation.

The different ratios between bead solution and sample alter the final polyethylene glycol concentration (present in the bead buffer), resulting in a size-specific capture. In general, lower amounts of bead solution will capture longer DNA molecules (Lundin, Stranneheim, Pettersson, Klevebring, & Lundeberg, 2010).

- It is important to note that the Illumina sequencer uses the first few sequencing cycles to call the different clusters and calibrate the base calling. For this reason, it is critical that the first bases to be sequenced are different between clusters. Hence, we recommend the use of four or more multiplexed adapters with barcode sequences as balanced as possible for each sequencing cycle (with respect to ratios of A, C, G, and T). If only one sample is processed, one can either split it between different barcodes (technical replicates) or use an alternative linker without barcode sequence (P5-1T, see Table 12.1). Otherwise, the clustering will be of poor quality resulting in a very low number of called clusters.
- We have applied this protocol to human and yeast samples, but in principle, any naturally or artificially polyadenylated RNA (e.g., prokaryotic mRNA) can be used as input.

3. SAMPLE PREPARATION

3.1. Extraction of total RNA

This protocol uses total RNA as input. Here, we provide a protocol suitable for *S. cerevisiae* RNA extraction, but any RNA extraction protocol that yields high-quality RNA is suitable.

1. Harvest 100 mL of *S. cerevisiae* culture at the desired OD_{600} (e.g., $OD_{600} \sim 1$) by centrifugation at room temperature.
2. Snap freeze the cell pellets in liquid N_2. Store the cells at $-80\,^{\circ}C$ or proceed directly to RNA extraction.
3. Prepare 2-mL screw-capped tubes containing 300 µL glass beads (425–600 µM, Sigma) and 200 µL phenol:chloroform:isoamyl alcohol (25:24:1).
4. Resuspend the cells in 200 µL of lysis buffer (50 mM Tris–HCl pH 7.5, 130 mM NaCl, 5 mM EDTA, and 5% SDS) and transfer them immediately to the 2-mL screw-capped tubes.
5. Disrupt the cells by vortexing the samples with a Fastprep-24 instrument (MP Biomedical). Perform two pulses of 10 s each at 4 °C using

Table 12.1 DNA oligos

Name	Sequence (5' to 3')
BioP7_dT16VN	CAAGCAGAAGACGGCATACGAGATCGGTCTCGGCATTCCTGCTGAACCGCTCTTCCGATCTTTTTTTTTTTTTTTVN
P5-1TF	AATGATACGGCGACCACCGAGATCTACACTCTTTCCCTACACGACGCTCTTCCGATCT
P5Mpx-1TF1	AATGATACGGCGACCACCGAGATCTACACTCTTTCCCTACACGACGCTCTTCCGATCT<u>CGTGATT</u>
P5Mpx-1TF2	AATGATACGGCGACCACCGAGATCTACACTCTTTCCCTACACGACGCTCTTCCGATCT<u>AAGCTAT</u>
P5Mpx-1TF3	AATGATACGGCGACCACCGAGATCTACACTCTTTCCCTACACGACGCTCTTCCGATCT<u>GTAGCCT</u>
P5Mpx-1TF4	AATGATACGGCGACCACCGAGATCTACACTCTTTCCCTACACGACGCTCTTCCGATCT<u>TACAAGT</u>
P5Mpx-1TF5	AATGATACGGCGACCACCGAGATCTACACTCTTTCCCTACACGACGCTCTTCCGATCT<u>ACATCGT</u>
P5Mpx-1TF6	AATGATACGGCGACCACCGAGATCTACACTCTTTCCCTACACGACGCTCTTCCGATCT<u>GCCTAAT</u>
P5Mpx-1TF7	AATGATACGGCGACCACCGAGATCTACACTCTTTCCCTACACGACGCTCTTCCGATCT<u>TGGTCAT</u>
P5Mpx-1TF8	AATGATACGGCGACCACCGAGATCTACACTCTTTCCCTACACGACGCTCTTCCGATCT<u>CACTGTT</u>
P5Mpx-1TF9	AATGATACGGCGACCACCGAGATCTACACTCTTTCCCTACACGACGCTCTTCCGATCT<u>ATTGGCT</u>

Continued

Table 12.1 DNA oligos—cont'd

Name	Sequence (5' to 3')
P5Mpx-1TF10	AATGATACGGCGACCACCGAGATCTACACTCTTTCCCTACACGACGCTCTTCCGATCT<u>GATCTGT</u>
P5Mpx-1TF11	AATGATACGGCGACCACCGAGATCTACACTCTTTCCCTACACGACGCTCTTCCGATCT<u>TTAATTT</u>
P5Mpx-1TF12	AATGATACGGCGACCACCGAGATCTACACTCTTTCCCTACACGACGCTCTTCCGATCT<u>CCTCCCT</u>
P5-1TR	[Phos]GATCGGAAGAGCGTCGTGTAGGGAAAGAGTGTAGATCTCGGTGGTCGCCGTATCATT[AmC7]
P5Mpx-1TR1	[Phos]<u>ATCACG</u>AGATCGGAAGAGCGTCGTGTAGGGAAAGAGTGTAGATCTCGGTGGTCGCCGTATCATT[AmC7]
P5Mpx-1TR2	[Phos]<u>TAGCTT</u>AGATCGGAAGAGCGTCGTGTAGGGAAAGAGTGTAGATCTCGGTGGTCGCCGTATCATT[AmC7]
P5Mpx-1TR3	[Phos]<u>GGCTAC</u>AGATCGGAAGAGCGTCGTGTAGGGAAAGAGTGTAGATCTCGGTGGTCGCCGTATCATT[AmC7]
P5Mpx-1TR4	[Phos]<u>CTTGTA</u>AGATCGGAAGAGCGTCGTGTAGGGAAAGAGTGTAGATCTCGGTGGTCGCCGTATCATT[AmC7]
P5Mpx-1TR5	[Phos]<u>CGATGT</u>AGATCGGAAGAGCGTCGTGTAGGGAAAGAGTGTAGATCTCGGTGGTCGCCGTATCATT[AmC7]
P5Mpx-1TR6	[Phos]<u>TTAGGC</u>AGATCGGAAGAGCGTCGTGTAGGGAAAGAGTGTAGATCTCGGTGGTCGCCGTATCATT[AmC7]

P5Mpx-1TR7	[Phos]TGACCAAGATCGGAAGAGCGTCGTGTAGGGAAAGAGTGTAGATCTCGGTGGTCGCCGTATCATT[AmC7]
P5Mpx-1TR8	[Phos]ACAGTGAGATCGGAAGAGCGTCGTGTAGGGAAAGAGTGTAGATCTCGGTGGTCGCCGTATCATT[AmC7]
P5Mpx-1TR9	[Phos]GCCAATAGATCGGAAGAGCGTCGTGTAGGGAAAGAGTGTAGATCTCGGTGGTCGCCGTATCATT[AmC7]
P5Mpx-1TR10	[Phos]CAGATCAGATCGGAAGAGCGTCGTGTAGGGAAAGAGTGTAGATCTCGGTGGTCGCCGTATCATT[AmC7]
P5Mpx-1TR11	[Phos]AATTAAAGATCGGAAGAGCGTCGTGTAGGGAAAGAGTGTAGATCTCGGTGGTCGCCGTATCATT[AmC7]
P5Mpx-1TR12	[Phos]GGGAGGAGATCGGAAGAGCGTCGTGTAGGGAAAGAGTGTAGATCTCGGTGGTCGCCGTATCATT[AmC7]
*PE 1.0 Illumina	AATGATACGGCGACCACCGAGATCTACACTCTTTCCCTACACGACGCTCTTCCGATCT
*PE 2.0 Illumina	CAAGCAGAAGACGGCATACGAGATCGGTCTCGGCATTCCTGCTGAACCGCTCTTCCGATCT

Sequences of the oligos used for this protocol. The multiplex barcode sequences are underlined. All the oligos are HPLC purified (Sigma), and modifications are marked as follows: [Biot] 5′Biotin, [Phos] 5′Phosphorilation, [AmC7] 3′Amine C7, B = (G + T + C), and V = (A + C + G).
*Oligonucleotide sequences © 2006–2012 Illumina, Inc.

a shaking velocity of 6 m/s. Alternatively, disrupt the cells using a conventional benchtop vortex with a tube adapter for 20 min at 4 °C at maximum velocity.
6. Centrifuge the tubes at 13,000 rpm in a benchtop microcentrifuge for 15 min at 4 °C.
7. Transfer the upper aqueous phase to Phase Lock Gel Heavy tubes (Eppendorf) containing 200 μL phenol:chloroform:isoamyl alcohol (25:24:1).
8. Shake the tubes vigorously and centrifuge at 13,000 rpm in a benchtop microcentrifuge for 10 min at 4 °C.
9. Transfer the upper aqueous phase to a new Phase Lock Gel Heavy tube containing 200 μL chloroform:isoamyl alcohol (24:1).
10. Shake the tubes vigorously and centrifuge at 13,000 rpm in a benchtop microcentrifuge for 10 min at 4 °C.
11. Take the upper aqueous layer (100–130 μL) and transfer to fresh, precooled 1.5-mL microtubes.
12. Precipitate the samples by adding 1/10 volume of 3 M sodium acetate (pH 4.2) and 2.5 volumes of absolute ethanol. Mix and incubate for a minimum of 30 min at -20 °C.
13. Centrifuge the tubes at 13,000 rpm in a benchtop microcentrifuge for 15 min at 4 °C and remove the supernatant.
14. Wash the pellet with 200 μL of 70% ethanol and spin 2 min at full speed.
15. Air-dry the pellet for 5–10 min.
16. Resuspend the pellet in 100 μL RNase-free water and store the RNA at -80 °C.
17. Determine the concentration of total RNA by UV spectrophotometry (e.g., NanoDrop from Thermo Scientific). The absorbance ratio $A_{260/280}$ should be approximately 2.
18. To check the RNA integrity, run a 2% agarose gel or use an RNA Bioanalyzer (Agilent Technologies). Discrete bands for tRNA and small and big subunits of ribosomal RNA should be visible.

3.2. Removal of DNA contamination

It is necessary to remove the DNA carryover from the RNA extraction in order to obtain good quality RNA. A standard procedure is to incubate the sample with RNase-free DNase. We use the TURBO DNA-free kit (Ambion), but any RNase-free DNase treatment should be suitable.

1. Mix 20 μg of total RNA with 5 μL of 10× TURBO DNA-free buffer, 3 μL TURBO DNase (Turbo DNA-free kit, Ambion), and RNase-free water up to 50 μL.
2. Incubate the sample for 30 min at 37 °C.
3. Add 5 μL of Turbo DNA-free inactivation reagent and mix.
4. Incubate the sample at room temperature for 5 min. Flick the tube to mix frequently during the incubation.
5. Centrifuge the tubes at 13,000 rpm in a benchtop microcentrifuge for 2 min and carefully recover the supernatant without touching the beads.
6. Measure the RNA concentration using Qubit (Invitrogen, Life Technologies) or any other method not influenced by the presence of free nucleotides. Alternatively, purify the sample by removing the free nucleotides and measure the RNA concentration using an UV spectrophotometer.

3.3. Fragmentation of total RNA

As this protocol is designed to sequence the terminal fragments of the RNA molecules, the first step involves the fragmentation of the total RNA, from which the 3' terminal mRNA fragments are selected at later steps.

1. Mix 10 μg of DNA-free total RNA with 4 μL of 5× RNA fragmentation buffer (200 mM Tris–acetate, pH 8.1, 500 mM KOAc, 150 mM MgOAc) in 0.2-mL PCR strips or a 96-well plate. Optionally, add 4 μL spike-in IVT stock solution for quality control (see below).
2. Adjust the total volume to 20 μL with RNase-free water and incubate the sample at 80 °C for 5 min.
3. Transfer the sample to ice immediately and proceed to the purification step.

3.4. Purification of the fragmented RNA

Use Agencourt RNAClean XP beads (Beckman Coulter Genomics) to purify the fragmented RNA according to the manufacturer's instructions.

1. Add 36 μL of Agencourt RNAClean XP beads (1.8× sample volume) and mix by pipetting.
2. Incubate the samples for 5 min at room temperature to allow binding.
3. Place the tubes on a magnetic stand for 5 min.
4. Discard the supernatant.
5. Add 200 μL of 70% EtOH to the tubes in the magnetic stand.
6. Wait for 30 s and remove the supernatant.

7. Repeat the wash by adding 200 μL of 70% EtOH to the tubes in the magnetic stand.
8. Wait for 30 s and remove the supernatant.
9. Air-dry the beads for 5 min.
10. Elute the fragmented RNA by resuspending the beads in 12 μL RNase-free water and mix thoroughly by pipetting up and down.
11. Place the tubes on a magnetic stand for 1 min and recover the supernatant.

4. LIBRARY CONSTRUCTION

4.1. Reverse transcription

The first step of the library preparation protocol consists of synthesizing biotinylated cDNA molecules by priming the poly(A) tail with a sequencing-compatible oligo dT primer (compatible with the PE 2.0, Illumina).

1. Mix 11.2 μL of the fragmented RNA with 1 μL biotinylated oligo dT primer (BioP7_dT16VN at 1 $\mu M = 2.5$ pmol, see Table 12.1) and 1 μL of 10 mM dNTP Mix. If less starting material is used, the amount of oligo dT primer should be adjusted accordingly.
2. Incubate the sample for 5 min at 65 °C to disrupt secondary structures and place on ice.
3. Add 4 μL of 5× First-Strand Buffer (Invitrogen), 2 μL DTT (0.1 M), 0.32 μL actinomycin D (1.25 mg/μL), and 0.5 μL RNasin Plus (Promega) to each sample.
4. Mix each tube and put in a thermocycler at 42 °C.
5. After 2 min, add 0.5 μL Superscript II (200 U/μL, Invitrogen) to each tube. It is critical to do so at 42 °C to prevent mispriming of the anchored oligo(dT) primer.
6. Incubate the samples at 42 °C for 50 min and then inactivate the enzyme by incubating the samples at 72 °C for 15 min.

4.2. Purification of the cDNA

A cleanup with Agencourt AMPure XP beads (Beckman Coulter Genomics) is performed to remove the excess of unused biotinylated oligo.

1. Add 30 μL of Agencourt AMPure XP beads (1.5 × sample volume) and mix by pipetting.
2. Incubate the samples for 5 min at room temperature to allow binding.
3. Place the tubes on a magnetic stand for 5 min.

4. Discard the supernatant.
5. Add 200 μL of 70% EtOH to the tubes in the magnetic stand.
6. Wait for 30 s and remove the supernatant.
7. Repeat the wash by adding 200 μL of 70% EtOH to the tubes in the magnetic stand.
8. Wait for 30 s and remove the supernatant.
9. Air-dry the beads for 1 min.
10. Elute the cDNA by resuspending the beads in 40 μL of 10 mM Tris–HCl pH 8.0 and mix thoroughly by pipetting up and down.
11. Place the tubes on a magnetic stand and recover the supernatant.

4.3. Second-strand synthesis

To generate the second strand of cDNA, we use RNase H, which nicks the RNA in the cDNA/RNA hybrid, and DNA polymerase I, which produces the second strand of the cDNA by nick translation.

1. Mix 40 μL cDNA from the previous step with 5 μL of 10× DNA polymerase I buffer (Fermentas) and 2.5 μL dNTPs (10 mM) on ice.
2. Add 0.5 μL RNaseH (5 U/μL, NEB) and 2 μL DNA polymerase I (10 U/μL, Fermentas).
3. Incubate the tubes at 16 °C for 2.5 h.

4.4. Purification of the double-stranded cDNA

A cleanup with Agencourt AMPure XP beads (Beckman Coulter Genomics) is performed to remove the enzymes and the short DNA fragments.

1. Add 45 μL of Agencourt AMPure XP beads (0.9 × sample volume) and mix by pipetting.
2. Incubate the samples for 5 min at room temperature to allow binding.
3. Place the tubes on a magnetic stand for 5 min.
4. Discard the supernatant.
5. Add 200 μL of 70% EtOH to the tubes in the magnetic stand.
6. Wait for 30 s and remove the supernatant.
7. Repeat the wash by adding 200 μL of 70% EtOH to the tubes in the magnetic stand.
8. Wait for 30 s and remove the supernatant.
9. Air-dry the beads for 1 min.
10. Elute the sample by resuspending the beads in 20 μL of 10 mM Tris–HCl pH 8.0 and mix thoroughly by pipetting up and down.
11. Place the tubes on a magnetic stand and recover the supernatant.

4.5. Capture of the 3′ terminal cDNA fragments

Once the cDNA fragments have been produced, it is necessary to select those molecules that contain the biotinylated sequencing primer at their 3′ ends. This is accomplished by using magnetic beads coated with streptavidin. It is important to remember that from this step until the enrichment PCR, the sample is bound to the magnetic beads, unlike in the Ampure bead purification where the sample is in the supernatant.

4.5.1 Preparation of streptavidin beads

First, it is necessary to prepare the magnetic beads according to the manufacturer's instructions.

1. Use 20 μL Dynabeads M-280 Streptavidin (Invitrogen) for each sample (good for binding 4 pmol of biotinylated oligo(dT)).
2. Transfer the tubes to a magnetic stand and remove the supernatant.
3. Take the tubes from the magnetic stand and resuspend the beads in 200 μL of 1× bind and wash buffer (1× B&W buffer, containing 5 mM Tris–HCl pH 7.5, 0.5 mM EDTA, and 1 M NaCl).
4. Transfer the tubes to a magnetic stand and remove the supernatant.
5. Repeat the wash with 200 μL of 1× B&W buffer.
6. Transfer the tubes to a magnetic stand and remove the supernatant.
7. Resuspend the beads in 20 μL of 2× B&W buffer (10 mM Tris–HCl pH 7.5, 1 mM EDTA, and 2 M NaCl) and keep them ready for the next step.

4.5.2 Binding the sample to streptavidin beads

Once the beads are ready, we use them to capture the biotinylated double-stranded cDNA.

1. Mix the 20 μL double-stranded cDNA with the 20 μL washed Dynabeads M-280 Streptavidin beads.
2. Incubate the samples at room temperature in a rotating wheel or a benchtop mixer (at 400 rpm) for 15–30 min to allow binding.
3. Transfer the tubes to a magnetic stand and remove the supernatant containing the unbiotinylated DNA.
4. Take the tubes from the magnetic stand and resuspend the beads in 200 μL of 1× B&W buffer.
5. Transfer the tubes to a magnetic stand and remove the supernatant.
6. Repeat the wash with 200 μL of 1× B&W buffer.
7. Transfer the tubes to a magnetic stand and remove the supernatant.

8. Take the tubes from the magnetic stand and resuspend the beads in 200 μL of 10 mM Tris–HCl pH 8.0 to remove the excess salt.
9. Transfer the tubes to a magnetic stand and remove the supernatant.
10. Resuspend the beads in 21 μL of 10 mM Tris–HCl pH 8.0.

4.6. End repair

Once the 3' terminal cDNA molecules have been captured, it is necessary to add a common sequencing adapter to the unbiotinylated end. For this, we follow the same procedure used in standard library constructions by Illumina, while making use of the fact that our samples are bound to the magnetic beads. The first step of this process is to produce blunt end molecules.

1. Mix the sample containing Dynabeads from the previous step with 2.5 μL of 10× End repair buffer and 1.25 End repair enzyme mix (NEBNext DNA Sample Prep Master Mix Set 1, NEB).
2. Incubate the tubes for 30 min at 20 °C.
3. Transfer the tubes to a magnetic stand and remove the supernatant.
4. Take the tubes from the magnetic stand and resuspend the beads in 200 μL of 1× B&W buffer.
5. Transfer the tubes to a magnetic stand and remove the supernatant.
6. Repeat the wash with 200 μL of 1× B&W buffer.
7. Transfer the tubes to a magnetic stand and remove the supernatant.
8. Take the tubes from the magnetic stand and resuspend the beads in 200 μL of 10 mM Tris–HCl pH 8.0 to remove the excess of salt.
9. Transfer the tubes to a magnetic stand and remove the supernatant.
10. Resuspend the beads in 21 μL of 10 mM Tris–HCl pH 8.0.

4.7. Addition of dA overhang

In order to produce sticky ends for the ligation of the sequencing adapter, it is necessary to introduce a common 3' adenine overhang.

1. Mix the sample containing Dynabeads from the previous step with 2.5 μL dA tailing buffer (10× NEBuffer 2 from NEB, supplemented with 0.2 mM dATP) and 1.5 μL Klenow fragment (3'→5' exo-) (5 U/μL, NEB).
2. Incubate the tubes for 30 min at 37 °C.
3. Transfer the tubes to a magnetic stand and remove the supernatant.
4. Take the tubes from the magnetic stand and resuspend the beads in 200 μL of 1× B&W buffer.
5. Transfer the tubes to a magnetic stand and remove the supernatant.
6. Repeat the wash with 200 μL of 1× B&W buffer.

7. Transfer the tubes to a magnetic stand and remove the supernatant.
8. Take the tubes from the magnetic stand and resuspend the beads in 200 μL of 10 mM Tris–HCl pH 8.0 to remove the excess of salt.
9. Transfer the tubes to a magnetic stand and remove the supernatant.
10. Resuspend the beads in 8 μL of 10 mM Tris–HCl pH 8.0.

4.8. Ligation of the sequencing adapter

The 3′ A-overhang produced in the previous step is now used to ligate the sequencing adapter (compatible with the PE 1.0 from Illumina) having a 3′ T-overhang. This sequencing adapter contains a barcode to allow sample multiplexing (see below).

1. Thaw the P5Mpx double-stranded linker on ice.
2. Mix the sample containing Dynabeads from the previous step with 12.5 μL of 2× Quick Ligation buffer (NEB), 2 μL P5Mpx double-stranded linker (2.5 μM), and 2.5 μL T4 DNA ligase (2000 U/μL, NEB). If less than 10 μg of starting material is used, the adapter concentration should be adapted accordingly to avoid adapter self-ligation.
3. Incubate the samples for 15–30 min at 20 °C.
4. Transfer the tubes to a magnetic stand and remove the supernatant.
5. Take the tubes from the magnetic stand and resuspend the beads in 200 μL of 1× B&W buffer.
6. Transfer the tubes to a magnetic stand and remove the supernatant.
7. Repeat the wash with 200 μL of 1× B&W buffer.
8. Transfer the tubes to a magnetic stand and remove the supernatant.
9. Resuspend the beads in 200 μL of 1× B&W and transfer the samples to new tubes.
10. Transfer the tubes to a magnetic stand and remove the supernatant.
11. Repeat the wash with 200 μL of 1× B&W buffer.
12. Transfer the tubes to a magnetic stand and remove the supernatant.
13. Take the tubes from the magnetic stand and resuspend the beads in 200 μL of 10 mM Tris–HCl pH 8.0 to remove the excess salt.
14. Transfer the tubes to a magnetic stand and remove the supernatant.
15. Resuspend the beads in 50 μL of 10 mM Tris–HCl pH 8.0.

4.9. Enrichment PCR

To enrich for molecules that contain the correct combination of sequencing adapters and to obtain fragments that are not bound to beads, the samples are amplified by PCR.

1. Since an excess of input DNA inhibits PCR amplification, we use only part of the beads (usually ½) for the PCR amplification. The optimal proportion and cycle number should be tested for each setup individually.
2. Mix 24 μL beads with DNA with 25 μL of 2 × Phusion Master Mix with HF buffer (Thermo Fisher Scientific) and 0.5 μL each of oligo PE1.0 and PE2.0 (10 μM, Illumina).
3. Incubate the sample in a thermocycler with the following program: 30 s at 98 °C, 18 cycles (10 s at 98 °C, 10 s at 65 °C, and 10 s at 72 °C) and a final extension of 5 min at 72 °C.
4. Transfer the PCR product to a magnetic stand and recover the 50 μL supernatant containing the PCR product.
5. The beads used for the PCR should be kept at 4 °C in case a reamplification is needed.

4.10. PCR purification

A cleanup with Agencourt AMPure XP beads (Beckman Coulter Genomics) is performed to remove the oligos and to have the sample in a low-salt buffer for the subsequent size selection. High ion concentration in the buffer would affect the migration of DNA in the gel.

1. Add 90 μL of Agencourt AMPure XP beads (1.8 × sample volume) and mix by pipetting.
2. Incubate the samples for 5 min at room temperature to allow binding.
3. Place the tubes on a magnetic stand for 5 min.
4. Discard the supernatant.
5. Add 200 μL of 70% EtOH to the tubes in the magnetic stand.
6. Wait for 30 s and remove the supernatant.
7. Repeat the wash by adding 200 μL of 70% EtOH to the tubes in the magnetic stand.
8. Wait for 30 s and remove the supernatant.
9. Air-dry the beads for 1 min.
10. Elute the sample in 25 μL of 10 mM Tris–HCl pH 8.0 and mix thoroughly by pipetting up and down.
11. Place the tubes on a magnetic stand and recover the supernatant.
12. Quantify the sample concentration using Qubit dsDNA High-Sensitivity reagent (Invitrogen, Life technologies) to ensure that the DNA was amplified and sufficient material (total of ≥ 150 ng) remains for size selection.

4.11. Size selection

Once the libraries have been produced, it is critical to perform a stringent size selection. It is important that the cDNA insert is long enough to map uniquely to the genome, but short enough to allow sequencing into the poly(A) tail. We recommend to perform a size selection of 200 bp for obtaining around 60 bp of cDNA using Illumina 105 bp single-end read sequencing (Fig. 12.1). In principle, this step can be performed using any gel purification method; however, we recommend E-Gel 2% SizeSelect (Invitrogen, Life technologies).

1. If multiplexed samples are used, samples can be pooled, merging equal amounts of DNA libraries.
2. Size select 200 bp libraries using the E-Gel. As this step is critical for the performance of the method, it may be advisable to recover additional, differently sized fractions in case the size estimation is not accurate. Note that a significant amount of the sample is lost during this step.
3. Add 45 μL of Agencourt AMPure XP beads (1.8 × sample volume) to the 25 μL of recovered library and mix by pipetting.
4. Incubate the samples for 5 min at room temperature to allow binding.
5. Place the tubes on a magnetic stand for 5 min.
6. Discard the supernatant.
7. Add 200 μL of 70% EtOH to the tubes in the magnetic stand.
8. Wait for 30 s and remove the supernatant.
9. Repeat the wash by adding again 200 μL of 70% EtOH to the tubes in the magnetic stand.
10. Wait for 30 s and remove the supernatant.
11. Air-dry the beads for 1 min.
12. Elute the sample in 8 μL of 10 mM Tris–HCl pH 8.0 with extensive mixing by pipetting up and down.
13. Place the tubes on a magnetic stand and recover the supernatant.
14. Quantify the sample concentration using Qubit dsDNA High-Sensitivity reagent (Invitrogen, Life technologies). The DNA concentration of the final library should be above 1 ng/μL to ensure accurate quantification for sequencing.
15. The samples are now ready for submitting to a sequencing facility for 105 bp single-end Illumina sequencing. However, it is recommended that additional quality controls be performed before sequencing (see below).

5. BIOINFORMATIC ANALYSIS OF POLYADENYLATION SITE SEQUENCING READS

Here, we describe general guidelines for analyzing the sequencing reads generated by the protocol described above.

5.1. Demultiplexing and recovering polyadenylated reads

If different samples have been multiplexed and sequenced in the same lane, the first step of data processing is to separate the reads into different files based on the sample barcodes. After that, one should keep those reads that run into the poly(A) tail. While in a transcript the poly(A) tail is at the very end, the sequencing read may run into adapters and/or other sequences added to the transcript during library preparation. Hence, reads with long internal adenosine homopolymers should be kept for downstream analyses. The yield of this trimming step depends on the accuracy of fragment size selection.

5.2. Alignment of trimmed reads to a reference genome

After the reads have been trimmed to remove experimentally added sequences (such as barcodes) and nongenomic sequences (such as poly(A) tails), they can be aligned to the reference genome using aligners such as Bowtie (Langmead, Trapnell, Pop, & Salzberg, 2009) or GSNAP (Wu & Nacu, 2010). Since GSNAP supports gapped alignments (useful for reads that cross splice junctions) and allows more randomly distributed mismatches, we recommend it for higher eukaryotes. However, this flexibility comes with a considerable increase in computing time; hence, the faster Bowtie may be suitable for rapid data exploration and performs well in species like yeast where splicing is infrequent.

5.3. Optional: Filtering for high-confidence reads

Since the protocol relies on priming from the poly(A) tail, it is possible that some transcripts or leftover genomic DNA with internal adenosine sequence stretches can cause mispriming. It is possible to filter these reads out by examining read adenosine content and genomic sequence. 3′UTR sequences are often AU-rich; therefore, it is important that the thresholds for these filters are not too stringent.

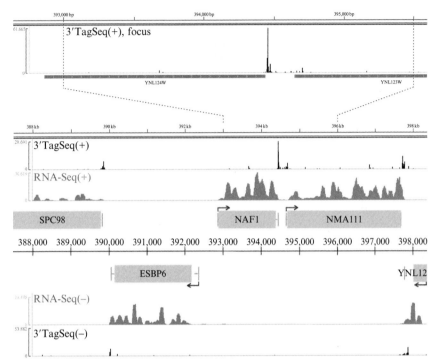

Figure 12.3 Example of polyadenylation site mapping. Example of strand-specific mapping obtained following this protocol for a region of *S. cerevisiae* chromosome 14. Tracks from a strand-specific total RNA-seq protocol (Parkhomchuk et al., 2009) are shown below.

5.4. Assignment of reads to a reference transcriptome and differential expression analysis

After the reads have been aligned to the reference genome, they can be assigned to a known reference transcriptome (available from, e.g., ENSEMBL) in order to ascertain what fraction of reads assign to known transcribed units and quantify their expression. In addition, it may be insightful to categorize read counts based on sequence type, for example, exon, intron, UTR, or protein-coding sequence. Most reads are expected to map to transcript ends (Fig. 12.3). It is worth noting that current annotations may not be complete in terms of mapping alternative transcript boundaries. Hence, one should consider extending the existing annotation boundaries by an appropriate length, for example, ±200 bp for *S. cerevisiae*, when assigning reads in order to account for some of the expected inaccuracies in current annotations.

In each read, the last base (after trimming) represents the last base of the transcript it came from. Thus, it is possible to collapse reads into single-base 3′ end calls and count their expression in a similar manner to counting expression levels of transcripts.

The python package HTSeq software (Anders, available online from http://www-huber.embl.de/users/anders/HTSeq/) is useful for the analyses described above, as it implements many functions for basic read manipulation (such as reading standard sequence file types and easy overlapping of read features with annotation features). Reproducibility of transcript expression level calls between replicates and differential expression analysis of samples profiled in different conditions can then be done; an effective tool for this is the R package DESeq (Anders & Huber, 2010).

5.5. Further steps for downstream biological analyses

After PAS calls have been established, it is important to address their reproducibility by examining the overlap between replicates. Confident calls can be used to estimate PAS usage for transcripts. If data from multiple conditions are available, differential calls could be identified using similar methods to those available for exon usage from count data (Anders, Reyes, & Huber, 2012).

6. QUALITY CONTROL

6.1. Library size

It is critical to check the library size with a Bioanalyzer chip (Agilent Technologies) or a similar device. A discrete peak around 200 bp should be observed (Fig. 12.4).

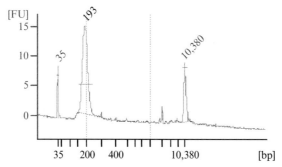

Figure 12.4 Library size distribution. Example of a Bioanalyzer profile from a final 3′RNA-seq library using a high-sensitivity DNA chip (Bioanalyzer 2100, Agilent Technologies). A main peak of approximately 200 bp can be observed.

6.2. Library cloning and Sanger sequencing

If using this protocol for the first time, it is recommended to confirm that the resulting libraries have the proper structure and that the cDNA fragments are sufficiently long to allow gene identification. For that, 1 µL of the final pooled library can be reamplified and cloned using the TOPO TA cloning kit (Invitrogen). Some of the cloned plasmid inserts (5–10) can be sequenced by standard Sanger sequencing. Note that the Phusion enzyme used for the PCR amplification has proofreading activity and thus does not incorporate the 3′ A-overhang necessary for T/A cloning.

6.3. Use of *in vitro* transcripts as internal controls

It is advisable to add *in vitro* transcripts (IVTs) with a poly(A) tail to the total RNA sample before processing it. These sequences can serve as a general quality control for the entire protocol. Additionally, the accuracy of the poly(A) tail mapping can be assessed using the known PAS of the IVT transcripts. In our case, more than 98% of the reads mapped to the IVTs exact PAS.

6.3.1 Preparation of internal controls

We use IVT spike-ins derived from *B. subtilis* (ATCC 87482 (Lys), ATCC 87483 (Phe), and ATCC 87484 (Thr)), but any other polyadenylated IVT of known sequence can be used.

1. Isolate the plasmid containing the IVT template following standard protocols.
2. Linearize the template using an appropriate digestion enzyme according to provider's instructions. In case of *B. subtilis* sequences, use *Not*I digestion.
3. Use the linearized DNA as a template for *in vitro* transcription. In the case of the *B. subtilis* sequences, mix 200 ng of DNA linearized template with 10 µL Transcription Optimized 5× buffer, 5 µL of 0.1 M DTT, 10 µL NTP mix (2.5 mM each), 0.5 µL RNasin Plus (Promega), 2 µL T3 RNA polymerase (Promega), and RNase-free water up to 50 µL.
4. Incubate the sample at 37 °C for 2 h.
5. Degrade the template by adding 5.8 µL of 10× TURBO DNA-free buffer and 2 µL TURBO DNase (TURBO DNA-free kit, Ambion).
6. Incubate the sample for 30 min at 37 °C.
7. Increase the volume of the sample to 200 µL with RNase-free water and transfer it to Phase Lock Gel Heavy tubes (Eppendorf) containing 200 µL phenol:chloroform:isoamyl alcohol (25:24:1).

8. Shake the tubes vigorously and centrifuge at 13,000 rpm in a standard benchtop microcentrifuge for 10 min at 4 °C.
9. Transfer the upper aqueous layer to fresh, precooled 1.5-mL microtubes.
10. Precipitate the samples by adding 1/10 volume of 3 M NaOAc (pH 4.2) and 2.5 volumes of absolute ethanol. Mix and incubate for a minimum of 30 min at -20 °C.
11. Centrifuge the tubes at 13,000 rpm in a standard benchtop microcentrifuge for 15 min at 4 °C and remove the supernatant.
12. Wash the pellet with 200 μL of 80% ethanol and spin 2 min at full speed.
13. Air-dry the pellet.
14. Resuspend in 50 μL RNase-free water and store the RNA at -80 °C.
15. Determine the concentration of the IVTs using Qubit (Invitrogen, Life technologies) and check their size by RNA Bioanalyzer (Agilent Technologies).

6.3.2 Preparation of spike-in IVT stock solution

After the quantification of IVTs, we prepare and aliquot a stock solution containing 5.66 pg/μL pGIBS-LYS (ATCC 87482), 15 pg/μL pGIBS-PHE (ATCC 87483), and 44.583 pg/μL pGIBS-THR (ATCC 87484). As the stock is very dilute, it is recommended to prepare a 100× stock and subsequently prepare the 1× stock by dilution. The RNA should be stored at -80 °C in nonsticky tubes.

7. PREPARATION OF DOUBLE-STRANDED LINKERS

To produce double-stranded DNA linkers, it is necessary to mix the forward and reverse oligos (e.g., P5Mpx-1TF1 and P5Mpx-1TR1, see Table 12.1 below) and subject them to heat denaturation followed by slow cooling to facilitate annealing of the double-stranded molecules.

1. Mix the oligos at 2.5 μM in the presence of 40 mM Tris–HCl pH 8.0 and 50 mM NaCl.
2. Incubate the sample for 5 min at 95 °C and let it cool slowly (-0.1 °C/s) to 65 °C.
3. Incubate the sample for 5 min at 65 °C during and let it cool slowly (-0.1 °C/s) to 4 °C.
4. Store the aliquots at -20 °C. Always thaw the linkers on ice to prevent denaturation.

5. Annealing efficiency can be checked using a 2% agarose gel comparing the migration of the annealed linkers to the individual oligos.

8. SUMMARY

The fate of a transcript strongly depends on the presence and length of its 3′UTR. As the 3′UTRs are underrepresented when sequencing with current RNA-seq methods, specialized methods are required to capture genome-wide PAS. Here, we describe a method for the accurate mapping of 3′ ends of polyadenylated transcripts that also allows quantification of transcript expression levels. When assessing this method in-house using spiked-in IVTs, we confirmed that the vast majority of the reads produced with our 3′ sequencing method map to the correct PAS. The protocol is relatively fast and simple, allowing for sequence library preparation within two working days. The use of barcodes for multiplexing and the capture of the libraries by magnetic beads make this protocol scalable for 96-well library preparation. In conclusion, this method for genome-wide mapping of PASs enables the identification of PAS at single-base resolution and quantification of their alternative usage in different conditions. Additionally, the integration of such data with chromatin features like nucleosome positioning and histone modifications will help elucidate the role of chromatin in shaping alternative 3′UTR usage.

ACKNOWLEDGMENTS

We thank Simon Anders, Vladimir Benes, and the EMBL GeneCore for advice and technical support during the development of this protocol, and to Raeka Aiyar for critical comments on the manuscript. This work was supported by grants to L. M. S. from the National Institutes of Health, the University of Luxembourg–Institute for Systems Biology Program, and the Deutsche Forschungsgemeinschaft. V. P. is supported by an EMBO postdoctoral Fellowship.

REFERENCES

Anders, S., available online. HTSeq: Analysing high-throughput sequencing data with Python. Software available from http://www-huber.embl.de/users/anders/HTSeq/.

Anders, S., & Huber, W. (2010). Differential expression analysis for sequence count data. *Genome Biology, 11*, R106.

Anders, S., Reyes, A., & Huber, W. (2012). Detecting differential usage of exons from RNA-Seq data. *Genome Research* [ahead of print].

Andreassi, C., & Riccio, A. (2009). To localize or not to localize: mRNA fate is in 3′UTR ends. *Trends in Cell Biology, 19*, 465–474.

Beck, A. H., Weng, Z., Witten, D. M., Zhu, S., Foley, J. W., Lacroute, P., et al. (2010). 3'-end sequencing for expression quantification (3SEQ) from archival tumor samples. *PLoS One*, 5, e8768.
Carninci, P. (2010). RNA dust: Where are the genes? *DNA Research*, 17, 51–59.
David, L., Huber, W., Granovskaia, M., Toedling, J., Palm, C. J., Bofkin, L., et al. (2006). A high-resolution map of transcription in the yeast genome. *Proceedings of the National Academy of Sciences of the United States of America*, 103, 5320–5325.
Derti, A., Garrett-Engele, P., MacIsaac, K. D., Stevens, R. C., Sriram1, S., Chen, R., et al. (2012). A quantitative atlas of polyadenylation in five mammals. *Genome Research*, 22, 1173–1183.
Di Giammartino, D. C., Nishida, K., & Manley, J. L. (2011). Mechanisms and consequences of alternative polyadenylation. *Molecular Cell*, 43, 853–866.
Fan, X., Moqtaderi, Z., Jin, Y., Zhang, Y., Liu, X. S., & Struhl, K. (2010). Nucleosome depletion at yeast terminators is not intrinsic and can occur by a transcriptional mechanism linked to 3'-end formation. *Proceedings of the National Academy of Sciences of the United States of America*, 107, 17945–17950.
Fox-Walsh, K., Davis-Turak, J., Zhou, Y., Li, H., & Fu, X. D. (2011). A multiplex RNA-seq strategy to profile poly(A+) RNA: Application to analysis of transcription response and 3' end formation. *Genomics*, 98, 266–271.
Jacquier, A. (2009). The complex eukaryotic transcriptome: Unexpected pervasive transcription and novel small RNAs. *Nature Reviews Genetics*, 10, 833–844.
Jan, C. H., Friedman, R. C., Ruby, J. G., & Bartel, D. P. (2011). Formation, regulation and evolution of *Caenorhabditis elegans* 3'UTRs. *Nature*, 469, 97–101.
Ji, Z., Lee, J. Y., Pan, Z., Jiang, B., & Tian, B. (2009). Progressive lengthening of 3' untranslated regions of mRNAs by alternative polyadenylation during mouse embryonic development. *Proceedings of the National Academy of Sciences of the United States of America*, 106, 7028–7033.
Langmead, B., Trapnell, C., Pop, M., & Salzberg, S. L. (2009). Ultrafast and memory-efficient alignment of short DNA sequences to the human genome. *Genome Biology*, 10, R25.
Li, B., Carey, M., & Workman, J. L. (2007). The role of chromatin during transcription. *Cell*, 128, 707–719.
Luco, R. F., Allo, M., Schor, I. E., Kornblihtt, A. R., & Misteli, T. (2011). Epigenetics in alternative pre-mRNA splicing. *Cell*, 144, 16–26.
Lundin, S., Stranneheim, H., Pettersson, E., Klevebring, D., & Lundeberg, J. (2010). Increased throughput by parallelization of library preparation for massive sequencing. *PLoS One*, 5, e10029.
Lutz, C. S., & Moreira, A. (2011). Alternative mRNA polyadenylation in eukaryotes: An effective regulator of gene expression. *Wiley Interdisciplinary Reviews: RNA*, 2, 23–31.
Mangone, M., Manoharan, A. P., Thierry-Mieg, D., Thierry-Mieg, J., Han, T., Mackowiak, S. D., et al. (2010). The landscape of C. elegans 3'UTRs. *Science*, 329, 432–435.
Mavrich, T. N., Ioshikhes, I. P., Venters, B. J., Jiang, C., Tomsho, L. P., Qi, J., et al. (2008). A barrier nucleosome model for statistical positioning of nucleosomes throughout the yeast genome. *Genome Research*, 18, 1073–1083.
Mayr, C., & Bartel, D. P. (2009). Widespread shortening of 3'UTRs by alternative cleavage and polyadenylation activates oncogenes in cancer cells. *Cell*, 138, 673–684.
Nagalakshmi, U., Wang, Z., Waern, K., Shou, C., Raha, D., Gerstein, M., et al. (2008). The transcriptional landscape of the yeast genome defined by RNA sequencing. *Science*, 320, 1344–1349.

Ozsolak, F., Kapranov, P., Foissac, S., Kim, S. W., Fishilevich, E., Monaghan, A. P., et al. (2010). Comprehensive polyadenylation site maps in yeast and human reveal pervasive alternative polyadenylation. *Cell, 143*, 1018–1029.

Parkhomchuk, D., Borodina, T., Amstislavskiy, V., Banaru, M., Hallen, L., Krobitsch, S., et al. (2009). Transcriptome analysis by strand-specific sequencing of complementary DNA. *Nucleic Acids Research, 37*, e123.

Proudfoot, N. J. (2011). Ending the message: Poly(A) signals then and now. *Genes & Development, 25*, 1770–1782.

Shepard, P. J., Choi, E. A., Lu, J., Flanagan, L. A., Hertel, K. J., & Shi, Y. (2011). Complex and dynamic landscape of RNA polyadenylation revealed by PAS-Seq. *RNA, 17*, 761–772.

Spies, N., Nielsen, C. B., Padgett, R. A., & Burge, C. B. (2009). Biased chromatin signatures around polyadenylation sites and exons. *Molecular Cell, 36*, 245–254.

Venters, B. J., & Pugh, B. F. (2009). How eukaryotic genes are transcribed. *Critical Reviews in Biochemistry and Molecular Biology, 44*, 117–141.

Wang, Z., Gerstein, M., & Snyder, M. (2009). RNA-Seq: A revolutionary tool for transcriptomics. *Nature Reviews Genetics, 10*, 57–63.

Wilkening, S., Pelechano, V., Järvelin, A.I., Tekkedil, M.M., Anders, S., Benes, V., et al. in preparation. *Efficient method for genome-wide polyadenylation site mapping and RNA quantification*.

Wu, T. D., & Nacu, S. (2010). Fast and SNP-tolerant detection of complex variants and splicing in short reads. *Bioinformatics, 26*, 873–881.

Xu, Z., Wei, W., Gagneur, J., Perocchi, F., Clauder-Munster, S., Camblong, J., et al. (2009). Bidirectional promoters generate pervasive transcription in yeast. *Nature, 457*, 1033–1037.

Yoon, O. K., & Brem, R. B. (2010). Noncanonical transcript forms in yeast and their regulation during environmental stress. *RNA, 16*, 1256–1267.

CHAPTER THIRTEEN

Genome-Wide Mapping of Nucleosome Occupancy, Histone Modifications, and Gene Expression Using Next-Generation Sequencing Technology

Gang Wei[*,†]**, Gangqing Hu**[*]**, Kairong Cui**[*]**, Keji Zhao**[*,1]

[*]Systems Biology Center, NHLBI, NIH, Bethesda, Maryland, USA
[†]Present address: CAS-MPG Partner Institute for Computational Biology, SIBS, CAS, Shanghai, China
[1]Corresponding author: e-mail address: zhaok@nhlbi.nih.gov

Contents

1. Introduction	298
2. Mapping Genome-Wide Nucleosome Occupancy and Positions by High-Throughput Next-Generation Sequencing	299
2.1 Preparation of mononucleosomes from cells	299
2.2 DNA purification from mononucleosomes	300
2.3 DNA end-repair	300
2.4 Add "A" overhangs to 3′ ends	300
2.5 Linker ligation	301
2.6 PCR amplification	301
2.7 Data analysis	302
3. Mapping Histone Modifications Using ChIP-Seq	302
4. Mapping Genome-Wide mRNA Profiles Using RNA-Seq	304
4.1 mRNA purification from total RNA	304
4.2 First-strand cDNA synthesis	305
4.3 Second-strand cDNA synthesis	306
4.4 Fragment the double-strand cDNA	306
4.5 Preparation of sequencing libraries for RNA-Seq	306
4.6 Data analysis	306
5. RNA-Seq Analysis with a Small Number of Cells	307
5.1 Synthesize the first-strand cDNA	308
5.2 Synthesize the second-strand cDNA	308
5.3 Amplification of the double-strand cDNA	310
5.4 Data analysis	311
Acknowledgment	311
References	311

Methods in Enzymology, Volume 513
ISSN 0076-6879
http://dx.doi.org/10.1016/B978-0-12-391938-0.00013-6

Abstract

Gene transcription can be regulated through alteration of chromatin structure, such as changes in nucleosome positioning and histone-modification patterns. Recent development of techniques based on the next-generation sequencing technology has allowed high-resolution analysis of genome-wide distribution of these chromatin features. In this chapter, we describe in detail the protocols of ChIP-Seq and MNase-Seq, which have been developed to detect the genome-wide profiles of transcription factor binding, histone modifications, and nucleosome occupancy. We also describe RNA-Seq protocols used to map global gene expression profiles.

1. INTRODUCTION

Recent global mapping of histone modifications has found that histone modifications are highly related to gene expression (Wang, Schones, & Zhao, 2009). Methylation of H3K4 and trimethylation of H3K36 are associated with gene activation whereas trimethylation of H3K27 and H3K9 associated with gene repression (Barski et al., 2007; Bernstein et al., 2006; Boyer et al., 2006; Kim et al., 2005; Roh, Cuddapah, Cui, & Zhao, 2006). Critical transcriptional regulatory elements are associated with various histone modifications. A large fraction of promoters are associated with seventeen different histone-modification backbones including H3K4me1, H3K4me2, H3K4me3, and H2A.Z. On the other hand, H3K4me2, H3K4me3, and H3K27ac are detected at active enhancers (He et al., 2010; Rada-Iglesias et al., 2011), whereas H3K4me1 can be associated with both active and inactive enhancers (Creyghton et al., 2010; Rada-Iglesias et al., 2011). Thus, identification of the distribution profiles of these histone modifications may predict regulatory elements of the underlying genes and help to understand their epigenetic regulation.

Another aspect of transcriptional regulation through chromatin structure is mediated by reorganization of nucleosome occupancy and positioning, which impact the accessibility of genomic target sites for transcription factors and transcription machinery to regulate gene expression (Henikoff, Furuyama, & Ahmad, 2004; Hu et al., 2011). Nucleosome phasing surrounding the transcription start sites is correlated with RNA polymerase II binding; nucleosome positioning at promoters and enhancers undergoes dynamic changes upon cellular signaling induction (Hu et al., 2011; Schones et al., 2008).

Traditionally, histone modifications at specific genomic loci can be determined by chromatin immunoprecipitation (ChIP) assays followed by

PCR assays to measure relative enrichment of ChIPed DNA at these genomic regions. In the past several years, genome-wide approaches have been developed to map histone modifications by combining ChIP assay with deep sequencing (Roh, Ngau, Cui, Landsman, & Zhao, 2004). More recently, with the advanced next-generation sequencing technologies, researchers have been able to generate genome-wide maps of histone modifications with high resolution and sensitivity (Barski et al., 2007; Mikkelsen et al., 2007). Similarly, genome-wide distribution of nucleosome occupancy and positioning can be determined by micrococcal nuclease (MNase) digestion coupled with the next-generation sequencing of the mononucleosome DNA fragments (Schones et al., 2008). In addition, next-generation sequencing technologies have been successfully applied to profile global gene expression by RNA-Seq with unprecedented sensitivity (Mortazavi, Williams, McCue, Schaeffer, & Wold, 2008; Nagalakshmi et al., 2008). In this chapter, we provide the detailed protocols of these genome-wide mapping assays.

2. MAPPING GENOME-WIDE NUCLEOSOME OCCUPANCY AND POSITIONS BY HIGH-THROUGHPUT NEXT-GENERATION SEQUENCING

Genome-wide nucleosome positioning has been mapped in resting and activated $CD4^+$ T cells (Schones et al., 2008). In this study, mononucleosomes are generated by MNase digestion, which is followed by high-throughput next-generation sequencing.

2.1. Preparation of mononucleosomes from cells

1. Harvest 50×10^6 cells by centrifugation at 1000 rpm for 5 min at 4 °C.
2. Wash the cells once with cold PBS.
3. Wash the cells once with cold digestion buffer. Resuspend the cells in 1 ml of the digestion buffer (50 mM Tris–HCl, pH 7.6, 1 mM CaCl$_2$, 0.2% Triton X-100, or NP-40) with 5 mM sodium butyrate, 1 × proteinase inhibitor cocktail, and 0.5 mM fresh PMSF.
4. Dispense 100 μl of aliquot in Eppendorf tubes and warm up to 37 °C for 2 min.
5. Add MNase in titration (0.5, 1, 2, 4, and 6 U) and incubate for 5 min at 37 °C.
6. Stop the reaction by adding 300 μl of stop buffer (10 mM Tris, pH 7.6, 5 mM EDTA).

7. Shear chromatin by sonication with four cycles of 30-s pulses at maximum setting with 30-s intervals in ice water bath.
8. Dialyze the fragmented chromatin against 200 ml of RIPA buffer (10 mM Tris, pH 7.6, 1 mM EDTA, 0.1% SDS, 0.1% Na-deoxycholate, 1% Triton X-100) for 2 h at 4 °C.
9. Centrifuge at full speed for 10 min at 4 °C. Transfer the supernatant to a new tube.
10. Transfer 50 μl of the supernatant to new tubes and purify the DNA by phenol–CHCl$_3$ extraction and ethanol precipitation.
11. Determine the size of digested chromatin by running the purified DNA on 2% agarose gel.
12. Select the optimal digestion condition that generates approximately 80% mononucleosomes and 20% dinucleosomes.

2.2. DNA purification from mononucleosomes

1. Purify the DNA from the chromatin of selected digestion condition by phenol–CHCl$_3$ extraction and ethanol precipitation.
2. Examine the DNA concentration by Nanodrop.
3. Run about 500 ng of purified DNA on a 2% agarose gel.
4. Excise the 100–200-bp fragments and purify the DNA using Qiagen Gel Extraction kit, eluting the DNA in 35 μl EB solution.

2.3. DNA end-repair

1. Mix the following reagents (Epicenter) together in the 1.5-ml tube:
   - 34 μl DNA (10–300 ng)
   - 5 μl of 10× end-repair buffer (330 mM Tris–acetate, pH 7.8, 660 mM potassium acetate, 100 mM magnesium acetate, 5 mM DTT)
   - 5 μl of 2.5 mM each dNTP
   - 5 μl of 10 mM ATP
   - 1 μl end-repair enzyme mix (T4 DNA Pol + T4 PNK)
2. Incubate for 45 min at room temperature.
3. Purify the DNA using Qiagen PCR purification kit and elute the DNA with 33 μl EB solution.

2.4. Add "A" overhangs to 3′ ends

1. Mix the following reagents together in the 1.5-ml tube:
   - 32 μl DNA, from above

- 5 μl of 10× Taq buffer
- 10 μl of 1 m*M* dATP
- 3 μl of 5 U/μl Taq DNA polymerase (New England Biolabs)
2. Incubate for 30 min at 70 °C.
3. Purify the DNA using Qiagen MinElute PCR purification kit and elute the DNA with 20 μl elution buffer.

2.5. Linker ligation

Note: The oligo mix should contain at least 10× molar excess of the DNA templates.

1. Mix the following reagents together in the 1.5-ml tube:
   - 19 μl DNA, from above
   - 2.5 μl of 10× T4 DNA ligase buffer (New England Biolabs)
   - 1 μl Adaptor oligo mix (Illumina) (reduce the volume if the amount of DNA is below 100 ng)
   - 2.5 μl T4 DNA ligase (400 units/μl) (New England Biolabs)
2. Incubate for 30 min at room temperature.
3. Run the linker-ligated DNA on a 2% precast agarose E-Gel (Invitrogen).
4. Cut the gel around 200–300 bp region.
5. Purify the DNA using Qiagen MinElute gel purification kit and elute the DNA with 20 μl elution buffer.

2.6. PCR amplification

Note: Amplify the DNA using Illumina primers and enzyme mix as described.

1. Mix the following reagents together in the 1.5-ml tube:
   - 10.5 μl DNA
   - 12.5 μl master mix (Phusion, Finnzymes)
   - 1 μl PCR primer 1 (2× diluted)
   - 1 μl PCR primer 2 (2× diluted)
     Total volume: 25 μl
2. Proceed with PCR as follows:
   Denature DNA at 98 °C for 30 s
   98 °C, 10 s; 65 °C, 30 s; 72 °C, 30 s
   Run 18–21 cycles
3. Run the amplified PCR products (total volume) on a 2% agarose 1× TBE gel.
4. Excise the 200–300 bp fragments and purify the DNA using Qiagen MinElute Gel Extraction kit, eluting the DNA in 10 μl EB solution.

5. Measure the DNA concentration using Qubit fluorometer.
6. Perform sequencing on Illumina Genome Analyzer.

2.7. Data analysis

The first step of data analysis is to inspect the quality of sequence reads (Kidder, Hu, & Zhao, 2011), for example, by using software such as FastQC (Babraham Institute). The sequence reads are then mapped to reference genomes using software such as Bowtie (Langmead, Trapnell, Pop, & Salzberg, 2009). Potential bias from PCR can be minimized by allowing only one sequence per position. Excluding sequences mapped to multiple genome positions are recommended if the focus is beyond highly repetitive regions. Recent methods for nucleosome positions calling from MNase-Seq data include NPS (Zhang, Shin, Song, Lei, & Liu, 2008) and NucleR (Flores & Orozco, 2011). It is important to realize that the genomic regions can be occupied by either well-positioned nucleosomes or poorly positioned nucleosomes, especially for mammalian genomes (Sadeh & Allis, 2011; Valouev et al., 2011). Tools for a genome-wide comparative analysis of nucleosome positioning have also been made available recently (Meyer, He, Brown, & Liu, 2011), though a systematic evaluation is lacking.

3. MAPPING HISTONE MODIFICATIONS USING ChIP-Seq

ChIP assays can be performed using cross-linked chromatin that is fragmented by sonication (Ren et al., 2000) or native chromatin digested with MNase (Barski et al., 2007). In this protocol, we use native chromatin that is digested with MNase to generate about 80% mononucleosomes (see Section 2.1). Using this method, many histone modifications have been mapped in resting CD4 T cells (Barski et al., 2007; Wang et al., 2008), activated CD4 T cells (Barski et al., 2007), human hematopoietic progenitor cells (Cui et al., 2009), murine helper T cells (Wei et al., 2009), and memory T cells (Araki et al., 2009). For each ChIP-Seq assay, preimmune or input sample from the same chromatin should be used as control. Having a high-quality antibody is critical to the success of ChIP-Seq assay (Kidder et al., 2011). Egelhofer et al. (2011) rigorously tested hundreds of histone-modification antibodies and posted the results to a public Website http://compbio.med.harvard.edu/antibodies/, which is valuable to the ChIP community for addressing the antibody quality issues.

1. Add 40 μl Dynabeads Protein A (Invitrogen) to each Eppendorf tube. Use 600 μl of 1 × PBS to wash the beads. Attach the beads to magnet for 1 min. Aspirate off 1 × PBS.
2. To each tube, add 100 μl of 1 × PBS containing 4 μg antibody or preimmune serum (rabbit IgG, negative control) to the beads. Tap occasionally to mix the beads and incubate for 40 min at RT to allow complete antibody binding to the beads.
3. Use magnet to remove supernatants (1 min) and wash the beads twice with 1 × PBS to remove free IgGs (0.6 ml and 5 min each time, rotating at RT).
4. Remove supernatant from the beads by magnet. Add chromatin extracts (from 5 to 10 million cells, in RIPA buffer) to the beads. Rotate at 4 °C overnight to reduce background.
5. Wash the beads (10 min each time at room temperature with rotation):
   - 2× with 1 ml of RIPA buffer, 10 min each time
   - 2× with 1 ml of RIPA buffer + 0.3 M NaCl, 10 min each time
   - 2×with 1 ml of LiCl buffer (0.25 M LiCl, 0.5% NP40, 0.5% NaDOC)
   - 1 × with 1 ml of 1 × TE + 0.2% Triton X-100
   - 1 × with 1 ml of 1 × TE
6. Resuspend the beads in 100 μl of 1 × TE. Add 3 μl of 10% SDS and 5 μl of 20 mg/ml proteinase K. Incubate at 65 °C overnight.
7. Vortex briefly and transfer the supernatant to a new tube using magnet. Wash the beads with 100 μl of 1 × TE + 0.5 M NaCl. Combine the supernatants.
8. Extract with 200 μl of phenol/chloroform twice.
9. Precipitate the DNA with 2 μl of 20 mg/ml glycogen, 20 μl of 3 M NaOAc, pH 5.3, and 500 μl of ethanol.
10. Wash the pellet once with 70% ethanol, air-dry briefly, and resuspend the pellet in 40 μl of 1×TE.
11. Use 2 μl or the same amount of DNA for PCR analysis. Make sure the ChIP has worked well before moving to preparation of sequencing library.
12. For sequencing library preparation, follow the protocol from 2.3 to 2.6.
13. *Data analysis*: see Section 2.7 for data quality inspection and sequences mapping. One should keep in mind that the tag enrichment of histone modification frequently occupies genomic regions of 100 bp. Therefore, tools designed to call peaks from ChIP-Seq for transcription factor, which binds to several genomic regions of several bps, are not

recommended (Kidder et al., 2011). Instead, we routinely use SICER to call tag enrichment regions from ChIP-Seq data for histone modifications. Other tools that can be used to address this issue include CCAT (Xu et al., 2010) and ZINBA (Rashid, Giresi, Ibrahim, Sun, & Lieb, 2011).

4. MAPPING GENOME-WIDE mRNA PROFILES USING RNA-Seq

Taking advantage of next-generation sequencing technologies, the RNA-Seq technique was developed to map genome-wide mRNA profiles. Compared to conventional RNA microarray assays, RNA-Seq not only offers higher sensitivity and broader linear detection range to quantify gene expression levels, but the sequencing information can also be used to detect alternative splicing of gene transcription novel transcript (Mortazavi et al., 2008; Nagalakshmi et al., 2008; Sultan et al., 2008) and genetic mutations in expressed exons (Chepelev, Wei, Tang, & Zhao, 2009). The RNA-Seq protocol in this section based on these pioneer studies requires 10 μg of total RNA.

4.1. mRNA purification from total RNA

1. Dilute 10 μg of total RNA with nuclease-free H_2O to 50 μl in a 1.5-ml RNase-free nonsticky Eppendorf tube.
2. Heat the sample at 65 °C for 5 min to disrupt the secondary structures and place the tube on ice.
3. Meanwhile, aliquot 100 μl of Dynal oligo(dT) beads (Invitrogen) into a 1.5-ml RNase-free nonsticky Eppendorf tube.
4. Wash the beads twice with 100 μl of binding buffer (20 mM Tris–HCl, pH 7.5, 1.0 M LiCl, and 2 mM EDTA) and remove the supernatant.
5. Resuspend the beads in 50 μl of binding buffer and add the 50 μl of total RNA sample from step 2; rotate the tube at RT for 5 min and remove the supernatant.
6. Wash the beads twice with 100 μl of washing buffer B (10 mM Tris–HCl, pH 7.5, 0.15 M LiCl, 1 mM EDTA).
7. Prepare for second round of oligo-dT purification by aliquoting 80 μl of binding buffer to a fresh 1.5-ml RNase-free nonsticky Eppendorf tube.
8. Remove the supernatant from the beads of step 6, add 20 μl of 10 mM Tris–HCl, and heat the beads at 80 °C for 2 min to elute mRNA. Immediately put the beads on the magnet stand and transfer the

supernatant (mRNA) to the tube from step 7. Add 100 μl of washing buffer B to the remaining beads.
9. Heat the mRNA sample from step 8 at 65 °C for 5 min to disrupt the secondary structures and place the tube on ice.
10. Meanwhile, wash the beads from step 8 twice with 100 μl of washing buffer B and remove the supernatant.
11. Add 100 μl of mRNA sample from step 9 to the beads; rotate the tube at RT for 5 min.
12. Remove the supernatant and wash the beads twice with 100 μl of washing buffer B.
13. Remove the supernatant from the beads, add 10 μl of 10 mM Tris–HCl, and heat the beads at 80 °C for 2 min to elute mRNA. Immediately put the beads on the magnet stand and transfer the supernatant (mRNA) to a fresh 200-μl thin-wall PCR tube, and there should be ∼9 μl of mRNA.

4.2. First-strand cDNA synthesis

1. Assemble the following reaction in a 200-μl thin-wall PCR tube:

Random hexamer primers (3 μg/μl, Invitrogen)	1 μl
mRNA (100 ng/μl)	10.5 μl

2. Incubate the sample in a PCR thermocycler at 65 °C for 5 min and then place the tube on ice.
3. Mix the following in order and make 10% extra reagent for multiple samples:

5 × first-strand buffer (Invitrogen)	4 μl
100 mM DTT (Invitrogen)	2 μl
dNTP mix (10 mM)	1 μl
RNaseOUT (40 U/μl) (Invitrogen)	0.5 μl

4. Add 7.5 μl of mixture to the tube, mix well, and heat the sample at 25 °C in a thermocycler for 2 min.
5. Add 1 μl SuperScript II (200 U/μl, Invitrogen) to the sample and incubate the sample in a thermocycler with the following program:

Step 1	25 °C	10 min
Step 2	42 °C	50 min
Step 3	70 °C	15 min
Step 4	4 °C	Hold

4.3. Second-strand cDNA synthesis

1. Put the tubes from Section, step 5 on ice.
2. Add 61 μl H_2O to the first-strand cDNA synthesis mix.
3. Add the following reagents:

10× second-strand buffer (500 mM Tris–HCl, pH 7.8, 50 mM $MgCl_2$, 10 mM DTT)	10 μl
dNTP mix (10 mM)	3 μl

4. Mix well, incubate on ice for 5 min or until well chilled, and add:

RNase H (2 U/μl, Invitrogen)	1 μl
DNA Pol I (10 U/μl, Invitrogen)	5 μl

5. Mix well and incubate at 16 °C in a thermocycler for 2.5 h.
6. Purify the DNA with a QIAquick PCR spin column and elute in 40 μl of EB solution.
7. Measure the DNA concentration.

4.4. Fragment the double-strand cDNA

1. Use Bioruptor to sonicate the double-stranded cDNA obtained above (300 ng DNA in 40 μl TE) for 30 min, put some ice every 10 min to keep the water cool.
2. Check the sonication efficiency by running a 2% agarose gel (load about 50 ng DNA). Make sure the size of sonicated DNA ranges from 100 to 500 bp.

4.5. Preparation of sequencing libraries for RNA-Seq

For sequencing library preparation, follow the protocol from 2.3 to 2.6.

4.6. Data analysis

The sequenced tags can be either mapped to whole genome or to annotated transcriptome. Similar to MNase-Seq and Chip-Seq data, sequences tags mapped to multiple genomic positions are routinely excluded from downstream analysis to minimize ambiguity. This would lead to an underestimation of expression for genes with multiple copies, but it would be overcome by mapping the tags to annotated transcriptome (Lee et al., 2011). Also note that since exon constitutes < 1% of the total genome, each genomic position would receive more than 1 tag, and therefore the "one sequence per position" rule is not applied to RNA-Seq data. The mRNA level of a gene is frequently measured by RPKM, reads per kilobase of exon per million reads (Mortazavi et al., 2008). This measure exhibits bias in terms of gene length and nucleotide composition, and the correction constitutes the theme in several publications (Oshlack & Wakefield, 2009; Zheng, Chung, & Zhao, 2011). Beyond a quantitative measure of gene expression, another major application of RNA-Seq is to define differentially expressed genes between two different cell types under different conditions. Software developed for this includes but is not limited to EdgeR (Robinson, McCarthy, & Smyth, 2010) and DegSeq (Wang, Feng, Wang, Wang, & Zhang, 2010). One limitation of these methods is the dependence of current gene annotation. RNA-Seq is also a useful technique for the discovery of new genes and/or new isoform of known genes and their expression dynamics across differentiation process, which constitutes the subject of the Cufflinks software (Trapnell et al., 2010).

5. RNA-Seq ANALYSIS WITH A SMALL NUMBER OF CELLS

The RNA-Seq protocol in Section 4 requires large number of cells. However, in many cases, the number of specific type of cells for study is limited. In order to map the transcriptome from the limited number of cells (100–1000 cells), modified RNA-Seq protocols have been developed, in which the reverse-transcribed cDNA has to be amplified before preparation of sequencing library (Tang et al., 2009).

The following primers are used for double-stranded cDNA synthesis and amplification.

Universal Primer 1 (UP1)	5′-ATA TGG ATC CGG CGC GCC GTC GAC TTT TTT TTT TTT TTT TTT TTT TTT-3′
Universal Primer 2 (UP2)	5′-ATA TCT CGA GGG CGC GCC GGA TCC TTT TTT TTT TTT TTT TTT TTT TTT-3′
NH2-Universal Primer 1 (NH2-UP1)	5′-NH2-ATA TGG ATC CGG CGC GCC GTC GAC TTT TTT TTT TTT TTT TTT TTT TTT-3′
NH2-Universal Primer 2 (NH2-UP2)	5′-NH2-ATA TCT CGA GGG CGC GCC GGA TCC TTT TTT TTT TTT TTT TTT TTT TTT-3′

5.1. Synthesize the first-strand cDNA

1. Prepare cell lysis buffer as followed. Scale up for multiple reactions.

	1×
10× PCR buffer II	0.45 μl
MgCl$_2$ (25 mM)	0.27 μl
NP-40 (5%)	0.45 μl
DTT (0.1 M)	0.225 μl
RNase inhibitor (40 U/μl)	0.09 μl
Primer (10 ng/μl)	0.09 μl
dNTP mix (2.5 mM each)	0.09 μl
Water, nuclease-free	2.84 μl
Total	4.5 μl

2. cDNA synthesis step should be performed within 2 h of adding the RNase inhibitor. Always keep the lysis buffer on ice.
3. Add 4.5 μl of the lysis buffer to the cell pellet in each tube.
4. Incubate the tubes at 70 °C for 90 s.
5. Place the tubes immediately on ice. Centrifuge briefly at room temperature. Back on ice for at least 5 min.
6. Prepare the following mixture of reagents for reverse transcription. Scale up for multiple reactions.

	1×
Superscript III	0.2 μl
RNase inhibitor	0.033 μl
Water, nuclease-free	0.067 μl
Total	0.3 μl

7. Chill the mixture on ice before using. Add 0.3 μl of RT mixture to each tube and tap the tube gently to mix.
8. Incubate at 50 °C for 35 min and heat inactive at 70 °C for 15 min.

5.2. Synthesize the second-strand cDNA

1. Prepare Exo I reaction mixture. Scale up for multiple reactions.

	1×
10× Exo1 buffer	0.1 μl
Water, nuclease-free	0.8 μl
Exonuclease I	0.1 μl
Total	1 μl

2. Add 1.0 μl of the Exo1 mix to the tubes. Mix thoroughly.
3. Incubate at 37 °C for 30 min and then inactivate at 80 °C for 25 min.
4. Place the tubes on ice for 1 min.
5. Prepare TdT reaction mixture as followed. Scale up for multiple reactions.

	1×
10× PCR buffer	0.6 μl
$MgCl_2$ (25 mM)	0.36 μl
dATP (100 mM)	0.18 μl
Water, nuclease-free	4.26 μl
RNase H	0.3 μl
TdT	0.3 μl
Total	6 μl

6. Add 6.0 μl of TdT master mix into each tube.
7. Incubate at 37 °C for 15 min and inactivate at 70 °C for 10 min.
8. Place on ice for 1 min.
9. Split the product above into four empty PCR tubes (3 μl each).
10. Prepare PCR Mix I to synthesize the second strand of cDNA. Scale up for multiple reactions.

	5×
10× ExTaq buffer	7.6
DNTP mix (2.5 mM each)	7.6
Primers (1 μg/μl)	1.52
Water, nuclease-free	58.52
ExTaq	0.76
Total	76

11. Add 19.0 µl of the PCR master mix I into each tube. Immediately put it on ice.
12. Perform one round of PCR for second-strand synthesis as follows: 95 °C for 3 min, 50 °C for 2 min, and 72 °C for 6 min.
13. Place the tubes on ice.

5.3. Amplification of the double-strand cDNA

1. Prepare PCR Mix II for the first round amplification. Scale up for multiple reactions.

	1×
10 × ExTaq buffer	7.6
DNTP mix (2.5 mM each)	7.6
Primers (1 µg/µl)	1.52
Water, nuclease-free	58.52
ExTaq	0.76
Total	76

2. Add 19.0 µl of the PCR master mix II into each tube. Immediately put it on ice.
3. Perform 20 rounds of PCR amplification as follows: 95 °C for 30 s, 67°°C for 1 min, and 72 °C for 6 min (plus 6 s more after each cycle). Place the tubes on ice.
4. Purify the PCR product with Qiagen PCR purification kit.
5. Prepare PCR Mix III for the second round amplification of cDNA. Scale up for multiple reactions.

	1×
Template from above	2 µl
10 × ExTaq buffer	5 µl
DNTP mix	5 µl
NH2_UP1 primers (1 µg/µl)	1 µl
NH2_UP2 primers	1 µl
Water, nuclease-free	35.5 µl
ExTaq	0.5 µl
Total	50 µl

6. Perform 12 PCR cycles: 95 °C for 30 s, 67 °C for 1 min, and 72 °C for 6 min.
7. Purify PCR product using Qiagen PCR purification kit.
8. Measure DNA concentration with Nanodrop.
9. Examine the size of amplified cDNA by running 50 ng DNA on a 2% agarose gel. Make sure that the size of cDNA ranges from 100 to 2000 bp.
10. Shear about 300 ng cDNA using Bioruptor as described in Section 4.4.
11. Prepare the sequencing libraries for RNA-Seq.

5.4. Data analysis

For data analysis, see Section 4.6.

ACKNOWLEDGMENT

This work was supported by the Division of Intramural Research Program of the NIH, National Heart, Lung, and Blood Institute.

REFERENCES

Araki, Y., Wang, Z., Zang, C., Wood, W. H., 3rd, Schones, D., Cui, K., et al. (2009). Genome-wide analysis of histone methylation reveals chromatin state-based regulation of gene transcription and function of memory CD8+ T cells. *Immunity, 30*, 912–925.
Barski, A., Cuddapah, S., Cui, K., Roh, T. Y., Schones, D. E., Wang, Z., et al. (2007). High-resolution profiling of histone methylations in the human genome. *Cell, 129*, 823–837.
Bernstein, B. E., Mikkelsen, T. S., Xie, X., Kamal, M., Huebert, D. J., Cuff, J., et al. (2006). A bivalent chromatin structure marks key developmental genes in embryonic stem cells. *Cell, 125*, 315–326.
Boyer, L. A., Plath, K., Zeitlinger, J., Brambrink, T., Medeiros, L. A., Lee, T. I., et al. (2006). Polycomb complexes repress developmental regulators in murine embryonic stem cells. *Nature, 441*, 349–353.
Chepelev, I., Wei, G., Tang, Q., & Zhao, K. (2009). Detection of single nucleotide variations in expressed exons of the human genome using RNA-Seq. *Nucleic Acids Research, 37*, e106.
Creyghton, M. P., Cheng, A. W., Welstead, G. G., Kooistra, T., Carey, B. W., Steine, E. J., et al. (2010). Histone H3K27ac separates active from poised enhancers and predicts developmental state. *Proceedings of the National Academy of Sciences of the United States of America, 107*, 21931–21936.
Cui, K., Zang, C., Roh, T. Y., Schones, D. E., Childs, R. W., Peng, W., et al. (2009). Chromatin signatures in multipotent human hematopoietic stem cells indicate the fate of bivalent genes during differentiation. *Cell Stem Cell, 4*, 80–93.
Egelhofer, T. A., Minoda, A., Klugman, S., Lee, K., Kolasinska-Zwierz, P., Alekseyenko, A. A., et al. (2011). An assessment of histone-modification antibody quality. *Nature Structural & Molecular Biology, 18*, 91–93.
Flores, O., & Orozco, M. (2011). nucleR: A package for non-parametric nucleosome positioning. *Bioinformatics (Oxford, England), 27*, 2149–2150.
He, H. H., Meyer, C. A., Shin, H., Bailey, S. T., Wei, G., Wang, Q., et al. (2010). Nucleosome dynamics define transcriptional enhancers. *Nature Genetics, 42*, 343–347.

Henikoff, S., Furuyama, T., & Ahmad, K. (2004). Histone variants, nucleosome assembly and epigenetic inheritance. *Trends in Genetics, 20*, 320–326.

Hu, G., Schones, D. E., Cui, K., Ybarra, R., Northrup, D., Tang, Q., et al. (2011). Regulation of nucleosome landscape and transcription factor targeting at tissue-specific enhancers by BRG1. *Genome Research, 21*, 1650–1658.

Kidder, B. L., Hu, G., & Zhao, K. (2011). ChIP-Seq: Technical considerations for obtaining high-quality data. *Nature Immunology, 12*, 918–922.

Kim, T. H., Barrera, L. O., Zheng, M., Qu, C., Singer, M. A., Richmond, T. A., et al. (2005). A high-resolution map of active promoters in the human genome. *Nature, 436*, 876–880.

Langmead, B., Trapnell, C., Pop, M., & Salzberg, S. L. (2009). Ultrafast and memory-efficient alignment of short DNA sequences to the human genome. *Genome Biology, 10*, R25.

Lee, S., Seo, C. H., Lim, B., Yang, J. O., Oh, J., Kim, M., et al. (2011). Accurate quantification of transcriptome from RNA-Seq data by effective length normalization. *Nucleic Acids Research, 39*, e9.

Meyer, C. A., He, H. H., Brown, M., & Liu, X. S. (2011). BINOCh: Binding inference from nucleosome occupancy changes. *Bioinformatics (Oxford, England), 27*, 1867–1868.

Mikkelsen, T. S., Ku, M., Jaffe, D. B., Issac, B., Lieberman, E., Giannoukos, G., et al. (2007). Genome-wide maps of chromatin state in pluripotent and lineage-committed cells. *Nature, 448*, 553–560.

Mortazavi, A., Williams, B. A., McCue, K., Schaeffer, L., & Wold, B. (2008). Mapping and quantifying mammalian transcriptomes by RNA-Seq. *Nature Methods, 5*, 621–628.

Nagalakshmi, U., Wang, Z., Waern, K., Shou, C., Raha, D., Gerstein, M., et al. (2008). The transcriptional landscape of the yeast genome defined by RNA sequencing. *Science, 320*, 1344–1349.

Oshlack, A., & Wakefield, M. J. (2009). Transcript length bias in RNA-seq data confounds systems biology. *Biology Direct, 4*, 14.

Rada-Iglesias, A., Bajpai, R., Swigut, T., Brugmann, S. A., Flynn, R. A., & Wysocka, J. (2011). A unique chromatin signature uncovers early developmental enhancers in humans. *Nature, 470*, 279–283.

Rashid, N. U., Giresi, P. G., Ibrahim, J. G., Sun, W., & Lieb, J. D. (2011). ZINBA integrates local covariates with DNA-seq data to identify broad and narrow regions of enrichment, even within amplified genomic regions. *Genome Biology, 12*, R67.

Ren, B., Robert, F., Wyrick, J. J., Aparicio, O., Jennings, E. G., Simon, I., et al. (2000). Genome-wide location and function of DNA binding proteins. *Science, 290*, 2306–2309.

Robinson, M. D., McCarthy, D. J., & Smyth, G. K. (2010). edgeR: A Bioconductor package for differential expression analysis of digital gene expression data. *Bioinformatics (Oxford, England), 26*, 139–140.

Roh, T. Y., Cuddapah, S., Cui, K., & Zhao, K. (2006). The genomic landscape of histone modifications in human T cells. *Proceedings of the National Academy of Sciences of the United States of America, 103*, 15782–15787.

Roh, T. Y., Ngau, W. C., Cui, K., Landsman, D., & Zhao, K. (2004). High-resolution genome-wide mapping of histone modifications. *Nature Biotechnology, 22*, 1013–1016.

Sadeh, R., & Allis, C. D. (2011). Genome-wide "re"-modeling of nucleosome positions. *Cell, 147*, 263–266.

Schones, D. E., Cui, K., Cuddapah, S., Roh, T. Y., Barski, A., Wang, Z., et al. (2008). Dynamic regulation of nucleosome positioning in the human genome. *Cell, 132*, 887–898.

Sultan, M., Schulz, M. H., Richard, H., Magen, A., Klingenhoff, A., Scherf, M., et al. (2008). A global view of gene activity and alternative splicing by deep sequencing of the human transcriptome. *Science, 321*, 956–960.

Tang, F., Barbaciorn, C., Wang, Y., Nordman, E., Lee, C., Xu, N., et al. (2009). mRNA-Seq whole-transcriptome analysis of a single cell. *Nature Methods, 6*, 377–382.

Trapnell, C., Williams, B. A., Pertea, G., Mortazavi, A., Kwan, G., van Baren, M. J., et al. (2010). Transcript assembly and quantification by RNA-Seq reveals unannotated transcripts and isoform switching during cell differentiation. *Nature Biotechnology, 28*, 511–515.

Valouev, A., Johnson, S. M., Boyd, S. D., Smith, C. L., Fire, A. Z., & Sidow, A. (2011). Determinants of nucleosome organization in primary human cells. *Nature, 474*, 516–520.

Wang, L., Feng, Z., Wang, X., Wang, X., & Zhang, X. (2010). DEGseq: An R package for identifying differentially expressed genes from RNA-seq data. *Bioinformatics (Oxford, England), 26*, 136–138.

Wang, Z., Schones, D. E., & Zhao, K. (2009). Characterization of human epigenomes. *Current Opinion in Genetics & Development, 19*, 127–134.

Wang, Z., Zang, C., Rosenfeld, J. A., Schones, D. E., Barski, A., Cuddapah, S., et al. (2008). Combinatorial patterns of histone acetylations and methylations in the human genome. *Nature Genetics, 40*, 897–903.

Wei, G., Wei, L., Zhu, J., Zang, C., Hu-Li, J., Yao, Z., et al. (2009). Global mapping of H3K4me3 and H3K27me3 reveals specificity and plasticity in lineage fate determination of differentiating CD4+ T cells. *Immunity, 30*, 155–167.

Xu, H., Handoko, L., Wei, X., Ye, C., Sheng, J., Wei, C. L., et al. (2010). A signal-noise model for significance analysis of ChIP-seq with negative control. *Bioinformatics (Oxford, England), 26*, 1199–1204.

Zhang, Y., Shin, H., Song, J. S., Lei, Y., & Liu, X. S. (2008). Identifying positioned nucleosomes with epigenetic marks in human from ChIP-Seq. *BMC Genomics, 9*, 537.

Zheng, W., Chung, L. M., & Zhao, H. (2011). Bias detection and correction in RNA-Sequencing data. *BMC Bioinformatics, 12*, 290.

CHAPTER FOURTEEN

A Chemical Approach to Mapping Nucleosomes at Base Pair Resolution in Yeast

Kristin R. Brogaard*,2, Liqun Xi†, Ji-Ping Wang†,1, Jonathan Widom*
*Department of Molecular Biosciences, Northwestern University, Evanston, Illinois, USA
†Department of Statistics, Northwestern University, Evanston, Illinois, USA
1Corresponding author: e-mail address: jzwang@northwestern.edu
2Co-Corresponding author: e-mail address: Kristin.Brogaard@systemsbiology.org

Contents

1. Introduction	316
2. Construction of H4S47C *S. cerevisiae* Strain	317
2.1 Genetic mutagenesis of the *S. cerevisiae* strain	317
3. Chemical Cleavage of Nucleosome Center Positions	318
3.1 H4S47C growth and permeabilization	319
3.2 Nucleosome labeling	320
3.3 Hydroxyl radical cleavage of nucleosomal DNA	321
3.4 Isolation of mapped DNA fragments and preparation for parallel sequencing	321
3.5 Preparing chemically mapped DNA fragments for parallel sequencing	323
4. Statistical Analysis of Chemical Mapping Data	324
4.1 Single-end versus paired-end sequencing	324
4.2 Example data	324
4.3 Primary and secondary site identification	325
4.4 Training of cleavage model	325
4.5 Nucleosome clustering and convolution of cleavage signals	327
4.6 Selection of unique and redundant nucleosome sets	329
4.7 Base composition-dependent cleavage patterns	329
4.8 Nucleosome occupancy scores	331
4.9 Sequencing depth	332
5. Summary	332
Acknowledgments	333
References	333

Abstract

Most eukaryotic DNA exists in DNA–protein complexes known as nucleosomes. The exact locations of nucleosomes along the genome play a critical role in chromosome functions and gene regulation. However, the current methods for nucleosome mapping do not provide the necessary accuracy to identify the precise nucleosome locations. Here we describe a new experimental approach that directly maps nucleosome center locations *in vivo* genome-wide at single base pair resolution.

1. INTRODUCTION

Nucleosomes distort and occlude the genomic DNA they bind from access to most DNA-binding proteins, and their exact positions affect the structure of the chromatin fiber (Richmond & Davey, 2003). Even single base pair shifts of nucleosome positions can change chromatin configurations (Koslover, Fuller, Straight, & Spakowitz, 2010) and protein binding kinetics (Li & Widom, 2004; Mao, Brown, Griesenbeck, & Boeger, 2011). A single base pair resolution map of nucleosome positions is necessary to fully understand a wide range of biological processes including RNA polymerase activity (Churchman & Weissman, 2011; Petesch & Lis, 2008), transcription factor binding kinetics (Li & Widom, 2004; Mao et al., 2011), DNA replication (Lipford & Bell, 2001), centromere structure (Cole, Howard, & Clark, 2011), and gene splicing (Schwartz, Meshorer, & Ast, 2009).

Currently, the most common method for nucleosome mapping relies on treatment of chromatin with micrococcal nuclease (MNase) to digest the exposed DNA sequences not protected by the nucleosome core. The nucleosome positions are indirectly inferred by the centers of the undigested DNA fragments. While this method has provided valuable insights into our understanding of nucleosomes, it is imprecise due to the transient unwrapping of nucleosomal DNA (Li, Levitus, Bustamante, & Widom, 2005), MNase sequence preferences (Dingwall, Lomonossoff, & Laskey, 1981), and the interference of other DNA-binding proteins, all of which lead to variable lengths in the undigested nucleosome DNA fragments. Thus, a different approach is required to measure nucleosome locations directly with greater accuracy.

Here we develop an approach for direct mapping of nucleosome centers at single base pair resolution. The chemical mapping method derives from previous work from Flaus, Luger, Tan, and Richmond (1996) and relies on hydroxyl radical cleavage to identify nucleosome center positions. For this method, the histone octamer has a modified histone H4 protein, where serine 47 is mutated to a cysteine (H4S47C). Importantly, this amino acid's position symmetrically flanks the nucleosome center axis and is in close proximity to the DNA backbone (see Fig. 14.1). By the covalent linkage of a sulfhydryl-binding, copper-chelating label (phenanthroline–iodoacetamide) to the cysteine, we can position a copper ion at this same position—symmetric around the center axis. Copper reacts with hydrogen peroxide creating reactive hydroxyl radicals. With the addition of hydrogen peroxide, a localized cloud of hydroxyl radicals cleave the DNA precisely at

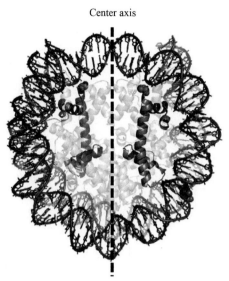

Center axis

Figure 14.1 A nucleosome structure highlighting histone H4 and residue serine 47 (sphere). Serine 47 is mutated to a cysteine and is the site where the sulfhydryl-reactive label covalently binds. Adapted from Brogaard et al., 2012. (See Color Insert.)

sites adjacent to the center. By identifying cleavage patterns, we can accurately determine the nucleosome center positions.

This "chemical method" for mapping nucleosome locations has been developed to map *in vivo* nucleosome positions in *S. cerevisiae* genome-wide (Brogaard, Xi, Wang, & Widom, 2012). The resulting map achieves unprecedented detail and accuracy in defining nucleosome center positions genome-wide. It reveals novel aspects of the *in vivo* nucleosome organization that are linked to transcription factor binding, RNA polymerase pausing, and the higher-order structure of the chromatin fiber. It will be interesting in future work to extend this protocol to more complex genomes, potentially revealing additional relationships between nucleosome positioning and aspects of chromosome function.

2. CONSTRUCTION OF H4S47C *S. CEREVISIAE* STRAIN
2.1. Genetic mutagenesis of the *S. cerevisiae* strain

A *S. cerevisiae* strain containing the H4S47C mutation is necessary to apply the chemical mapping technique *in vivo*. *S. cerevisiae* has two gene copies of the H4 histone, *HHF1* and *HHF2*. The strain is created using the MIRAGE (mutagenic inverted repeat assisted genome engineering) system (Nair & Zhao, 2009) to mutate the endogenous *HHF1* locus to contain the cysteine mutation at position 47. The *HHF2* gene is replaced by a *URA3* gene using a standard yeast integration method (Amberg, Burke, & Strathern, 2005). *S. cerevisiae*'s histone octamer has no endogenous cysteines. The H4S47C/ *ura3* strain is viable but maintains a growth phenotype and is temperature sensitive.

1. In one copy of the histone H4 gene, modify the serine at position 47 in *S. cerevisiae* to a cysteine by the mutagenesis of a single cytosine at position 143 to a guanine using the established MIRAGE system (Nair & Zhao, 2009).
2. Replace the second copy of the histone H4 gene by a *URA3* gene using a standard homologous recombination gene replacement technique for yeast (Amberg et al., 2005).
3. Confirm mutagenesis and *URA3* integration by sequencing.

3. CHEMICAL CLEAVAGE OF NUCLEOSOME CENTER POSITIONS

The sulfhydryl-reactive, copper-chelating label (N(1,10-phenan-throline-5-yl)iodoacetamide) covalently binds to the cysteine present in histone H4 and is used to anchor copper at locations that are in close proximity to the DNA backbone and symmetrically flanking the nucleosome center axis. Removal of excess label is crucial to minimize any background cleavage. With the addition of hydrogen peroxide, the anchored copper becomes the site of hydroxyl radical production. The localized cloud of hydroxyl radicals cleaves the DNA backbone at very specific positions adjacent to the nucleosome center position. On an agarose gel, the resulting cleaved DNA products produce a banding pattern where the smallest fragment corresponds to DNAs between the centers of two closely spaced neighboring nucleosomes (see Fig. 14.2).

There are two important controls for this experiment. The first is a "no mapping control" where a small fraction of yeast culture (~1/10th the total) is put aside after the initial harvesting and is purified. This negative control will show, on an agarose gel, the actual size distribution of whole genomic DNA, without any hydroxyl radical cleavage. The second important control that should be done in parallel with the H4S47C mapping is the chemical mapping of wild-type *S. cerevisiae*. The resulting DNA, after this reaction, should show no observable background cleavage (it should resemble the

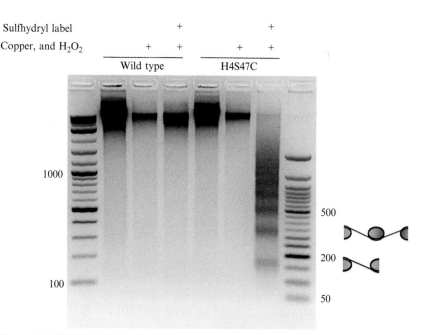

Figure 14.2 Ethidium bromide-stained agarose gel showing the chemical mapping results in a DNA banding pattern. Mapping (and observable DNA cleavage) only occurs when the reaction includes (indicated by "+") the sulfhydryl-reactive label, copper, H_2O_2, and the H4S47C mutant yeast. The cartoons adjacent to the agarose gel illustrate that the banding pattern is produced from mapping successive nucleosomes' centers. Adapted from Brogaard et al., 2012. (See Color Insert.)

"no mapping control"). If background cleavage is observed, additional washes are needed to remove the excess label from the cells.

The following protocol is tailored for 1 L of log-phase *S. cerevisiae* culture. The resulting chemically cleaved DNA concentration is more than sufficient for further processing and parallel sequencing. All steps are scalable to different volumes of culture.

3.1. H4S47C growth and permeabilization

The H4S47C strain has a growth phenotype, taking it longer to reach exponential growth as compared to wild type. It is recommended to make a growth curve prior to the chemical mapping experiment. It is important to immediately permeabilize and wash the yeast after harvesting. This depletes endogenous ATP and decreases the salt concentration to below physiological concentration, reducing any chances of nucleosome shifting after harvesting and during the subsequent steps (Teif & Rippe, 2009; Yager, McMurray, & van Holde, 1989).

1. Inoculate 15 mL YPD (10% yeast extract, 20% dextrose, 20% peptone) with a fresh colony of the H4S47C strain.
2. Grow the culture overnight (~12–16 h) at 30 °C with shacking.
3. Inoculate 1 L YPD with the 15-mL culture.
4. Grow at 30 °C with shacking until cells reach a density of $\sim 1 \times 10^7$ cells/mL.
5. Pellet the cells by spinning at $3000 \times g$ for 5 min.
6. Decant the supernatant.
7. Resuspend the cells with 10 mL of 1 M sorbitol and transfer to a 15-mL tube.
8. Spin again at $3000 \times g$ for 5 min.
9. Resuspend the pellet fully with 10 mL of Permeabilizing Buffer (1 M sorbitol, 5 mM BME, and 1 mg of lyticase—Sigma, cat# L5263-200KU).
10. Rotate the solution at room temperature for 5 min.
11. Spin at $3000 \times g$ for 5 min and decant supernatant.
12. Wash the cells with 10 mL of 1 M sorbitol with 0.1% NP-40, resuspending cells completely.
13. Spin at $3000 \times g$ for 5 min and decant supernatant.
14. Repeat the wash step.
15. Resuspend the pellet in 4 mL of Labeling Buffer (1 M sorbitol, 50 mM NaCl, 10 mM Tris–HCl pH 7.5, 5 mM MgCl$_2$, 0.5 mM spermidine, 0.15 mM spermine, 0.1% NP-40).

3.2. Nucleosome labeling

The phenanthroline–iodoacetamide label is light sensitive and soluble in DMSO. DMSO is used in high concentration to help facilitate the entry of the label into the cells and nuclei. Previous publications indicate that DMSO at this concentration does not affect protein–protein and protein–DNA interactions (Ou, Park, & Zhou, 2002; Sidorova, Muradymov, & Rau, 2005). Additionally, we are not concerned about DMSO-induced DNA melting because of the low temperature used during the labeling process (Escara & Hutton, 1980).

1. Make (N(1,10-phenanthroline-5-yl)iodoacetamide) label (Biotium, cat# 92015) stock by dissolving the lyophilized label to a concentration of 7 mM.
2. Add 1 mL of 7 mM label to the cells for a final concentration of 1.4 mM of label and 20% total volume of DMSO.

3. Incubated for 2 h with rotation at room temperature.
4. Continue incubation overnight with rotation at 4 °C.
5. Wash the cells to get rid of excess label with 10 mL of 1 M sorbitol with 0.1% NP-40, resuspend completely. The NP-40 facilitates the removal of the excess label.
6. Repeat the wash two more times (three washes total). These washes are crucial, as excess label will cause background DNA cleavage.
7. Resuspend the pellet in 3 mL of Mapping Buffer (1 M sorbitol, 2.5 mM NaCl, 50 mM Tris–HCl pH 7.5, 5 mM MgCl$_2$, 0.5 mM spermidine, 0.15 mM spermine, and 0.1% NP-40).

3.3. Hydroxyl radical cleavage of nucleosomal DNA

1. Add CuCl$_2$ to the cells in a final concentration of 150 μM.
2. Incubate for 2 min at room temperature.
3. Spin the cells at 3000 × g for 5 min.
4. Decant the supernatant.
5. Wash with 10 mL of Mapping Buffer and resuspend completely.
6. Repeat the spin and wash two more times (three washes total).
7. After the final wash, resuspend pellet in 3 mL of Mapping Buffer.
8. Add MPA (3-mercaptopropionic acid) to a final concentration of 6 mM. MPA will reduce the copper from copper II to copper I. Copper I is necessary to produce the hydroxyl radicals.
9. To initiate the mapping reaction, add H$_2$O$_2$ (hydrogen peroxide) to a final concentration of 6 mM.
10. Incubate the reaction for 20 min at room temperature. The reaction is quick, with most cleavage occurring after 1.5 min.
11. To quench the reaction, add Neocuproine to a final concentration of 2.8 mM. Neocuproine is soluble in DMSO. After quenching the reaction, the protocol is no longer light sensitive.
12. Spin at 3000 × g for 5 min.
13. Resuspend the cells in 2 mL of 1 × TE (10 mM Tris–HCl pH 7.5 and 1 mM EDTA pH 8.0).

3.4. Isolation of mapped DNA fragments and preparation for parallel sequencing

Salt and detergent are added to the cells to disrupt the DNA–histone contacts. If these contacts are not disrupted, DNA can be lost in the subsequent phenol–chloroform purification.

1. Add SDS (sodium dodecyl sulfate) to a final concentration of 2%.
2. Add NaCl (sodium chloride) to a final concentration of 2 M.
3. Incubate for 30 min at room temperature.
4. Dilute the cells with water to reduce NaCl concentration to below 1 M.
5. In a hood, add equal volume of Tris-saturated phenol to the cells.
6. Mix well by vortexing the sample.
7. Spin at 3000 × g for 5 min.
8. Remove the aqueous layer, containing the DNA, and add it to a new tube.
9. If residual contaminants from the interface remain in the aqueous layer, the phenol wash can be repeated until sample is clean.
10. To the aqueous phase, add an equal volume of chloroform.
11. Mix well by vortexing the sample.
12. Spin at 3000 × g for 5 min.
13. Remove the aqueous layer, containing the DNA, and add it to a new tube.
14. Ethanol precipitate the sample by adding 2× volume of ice cold 100% ethanol.
15. Place in the −80 freezer for 30 min.
16. Spin at 5000 × g for 10 min.
17. Remove the ethanol carefully without disrupting the DNA pellet.
18. Fill the tube with 75% ethanol.
19. Place the tube in centrifuge with the opposite side of the tube facing out—forcing the pellet to move to the other side of the tube.
20. Spin at 5000 × g for 10 min.
21. Carefully remove the ethanol without disrupting the DNA pellet.
22. Resuspend the pellet completely in 2 mL of 1× TE by repeatedly washing the walls of the tube.
23. Add RNase to 1 mg/mL final concentration.
24. Incubate overnight at 42 °C.
25. DNA can be purified further, away from RNA and RNase (using various commercial DNA purification kits or by repeating the phenol–chloroform/ethanol precipitation) or otherwise can be added directly to the agarose gel.
26. Make a 2–2.5% agarose gel with 0.5 μg/mL ethidium bromide in both the gel and the running buffer.
27. For standard 0.75–1 cm well sizes, do not exceed 1/3 of the purified DNA for each well.

28. Run the gel until there is a clear separation of the mapping fragments (see Fig. 14.2)—creating a distinct banding pattern.
29. Cut out the shortest DNA band (150–200 bp) from the agarose gel. This band represents the DNAs that span two closely spaced, neighboring nucleosomes.
30. Extract the DNA from agarose gel slab.
31. Determine the final DNA concentration.

3.5. Preparing chemically mapped DNA fragments for parallel sequencing

The order of events in the blunting and phosphorylation of the mapped DNA ends is crucial for the subsequent analysis of the mapped products, as the ends are the sites of cleavage. Initial treatment with Klenow polymerase large fragment first removes 3′ overhangs and fills in 5′ overhangs to create blunted DNA fragments. The use of a single polymerase, such as Klenow, creates a more homogenous set of DNA ends. Lucigen's DNATerminator End Repair Kit was used primarily for its phosphorylation abilities. However, any possible residual overhanging DNA ends would be blunted by this treatment (as the kit blunts all single-stranded DNA overhangs). The blunting and phosphorylation are done sequentially without any purification. This reduces the loss of DNA during the purification steps.

Following the blunting and phosphorylation steps, the chemically mapped DNA fragments are ready to be adapted to any high-throughput parallel sequencing platforms.

1. Remove the overhangs using NEB's Klenow large fragment (NEB, cat# MO210). 5 U of Klenow per 1 μg of DNA, 33 μM dNTPs, 1 × NEB2, 25 μL final volume.
2. Incubate the reaction for 15 min at room temperature.
3. Heat inactivate the reaction for 10 min at 72 °C.
4. Without any purification, phosphorylate the chemically mapped ends using Lucigen's DNATerminator End Repair Kit (cat# 40035-3). 1 μg of DNA terminator/1 μg of DNA, 1 × DNA terminator buffer, 100 μL total volume.
5. For >1 μg of DNA, incubate for 30 min at room temperature. Incubation times can be decreased based on DNA concentration.
6. Quench the reaction with 5 × volume of Qiagen's PB buffer from the MinElute Reaction Cleanup kit (Qiagen, cat# 28204).
7. Purify the DNA using Qiagen's MinElute Reaction Cleanup kit.

8. Elute the DNA in 50 μL of 1 × TE.
9. Determine the DNA concentration.
10. The chemical mapping DNAs are ready to be adapted and processed further for all high-throughput parallel sequencing technologies.

4. STATISTICAL ANALYSIS OF CHEMICAL MAPPING DATA

In this section, we discuss the computational aspects of chemical mapping data. The goal is to define the nucleosome center positions genome-wide based on the chemical cleavage pattern. We illustrate a pipeline developed for yeast data generated using the protocols described above (Brogaard et al., 2012). It should be noted that in practice, the characteristics of chemical cleavages around nucleosome center may depend upon the protocol details used in the experiments. However, we believe this pipeline will provide general guidelines that can be readily adapted to similar data sets.

4.1. Single-end versus paired-end sequencing

The 5′ and 3′ ends of the isolated DNA fragments from chemical mapping approach represent cleavage sites in two consecutive nucleosomes (see Fig. 14.2). Each end of the DNA fragment alone provides sufficient information to define the center of one nucleosome. Thus, under the same sequencing effort, we expect that single-end sequencing provides equivalent mapping accuracy compared to the paired-end sequencing (a feature that is unique to this new mapping approach). For paired-end sequencing, the DNA ends (cleavage sites) can be treated independently in the data analysis. For higher organisms where mappability of short reads becomes an issue due to genome repeats, paired-end sequencing can be more advantageous in chemical mapping if nucleosome positioning in such repeats regions is of interest.

4.2. Example data

The data to be used for illustration here are from Brogaard et al. (2012), which contain two paired-end and four single-end data sets obtained from six independent chemical mapping experiments. Only reads that map to the genome uniquely are kept for the following analyses to avoid ambiguity. The pooled data give 105 million cleavages in total for each strand.

4.3. Primary and secondary site identification

The first step to define the chemical cleavage pattern is to identify the major cleavage sites around a nucleosome center (analogous to the primary and secondary cleavage sites described in the earlier work of Flaus et al., 1996). If the cleavages do occur dominantly at a few specific sites, then the distance between the cleavage sites on different strands will show a distinct pattern because of the mirror symmetry of the two strands (see Fig. 14.3).

We first calculate the cleavage frequency at every genomic location on each strand. We define local cleavage frequency peaks on each strand requiring that peaks are local maxima within ± 73 bp and peaks are unique within every 147 bp. The selected peaks can be regarded as the dominant cleavage sites within nucleosomes. Figure 14.4 plots the frequency of peak–peak distance between the Crick and Watson strands, showing two dominant distances equal to 2 and -5-nucleotides, respectively (with the 5′ to 3′ direction defined as positive). These distances can be explained if the major cleavage sites are at position -1 and $+6$ relative to nucleosome center (dyad defined as position 0). These two positions are referred to as primary and secondary sites.

4.4. Training of cleavage model

We assume that given a nucleosome centered at position 0, cleavages may occur at multiple positions around the two major sites. This is evidenced by patterned cleavage clusters centered around the local cleavage frequency peaks on both strands (see Fig. 14.5). We define a weight template that contains four positions including $(-2, -1, 0, 1)$ around the primary site (-1)

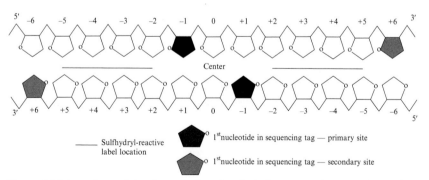

Figure 14.3 Locations of dominate hydroxyl radical cleavages relative to the nucleosome center (base pair 0). The cleavage sites are the first nucleotide in the sequencing outputs. Adapted from Brogaard et al., 2012.

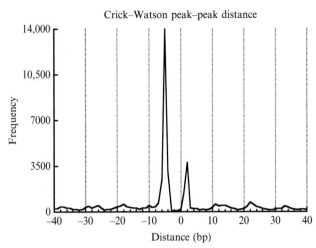

Figure 14.4 Crick–Watson cleavage peak–peak distances showing two dominant distances: $+2$ and -5 nucleotides.

and the other four positions including (4, 5, 6, 7) around the secondary site ($+6$), to represent the cleavage pattern (more complicated templates can be defined, while the resulting changes in nucleosome map are very minor based on our experience). These positions are indexed as $J=(-2,-1,0,1,4,5,6,7)$. We initialize the weight template as $P=\{p_j:j\in J\}=(1/8,1/4,1/8,0,0,1/8,1/4,1/8)$. Let $I=\{1,\ldots,L\}$ be the genomic locations, and $W=\{W_i:i\in I\}, C=\{C_i:i\in I\}$ be the observed cut frequency on Watson and Crick, respectively. At each position $i\in I$, we calculate the template-weighted score as

$$S_i = \sum_{j\in J}(p_j W_{i+j}+p_j C_{i-j}) \qquad (14.1)$$

We select peaks of the template-weighted score genome-wide based on their magnitude sequentially as putative nucleosome centers by requiring no overlap between nucleosomes. The template is updated based on the selected nucleosomes. These two steps are repeated iteratively until convergence. The converged template is (0.06, 0.26, 0.16, 0.056, 0.07, 0.14, 0.17, 0.08) (see Fig. 14.6), suggesting that the primary site tends to have more cleavages than the secondary site.

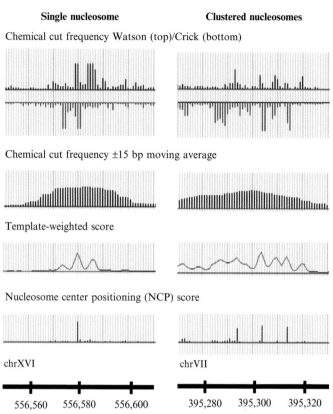

Figure 14.5 A well-positioned nucleosome results in a three-peak template-weighted score pattern with the middle highest peak corresponding to the nucleosome center position. A single, well-positioned nucleosome also results in a dominant NCP score peak indicating the nucleosome center. Clustered nucleosomes produce many peaks in the template-weighted score, causing difficulty in identification of true nucleosome centers. The deconvolution algorithm is able to identify three clear major NCP score peaks, defining three overlapping nucleosomes, separated by 10 bp. Adapted from Brogaard et al., 2012. (See Color Insert.)

4.5. Nucleosome clustering and convolution of cleavage signals

Nucleosomes may be positioned in an overlapping manner due to population or dynamic average. Closely distanced nucleosomes may cause convolution of cleavage signals. The template-weighted score approach can effectively identify nucleosomes that show clear primary–secondary configurations, while convolution of signals makes it unable to accurately define clustered nucleosomes (see Fig. 14.5). To deconvolute the signals, we hypothesize that

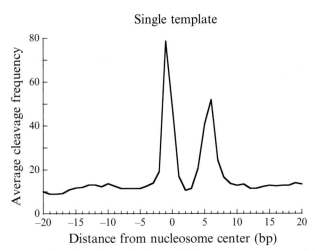

Figure 14.6 A template describing the chemical cleavage pattern around the nucleosome centers. The plot shows the average cleavage frequency from position −20 to +20 bp (5′ to 3′ direction) relative to the nucleosome center. Adapted from Brogaard et al., 2012.

every position can, in principle, be the nucleosome center. Suppose the average cut frequency of the eight positions defined in the template model is $\Lambda = \{\lambda^j : j \in J\}$ where $J = (-2, -1, 0, 1, 4, 5, 6, 7)$ (This is estimated based on nucleosomes defined from the template-weighted score method above). A nucleosome positioned at i incurs cleavages at Watson positions $i+j$ and at Crick positions $i-j$ for $j \in J$. We denote the cuts as $X_i = \{X_i^j : j \in J\}$ and $Y_i = \{Y_i^{-j} : j \in J\}$, respectively. We assume that X_i^j and Y_i^{-j} follow a Poisson distribution with mean parameter $k_{i1}\lambda^j$ and $k_{i2}\lambda^j$, respectively, where k_{i1} and k_{i2} are defined as *n*ucleosome *c*enter *p*ositioning (NCP) scores from the Watson and Crick strands, respectively, at position i, measuring the nucleosome positioning signal strength relative to the average template. We denote these parameters of interest as $K = \{k_{ij} : i \in I, j = 1, 2\}$.

Due to the existence of clustered nucleosomes, X_i and Y_i are not observed, but instead, the cut frequency W_i and C_i on the Watson and Crick strands as a summation of all $X_{i'}^j$ where $i' + j = i$ and all $Y_{i'}^{-j}$ where $i' - j = i$ for $j \in J$, respectively. The following is the convolution model:

$$W_i = \sum_{j \in J} X_{i-j}^j + \varepsilon_{i1}, \quad C_i = \sum_{j \in J} Y_{i+j}^{-j} + \varepsilon_{i2}, \quad (14.2)$$

where $\varepsilon_{i1}, \varepsilon_{i2}$ are the added noise terms, also assumed to follow a Poisson distribution. The noise term is defined to account for the background cleavages, nonspecific to the targeted nucleosome centers. We shall use it as a control to measure the nucleosome positioning signal strength to select nucleosomes. To simplify calculation in the following, we use the average of the lower half of the observed cut frequency in the region of $i\pm73$ of each strand as the mean parameter of the Poisson noise, denoted as $\lambda_{i1}^0, \lambda_{i2}^0$ henceforth. Assuming that all terms in the model are independent, then W_i and C_i have the following distributions:

$$W_i \sim \text{Poisson}\left(\sum_{j\in J} k_{(i-j)1}\lambda^j + \lambda_{i1}^0\right),$$

$$C_i \sim \text{Poisson}\left(\sum_{j\in J} k_{(i+j)2}\lambda^j + \lambda_{i2}^0\right).$$

Algorithmic details for estimation of $K = \{k_{ij}: i \in I, j=1,2\}$ are referred to Brogaard et al. (2012) and http://bioinfo.stats.northwestern.edu/~jzwang/.

4.6. Selection of unique and redundant nucleosome sets

We define the NCP score at position i as the average of the Watson and Crick NCP scores, that is, $\bar{k}_i = (\bar{k}_{i1} + \bar{k}_{i2})/2$. Likewise, we define the average noise at position i as $\lambda_i^0 = (\lambda_{i1}^0 + \lambda_{i2}^0)/2$. Nucleosomes can be called using $\bar{k}_i/\lambda_i^0$ as the signal/noise index. If the value of λ_i^0 is <0.5, we set it as 0.5 to avoid overinflation. In a unique map, we sequentially define nucleosomes in descending order of $\bar{k}_i/\lambda_i^0$ genome-wide by requiring that two neighboring nucleosomes not to overlap or to overlap by no more than a threshold value base pair. For example, in Brogaard et al. (2012), they allowed a maximum of 40 bp overlap between two neighboring nucleosomes. This could help reduce possible miscalls due to miscalls from previous rounds in some regions.

A redundant map can be defined by selecting all positions whose NCP score/noise ratios exceed a given threshold value. Such a redundant map can provide insight into competing preferential nucleosome positions in local regions. The choice of threshold value is arbitrary. One sensible choice could be the smallest NCP score/noise ratio value obtained from the unique map defined above.

4.7. Base composition-dependent cleavage patterns

The cleavage pattern described by the template model may depend on the base composition around the dyad. In the data from Brogaard et al. (2012), they found four substantially different cleavage patterns related to the presence/absence of a nucleotide "A" or "T" at -3 and $+3$, respectively

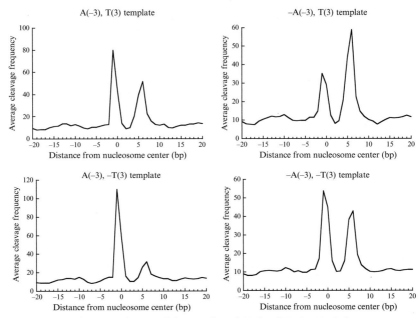

Figure 14.7 A four-template model trained to describe the chemical cleavage pattern around the nucleosome center. Each plot shows the average cleavage frequency from position −20 to +20 bp (5′ to 3′ direction) relative to the nucleosome center. The four-template model contains four separate templates specified by whether having an A at position −3 or a T at position +3 (a "−" in front of A or T stands for absence of that letter). Adapted from Brogaard et al., 2012.

(see Fig. 14.7). The difference could be due to a bias arising from the chemical process. If so, then the resulting NCP score could be biased under the single-template model. To account for this possible bias, one could further train four individual templates using the nucleosomes called from the deconvolution algorithm based on single template. The average frequency at the eight positions of the four models is denoted as:

$$\Lambda_m = \{\lambda_m^j : j \in J\}, m = 1, 2, 3, 4,$$

where J is the eight positions defined above. At each genomic location, the applicable template model is predetermined on each strand based on whether there is an "A" nucleotide at position −3 or a "T" at position +3. The AA/TT/AT/TA signal from the unique map of four-template model is slightly but consistently better than that from the single-template model (Brogaard et al., 2012), suggesting that the four-template model does provide better accuracy in the nucleosome map.

4.8. Nucleosome occupancy scores

The NCP score $\bar{k}_i$ defined above provides a measure of the relative amount of nucleosomes centered at position i. Hence, the nucleosome occupancy at position i, defined as the relative coverage by all possible nucleosomes, can be written as

$$O_i = \sum_{j=i-73}^{i+73} \bar{k}_j. \qquad (14.3)$$

The defined nucleosome occupancy is an evidence-based relative measure of the amount of nucleosome coverage at each position. This is analogous to the reads occupancy commonly used in the MNase mapping, in which the nucleosome occupancy is measured by the number of reads that cover each genomic position. This occupancy is only meaningful in a relative sense when comparing different positions.

There are two issues in this definition. First, the occupancy can be affected by the bias arising from the gel extraction of the mapped genomic DNA fragments. In some regions, particularly those with long linkers (e.g., upstream of TSS), the cleavage cut frequency on one strand can be substantially lower than the other. As a result, the NCP score on the strand with low-cleavage frequency tends to be underestimated. Let δ_i denote the ratio of cleavage total of the two strands in the region $i \pm 73$. We can set the rule that if $|\log_{10}(\delta_i)|$ exceeds two standard deviations from 0, then the corrected NCP score is defined as $\bar{k}_i' = \max(\bar{k}_{i1}, \bar{k}_{i2})$, otherwise $\bar{k}_i' = (\bar{k}_{i1} + \bar{k}_{i2})/2$.

The second issue is the noise. The definition in Eq. (14.3) does not account for possible noise component remaining in $\bar{k}_i'$ due to conservative noise level used in the deconvolution (i.e., the average of the lower half of cleavage frequency within every 147-bp window). To control the noise in the occupancy, one could set the same NCP score/noise ratio criterion as used in defining the redundant map (denoted R) (Brogaard et al., 2012). The bias corrected nucleosome occupancy score based on the redundant map thus is defined as

$$O_i = \sum_{|j-i|\leq 73, j\in R} \bar{k}_j'. \qquad (14.4)$$

4.9. Sequencing depth

One important practical issue is the sequencing depth needed for the chemical approach. In MNase mapping, increasing sequencing coverage does not necessarily improve the mapping accuracy because of systematic sequence preferences of MNase. The chemical mapping relies on the cleavage signal to accurately identify the nucleosome center. Thus, more cleavages (or larger sample size) provide larger power to distinguish the nucleosome positioning signal from noise. An interesting question is how many cleavages are needed to generate accurate nucleosome map? We did a simulation study by randomly sampling a fixed fraction of sequences from the original data by Brogaard et al. (2012). At each fraction equal to 0.2, 0.5, and 0.8, we generated the unique map and compared the AA/TT/TA/AT dinucleotide frequency in the nucleosome DNAs with the original map from the full data set. At fraction equal to 0.2 ($\sim$2 cleavages/bp), about 76% nucleosomes from the unique map were identical to the original map, while the AA/TT/TA/AT signal shows an obvious degradation (results not shown). In contrast at fraction 0.5 and 0.8 ($\sim$5 and 8 cleavages/bp), about 82% and 92% of the nucleosomes were identical to the original map, respectively, and the AA/TT/TA/AT plot almost overlapped with the original plot. This simulation shows that the mapping accuracy does improve with increased sequencing coverage (as reflected by the dinucleotide plot). It further suggests that an average of 5 cleavages per base pair would probably provide reasonable accuracy in chemical mapping experiments.

5. SUMMARY

The development of the chemical mapping technique provides a new approach for mapping nucleosomes *in vivo* with unprecedented accuracy and detail. The chemical map from *S. cerevisiae* has revealed new aspects regarding the role of nucleosomes in chromosome functions, gene regulation, and higher-order chromatin structure. It should be noted that change of experimental protocols may result in alteration of chemical cleavage patterns, or even details of nucleosome positioning. The pipeline we describe here provides a general framework that can be readily adapted to analyze new chemical mapping data in future studies. We expect that chemical map technique can be applied to higher organisms to advance our understanding of nucleosomes.

ACKNOWLEDGMENTS

We would like to acknowledge the life and achievements of Jonathan Widom who passed away during the development of this protocol. We are grateful to Northwestern University's Genomic Core for all sequencing completed for this project. The work was supported by NIH grants T32GM00806 (K.R.B.), R01GM058617 (J.W.), R01GM075313 (J-.P.W.), and U54CA143869 (J.W.).

REFERENCES

Amberg, D. C., Burke, D. J., & Strathern, J. N. (2005). Methods in yeast genetics: A Cold Harbor Laboratory course manual (5th ed.), Cold Spring Harbor Laboratory Press.

Brogaard, K. R., Xi, L., Wang, J. P., & Widom, J. (2012). A map of nucleosome positions in yeast at base-pair resolution. *Nature, 486,* 496–501.

Churchman, L. S., & Weissman, J. S. (2011). Nascent transcript sequencing visualizes transcription at nucleotide resolution. *Nature, 469,* 368–373.

Cole, H. A., Howard, B. H., & Clark, D. J. (2011). The centromeric nucleosome of budding yeast is perfectly positioned and covers the entire centromere. *Proceedings of the National Academy of Sciences of the United States of America, 108,* 12687–12692.

Dingwall, C., Lomonossoff, G. P., & Laskey, R. A. (1981). High sequence specificity of micrococcal nuclease. *Nucleic Acids Research, 9,* 2659–2673.

Escara, J. F., & Hutton, J. R. (1980). Thermal stability and renaturation of DNA in dimethyl sulfoxide solutions: Acceleration of the renaturation rate. *Biopolymers, 19,* 1315–1327.

Flaus, A., Luger, K., Tan, S., & Richmond, T. J. (1996). Mapping nucleosome position at single base-pair resolution by using site-directed hydroxyl radicals. *Proceedings of the National Academy of Sciences of the United States of America, 93,* 1370–1375.

Koslover, E. F., Fuller, C. J., Straight, A. F., & Spakowitz, A. J. (2010). Local geometry and elasticity in compact chromatin structure. *Biophysical Journal, 99,* 3941–3950.

Li, G., Levitus, M., Bustamante, C., & Widom, J. (2005). Rapid spontaneous accessibility of nucleosomal DNA. *Nature Structural & Molecular Biology, 12,* 46–53.

Li, G., & Widom, J. (2004). Nucleosomes facilitate their own invasion. *Nature Structural & Molecular Biology, 11,* 763–769.

Lipford, J. R., & Bell, S. P. (2001). Nucleosomes positioned by ORC facilitate the initiation of DNA replication. *Molecular Cell, 7,* 21–30.

Mao, C., Brown, C. R., Griesenbeck, J., & Boeger, H. (2011). Occlusion of regulatory sequences by promoter nucleosomes in vivo. *PLoS One, 6,* e17521.

Nair, N. U., & Zhao, H. (2009). Mutagenic inverted repeat assisted genome engineering (MIRAGE). *Nucleic Acids Research, 37,* e9.

Ou, W. B., Park, Y. D., & Zhou, H. M. (2002). Effect of osmolytes as folding aids on creatine kinase refolding pathway. *The International Journal of Biochemistry & Cell Biology, 34,* 136–147.

Petesch, S. J., & Lis, J. T. (2008). Rapid, transcription-independent loss of nucleosomes over a large chromatin domain at Hsp70 loci. *Cell, 134,* 74–84.

Richmond, T. J., & Davey, C. A. (2003). The structure of DNA in the nucleosome core. *Nature, 423,* 145–150.

Schwartz, S., Meshorer, E., & Ast, G. (2009). Chromatin organization marks exon-intron structure. *Nature Structural & Molecular Biology, 16,* 990–995.

Sidorova, N. Y., Muradymov, S., & Rau, D. C. (2005). Trapping DNA-protein binding reactions with neutral osmolytes for the analysis by gel mobility shift and self-cleavage assays. *Nucleic Acids Research, 33,* 5145–5155.

Teif, V. B., & Rippe, K. (2009). Predicting nucleosome positions on the DNA: Combining intrinsic sequence preferences and remodeler activities. *Nucleic Acids Research*, *37*, 5641–5655.

Yager, T. D., McMurray, C. T., & van Holde, K. E. (1989). Salt-induced release of DNA from nucleosome core particles. *Biochemistry*, *28*, 2271–2281.

AUTHOR INDEX

Note: Page numbers followed by "*f*" indicate figures, and "*t*" indicate tables.

A

Abbondanzieri, E. A., 5–6, 9–10
Abraham, L., 252
Adams, S. R., 64–65, 147–148, 166, 170–171
Ahmad, K., 170–171, 298
Ahmadiyeh, N., 115–116
Ai, L., 186
Alani, E., 30–33
Alber, F., 116
Albert, I., 147, 166, 233–234
Alekseyenko, A. A., 302–304
Alfieri, J. A., 164–165
Allis, C. D., 146, 302
Allo, M., 272
Almouzni, G., 30
Altenburger, W., 165–166
Altheim, B. A., 207–211
Altschul, S. F., 196
Altschuler, S. J., 147, 206–207
Amberg, D. C., 317–318
Amitani, I., 5
Amstislavskiy, V., 290*f*
Ananda, G., 134–135
Anders, S., 273, 290*f*, 291
Anderson, J. D., 233–234
Andreassi, C., 272–273
Andreu-Vieyra, C., 186
Andrews, A. J., 30
Andrieu-Soler, C., 91–92
Andronescu, M., 116
Antos, J. M., 64–65
Aparicio, O., 302–304
Araki, Y., 302–304
Arcangeli, M., 91–92
Arnaout, R. A., 91, 114–115, 116, 118–119
Ashkin, A., 39–41
Ast, G., 316
Axelrod, D., 71
Ayton, P. M., 119–121
Azari, H., 197–199

B

Baban, S., 119–121
Babcock, H. P., 60, 69
Badis, G., 229–230
Bai, L., 30–35, 32*f*, 36–37, 42–43, 46–49, 47*f*, 50–53, 51*f*, 52*f*, 54
Bailey, S. T., 298
Bajpai, R., 298
Balci, H., 30–31
Balwierz, P. J., 119–121
Banaru, M., 290*f*
Banfalvi, G., 253–254
Bannister, A. J., 30
Bantignies, F., 91–92
Bao, Y. H., 36–37, 66–69, 233–234
Barbacioru, C., 307
Barrera, L. O., 298
Barski, A., 206–207, 298–299, 302–304
Bartel, D. P., 272–273
Bartholomew, B., 60–62, 186, 190
Bartley, L. E., 60
Barton, G. J., 4–5
Baruah, H., 64–65
Baskin, J. M., 64–65
Baskin, R. J., 5
Bau, D., 114–115, 118–119, 138
Beck, A. H., 273
Becker, P. B., 4–5, 30, 60–62, 206–207, 224–226, 234
Bedei, M., 253–254
Bell, O., 30, 206–207
Bell, S. P., 316
Belsky, J. A., 147–148, 166, 170–171
Beltramini, C., 119–121
Benes, V., 273, 290*f*
Benesch, R. E., 73–74
Bentley, D. R., 170–171
Berman, B. P., 186
Bernardi, G., 211–214
Bernaschek, G., 253
Bernstein, B. E., 298
Bertozzi, C. R., 64–65

335

Bhat, W., 171
Bird, Ch 17
Birol, I., 110
Biserni, G., 119–121
Bjorkholm, J. E., 39–41
Blanchette, M., 114–115, 118–122, 125, 137
Blankenberg, D., 134–135
Blaschke, D., 206–211, 227
Blobel, G. A., 252
Block, S. M., 5–6, 9–10, 41, 43–44
Blom, M., 91–92
Blosser, T. R., 60–62
Bockelmann, U., 33, 45–46
Boeger, H., 4–5, 316
Bofkin, L., 272
Boitano, M., 188
Bonaldi, T., 211
Bonnet, J., 91–92
Bonte, E. J., 4–5, 60–62
Bonthron, D. T., 186, 197–199
Borodina, T., 290f
Botstein, D., 216
Bouazoune, K., 186, 190
Bowman, G. D., 4–5, 30
Boyd, S. D., 206–207, 302
Boyer, L. A., 298
Brambrink, T., 298
Branco, M. R., 115–116
Braunschweig, U., 107–109
Bray, N., 147, 206–207
Brehm, A., 211
Brem, R. B., 272–273
Brogaard, K. R., 317, 324, 329–330, 331, 332
Brower-Toland, B. D., 23, 30–31, 39–41, 53
Brown, C. R., 316
Brown, G. D., 115–116
Brown, K. D., 186
Brown, M., 302
Brugmann, S. A., 298
Bryne, J. C., 91–92
Buchberg, A. M., 119–121
Buranachai, C., 30–31
Burge, C. B., 272–273
Burhans, R., 134–135

Burke, D. J., 317–318
Bussemaker, H. J., 107–109
Bustamante, C., 5–10, 12, 17, 30–31, 62–63, 316
Byron, M., 114–115, 118–119, 138
Byun, H. M., 186

C

Cai, Y. F., 64–65
Cairns, B. R., 4–5, 9–10, 13, 14–15, 14f, 19f, 20, 30, 60–62, 146, 229–230
Camblong, J., 272
Campos, E. I., 30
Cao, H., 198f
Capriotti, E., 114–115, 118–119, 138
Carey, B. W., 298
Carey, M., 272
Carninci, P., 272
Carr, I. M., 186, 197–199
Carragher, B. O., 65–66
Carroll, J. S., 115–116
Chakravarthy, S., 36–37, 66–69
Chambon, P., 214–215, 217–218, 223
Chan, E. T., 229–230
Chartrand, P., 116
Chatterjee, N., 186
Chemla, D. S., 60
Chemla, Y. R., 5–10, 12, 16, 30–31
Chen, H. L., 116
Chen, L., 116
Cheng, A. W., 298
Cheng, W., 5–6, 69
Chepelev, I., 304
Chien, F. T., 25f
Childs, R. W., 302–304
Choi, E. A., 273
Choi, Y. A., 64–65
Chou, M. M., 252
Chu, S., 39–41
Chung, H., 165–166
Chung, L. M., 306–307
Churchill, M. E., 30
Churchman, L. S., 316
Clapier, C. R., 4–5, 9–10, 13, 14–15, 14f, 19f, 30, 60–62
Clark, D. J., 147–148, 149–150, 158–159, 160f, 161–166, 163f, 316

Clark, J., 192–193
Clark, S. J., 196
Clark, T. A., 188
Clarklewis, I., 64–65
Clauder-Munster, S., 272
Cleary, M. L., 119–121
Cline, M. S., 198f
Cobb, B. S., 90
Cole, H. A., 147–148, 158–159, 160f, 161–165, 163f, 316
Cole, P. A., 64–65
Comet, I., 91–92
Connors, J., 110
Coraor, N., 134–135
Corces, V. G., 91
Cornish, V. W., 64–65
Corona, D. F. V., 4–5, 60–62
Cortez, C. C., 186
Coulter, B., 253–254, 255f, 257f, 259f
Cox, A. J., 133–134
Cox, T., 216
Crabtree, G. R., 4–5
Cremer, M., 115–116
Cremer, T., 115–116
Creyghton, M. P., 298
Cuddapah, S., 206–207, 298–299, 302–304
Cuff, J., 298
Cui, K., 178–179, 206–207, 298–299, 302–304
Cui, Y., 17

D

Darst, R. P., 186, 190, 197–199, 201–202
Das, C., 30
Daub, C. O., 119–121
Davey, C. A., 53, 233–234, 316
Davey, M., 129
David, L., 272
Davis, R. W., 147, 206–207
Davis-Turak, J., 273
Dawson, P. E., 64–65
de Boer, E., 91–92
De Gobbi, M., 91–92
de Laat, W., 90, 91–92, 115–116
de Rooi, J. J., 91–92
de Wit, E., 91–92, 101, 107–109, 116
Deal, R. B., 170–171

Dean, A., 165–166
Dechassa, M. L., 30–35, 36–37, 42, 48–49, 49f, 50, 186
Dekker, J., 90, 91, 114–115, 116, 118, 121–122, 125, 134–137
Dekker, M., 90, 114–115, 116
Deleyrolle, L. P., 197–199
Delmas, A. L., 186, 199, 201–202
Delrow, J., 234
Dennehey, B. K., 30
Di Giammartino, D. C., 272–273
Dingwall, C., 165–166, 316
Dion, M. F., 147, 206–207
Dirac-Svejstrup, A. B., 206–211, 227
Doddapaneni, H., 115–116
Donnenberg, A. D., 253
Dorin, M. (1994), Ch 11
Dostie, J., 91, 114–115, 116, 118–119, 121–122, 125, 137
Dowen, R. H., 186
Dragowska, W., 119–121
Dreszer, T. R., 198f
Drew, H. R., 233–234
Drissen, R., 90
Duan, Z., 116
Duboule, D., 91–92
Duda, S., 25f
Dungrawala, H., 252
Dunkel, I., 165–166
Dunlap, D., 60–62
Durbin, R., 133–134, 158, 178–179, 180
Dyer, L. M., 186, 199, 201–202
Dyer, P. N., 36–37, 66–69
Dziedzic, J. M., 39–41

E

Ebright, R. H., 64–65
Ebright, Y. W., 64–65
Edayathumangalam, R. S., 36–37, 66–69
Edgar, R. C., 196
Egelhofer, T. A., 302–304
Egger, G., 186, 199
Ehrenhofer-Murray, A. E., 165–166
Ekwall, K., 206–207
Elgin, S. C., 211
Elnitski, L., 134–135
Emelyanov, A. V., 4–5

Enderle, T., 60
Ertel, F., 206–211, 227
Escara, J. F., 320–321
Essevaz-Roulet, B., 33, 45–46
Ethier, S. D., 114–115
Euskirchen, G., 30, 147, 206–207, 207f, 214–215
Evans, E., 41

F

Fan, H. Y., 60–62
Fan, X., 272–273
Fatemi, M., 186
Faulhaber, I., 211–214
Felsenfeld, G., 165–166, 211–214
Feng, Z., 306–307
Fennell, T., 158, 159–160
Ferdinand, M. B., 30–31
Fernandez-Suarez, M., 64–65
Ferraiuolo, M. A., 114–115, 118–122, 125, 137
Ferreira, H., 60–62
Field, Y., 30, 147–148, 165–166, 207f, 214–215
Fink, G. R., 216
Fire, A. Z., 206–207, 302
Fishel, R., 41–42
Fisher, C., 91–92
Fishilevich, E., 273
Flanagan, J. F., 20
Flanagan, L. A., 273
Flaus, A., 4–5, 60–62, 316–317, 325
Flavell, R. A., 90, 115–116
Flores, O., 302
Flusberg, B. A., 188
Flynn, R. A., 298
Foissac, S., 273
Foley, J. W., 273
Fondufe-Mittendorf, Y., 30, 147–148, 165–166, 207f, 214–215
Forrest, A. R., 119–121
Forster, T., 60
Forties, R. A., 30–31, 41–42
Fox, M. H., 252
Fox-Walsh, K., 273
Fragoso, G., 164–165, 166, 234–235
Fraser, J., 114–115, 118–122, 125, 137

Fraser, P., 90
Frenkel, B., 186
Friedli, M., 91–92
Friedman, R. C., 273
Fritsch, E. F., 226, 227
Fu, X. D., 273
Fujita, P. A., 198f
Fulbright, R. M., 30–35, 32f, 36–37, 42–43, 46–49, 47f, 50–53, 51f, 52f, 54
Fuller, C. J., 316
Fullwood, M. J., 91
Furey, T. S., 103–104
Furuyama, T., 298
Fyodorov, D. V., 36–37

G

Gagneur, J., 272
Galjaard, R. J., 91–92
Gal-Yam, E. N., 186, 199
Gangaraju, V. K., 186, 190
Gansen, A., 38–39
Gao, Y., 5, 9–10, 13, 14–15, 14f, 19f, 60–62
Gardner, K. E., 146
Garinther, W. I., 207–211
Gascoyne, R., 110
Gasser, S. M., 115–116
Gauss, G. H., 69
Gdula, D. A., 4–5, 60–62
Gehrig, C., 91–92
Gelles, J., 35, 41, 43–44
Gemici, Z., 30–33
Gemmen, G. J., 30–31
Germond, J. E., 214–215, 217–218, 223
Gerstein, M., 272, 298–299, 304
Getz, G., 109
Gheldof, N., 116
Giannoukos, G., 298–299
Giardine, B., 134–135
Gillemans, N., 90
Giresi, P. G., 94–96, 303
Gish, W., 196
Gittes, F., 6–9
Gkikopoulos, T., 4–5, 229–230, 234
Goecks, J., 134–135
Gondor, A., 91, 116
Gossett, A. J., 30, 147, 207f, 214–215

Granovskaia, M., 272
Green, M. R., 146
Greenleaf, W. J., 5–6, 9–10
Greil, F., 107–109
Griesenbeck, J., 4–5, 316
Griffin, B. A., 64–65
Grill, S. W., 5, 9–10, 18–20, 23–24, 60–62
Gross-Bellard, M., 217–218
Gross-Bellark, M., 214–215, 217–218, 223
Grosveld, F., 90, 91–92, 115–116
Grotenbreg, G. M., 64–65
Guipponi, M., 91–92
Gustafson, T. L., 186
Guttman, M., 109

H

Ha, T., 30–31, 60, 62, 69, 73–74, 82–83
Hadjur, S., 90
Hager, G. L., 164–165, 166, 234–235
Hakim, O., 91–92
Hall, M. A., 30–35, 32f, 36–37, 42–43, 46–49, 47f, 50–53, 51f, 52f, 54
Hallen, L., 290f
Hamiche, A., 4–5, 60–62
Han, H. F., 186
Han, T., 272–273
Handoko, L., 303
Handsaker, B., 158, 159–160
Hardison, R. C., 134–135
Hargreaves, D. C., 4–5
Harkness, T. A., 207–211
Hart, C. L., 65–66
Hartley, P. D., 229–230, 234
Hashimoto, K., 192
Hauger, F., 38–39
Haugland, R. P., 60
Haushalter, K. A., 30–31
Hawkins, R. D., 186
Hayashizaki, Y., 114–115, 118–121
Hayes, J. J., 171
He, H. H., 298, 302
Heath, H., 90, 115–116
Heise, F., 165–166
Henderson, S., 91–92
Hengstschlager, M., 253

Hengstschlager-Ottnad, E., 253
Henikoff, J. G., 170–171, 178–179
Henikoff, S., 147–148, 166, 170–171, 298
Hennikoff, J. G., 147–148, 166
Henry, N. M., 60–62
Herman, P., 115–116
Herschlag, D., 60
Hertel, C. B., 206–211, 214–215, 227
Hertel, K. J., 273
Heslot, F., 33, 45–46
Heyduk, T., 64–65
Hillen, W., 13
Hinrichs, A. S., 198f
Hinrichs, W., 13
Hinz, W., 129
Hirt, B., 214–215, 217–218, 223
Hochheimer, A., 206–211, 227
Hockman, D. J., 207–211
Hoffman, C. J., 186, 197–199
Hohng, S., 62
Homer, N., 158, 159–160
Homminga, I., 91–92
Hon, G., 186
Honan, T. A., 91, 114–115, 116, 118–119
Hong, S. H., 138
Hoose, S. A., 186, 190
Horowitz-Scherer, R., 20
Horsman, D., 110
Hörz, W., 60–62, 163f, 165–166, 206–211, 214–215
Hough, M. R., 119–121
Howard, B. H., 147–148, 158–159, 160f, 161–165, 163f, 316
Hu, G., 298, 302, 303
Hu, J. F., 116
Huber, W., 272, 291
Huebert, D. J., 298
Huff, J. T., 229–230
Hughes, J. R., 91–92
Hughes, T. R., 147, 165–166, 206–207, 233–234
Hu-Li, J., 302–304
Humphray, S., 216
Hurtado, A., 115–116
Hutton, J. R., 320–321
Huynh, V. A., 214–215

I

Ibrahim, J. G., 303
Ichikawa, J., 147, 206–207
Imakaev, M., 91, 94–96, 116
Imbalzano, A. N., 146
Imhof, A., 211
Ioshikhes, I. P., 206–207, 272–273
Irvine, K. M., 119–121
Ishitsuka, Y., 30–31
Israel, L., 206–211, 227
Issac, B., 298–299
Itoi, E., 192
Iyer, V. R., 94–96, 233–234
Izhaky, D., 6–10, 12
Izumchenko, E. G., 186, 199, 201–202

J

Jackson, V., 214–215
Jacques, P. E., 171
Jacquier, A., 272
Jaffe, D. B., 298–299
Jan, C. H., 273
Jantzen, B. C., 30–33, 39–41, 43
Järvelin, A. I., 273, 290f
Javaid, S., 41–42
Jayathilaka, N., 116
Jennings, E. G., 302–304
Jeong, S., 186, 199
Jessen, W. J., 186
Ji, Z., 272–273
Jia, L., 115–116, 186
Jiang, B., 272–273
Jiang, C., 201, 206–207, 272–273
Jiang, J., 30–33
Jin, J., 30–33, 36–37, 48, 54
Jin, Y., 272–273
John, S., 91–92, 164–165, 166
Johnson, A., 4–5
Johnson, B. E., 198f
Johnson, D. S., 30–33, 36–37, 48, 50–53, 54
Johnson, E. M., 198f
Johnson, S. M., 206–207, 302
Jones, G. M., 216
Jones, P. A., 186, 190, 199
Joo, C., 30–31, 82–83
Joye, E., 91–92

K

Kadauke, S., 252
Kadonaga, J. T., 30–31, 36–37, 147, 206–207, 207f, 214–215
Kaestner, K. H., 159
Kalhor, R., 116
Kamal, M., 298
Kapanidis, A. N., 64–65
Kaplan, N., 30, 147–148, 165–166, 207f, 214–215
Kapranov, P., 273
Karolchik, D., 198f
Kashlev, M., 30–33, 36–37, 48, 54
Kassabov, S. R., 60–62, 186
Kauffman, M. G., 253
Ke, P. C., 30–31
Kelly, T. J., 253
Kelly, T. K., 186, 190
Kent, N. A., 147–148, 166, 170–171
Kent, S. B. H., 64–65
Kent, W. J., 103–104
Kerkhoven, R. M., 91–92
Kidder, B. L., 302–304
Kilgore, J. A., 186
Killian, J. L., 30–31, 56
Kim, J., 94–96
Kim, M., 306–307
Kim, S. W., 273
Kim, T. H., 298
Kim, Y., 149–150, 164–165
Kim, Y. J., 116
Kingston, R. E., 60–62, 146, 186
Kirchmaier, A. L., 252
Kireeva, M. L., 30–33, 36–37, 48, 54
Kislyuk, A. O., 188
Kladde, M. P., 186, 190
Kleckner, N., 90, 114–115, 116
Klevebring, D., 276
Klingenhoff, A., 304
Klous, P., 90, 91–92, 101, 107–109, 115–116
Klugman, S., 302–304
Knight, J., 64–65
Koch, S. J., 30–33, 39–41, 42, 43
Kokubun, S., 192
Kolasinska-Zwierz, P., 302–304
Konev, A. Y., 4–5

Kooistra, T., 298
Kooren, J., 90, 115–116
Korber, P., 30, 206–211, 207f, 214–215, 217–218, 227, 234
Kornberg, R. D., 4–5, 60–62
Kornblihtt, A. R., 272
Kortkhonjia, E., 64–65
Koslover, E. F., 316
Kouzarides, T., 30
Kowalczykowski, S. C., 5
Kozarewa, I., 178–179, 180
Krajewski, W. A., 224–226
Krassovsky, K., 147–148, 166, 170–171
Kraus, W. L., 53
Krobitsch, S., 165–166, 290f
Kroon, E., 119–121
Krosl, J., 119–121
Krstulovic, B. S., 206–211
Kruithof, M., 25f
Krzywinski, M., 110
Ku, M., 298–299
Kuehn, M. S., 198f
Kuhn, R. M., 198f
Kunert, N., 211
Kwan, G., 306–307
Kwon, H., 146

L

Lacroute, P., 273
Lai, J., 186
Lajoie, B. R., 114–115, 118–119, 122, 134–137, 138
Lamparska-Kupsik, K., 192–193
Landel, C. C., 60–62
Lander, E. S., 109
Landick, R., 9–10, 35, 41, 43–44
Landry, J., 4–5
Landry, M. P., 16
Landsman, D., 298–299
Langerak, A. W., 91–92
Langmead, B., 133–134, 289, 302
Langowski, J., 38–39, 100–101
Längst, G., 4–5, 60–62, 206–211, 214–215, 234
Lansdorp, P. M., 119–121
Lantermann, A. B., 206–207
Lasken, R. S., 195

Laskey, R. A., 165–166, 217–218, 316
Lau, A. W., 252
Lawrence, H. J., 119–121
Lawrence, J. B., 114–115, 118–119, 138
Lazarus, R., 134–135
Leblanc, B. P., 91–92, 164–165
Lee, A. S., 186, 199
Lee, C., 116, 307
Lee, J. H., 188
Lee, J. Y., 272–273
Lee, K. M., 67–69, 302–304
Lee, S., 306–307
Lee, T. I., 298
Lee, W. L., 91, 114–115, 116, 118–119, 147, 206–207
Lei, Y., 302
Leleu, M., 91–92
Lemieux, J., 178–179
Levitus, M., 62–63, 316
Li, B., 272
Li, G., 62–63, 316
Li, H., 133–134, 158, 159–160, 273
Li, M., 30–35, 36–37, 42, 48–49, 50, 56
Li, T., 116
Li, X., 206–207
Li, Z., 159
Lia, G., 60–62
Liang, G., 186
Lieb, J. D., 94–96, 165–166, 303
Lieberman, E., 298–299
Lieberman-Aiden, E., 91, 94–96, 116
Lim, B., 306–307
Lim, C. Y., 4–5
Lin, A. W., 65–66
Lin, J. C., 186
Lindahl (1948), Ch 11
Lindstrom, K., 164–165
Ling, J. Q., 116
Linke, C., 165–166
Lipford, J. R., 316
Lipman, D. J., 196
Lis, J. T., 23, 30–35, 32f, 36–37, 42–43, 46–49, 47f, 50–53, 51f, 52f, 54, 55f, 60–62, 316
Lister, R., 186
Liu, J., 91
Liu, M. H., 91

Liu, X. S., 272–273, 302
Liu, Y. J., 147, 206–207
Locatelli, F., 119–121
Loeb, G. B., 170–171
Logie, C., 25f
Lohman, T. M., 69
Lomonossoff, G. P., 165–166, 316
Lopez-Barragan, M. J., 178–179
Lopez-Jones, M., 91–92
Lorch, Y., 60–62
Lowary, P. T., 19–20, 33–35, 36–37
Lower, K. M., 91–92
Lu, J., 273
Lubling, Y., 147–148, 165–166
Luco, R. F., 272
Luger, K., 25f, 30–35, 36–37, 42, 48–49, 50, 52f, 53, 66–69, 146, 217–218, 233–234, 316–317, 325
Lundeberg, J., 276
Lundin, S., 276
Lusser, A., 206–207
Lutz, C. S., 272–273

M

MacAlpine, D. M., 147–148, 166, 170–171
Mackowiak, S. D., 272–273
Mäder, A. W., 25f, 30, 52f, 53, 146, 217–218
Madhani, H. D., 229–230, 234
Madrid, T. S., 65–66
Maeder, A. W., 53
Magen, A., 304
Mangan, M., 134–135
Mangone, M., 272–273
Maniatis, T., 226, 227
Manley, J. L., 272–273
Manohar, M., 30–31
Manoharan, A. P., 272–273
Manukyan, A., 252
Mao, C., 316
Mariano, P., 91, 116
Marin, M., 91–92
Markham, A. F., 186, 197–199
Martin, D. M., 4–5
Martinez-Hernandez, L., 64–65
Martino, V., 119–121
Mascrez, B., 91–92

Mavrich, T. N., 147, 166, 206–207, 233–234, 272–273
Mayr, C., 272–273
McCall, P. M., 16
McCarthy, D. J., 306–307
McCue, K., 298–299, 304
McDaniell, R. M., 94–96
McGhee, J. D., 165–166
McIlwain, S., 116
McKinney, S. A., 73–74, 82–83
McLaughlin, N., 164–165
McMurray, C. T., 319–320
Medeiros, L. A., 298
Meister, P., 115–116
Mejia, Y. X., 5–6
Mekler, V., 64–65
Mellor, J., 4–5, 229–230, 234
Mendel, D., 64–65
Meshorer, E., 316
Meyer, C. A., 298, 302
Miele, A., 116
Mihardja, S., 5, 9–10, 18–20, 23–24, 30–31, 60–62
Mikkelsen, T. S., 298–299
Mileski, W., 129
Miller, L. W., 64–65
Miller, W., 100–101, 196
Minoda, A., 302–304
Miranda, T. B., 186, 190
Misteli, T., 272
Miura, H., 114–115
Miyamoto, C., 119–121
Mizuguchi, G., 4–5
Moazed, D., 4–5
Moffitt, J. R., 5–10, 12, 30–31
Mohamed, Y. B., 91
Monaghan, A. P., 273
Montavon, T., 91–92
Moore, I. K., 30, 147–148, 165–166, 207f, 214–215
Moorhouse, A., 147–148, 166, 170–171
Moqtaderi, Z., 30, 147, 206–207, 207f, 214–215, 272–273
Moreira, A., 272–273
Morgan, R. D., 188
Morse, R. H., 147, 206–207
Mortazavi, A., 298–299, 304, 306–307

Moshkin, Y., 91–92, 101, 107–109, 116
Muir, T. W., 64–65
Mukhopadhyay, J., 64–65
Mullikin, J. C., 133–134
Munson, K., 192–193
Muradymov, S., 320–321
Murray, I. A., 188
Musladin, S., 206–211
Muthurajan, U. M., 36–37, 66–69
Myers, E. W., 196

N
Naber, N., 65–66
Nabilsi, N. H., 186, 190
Nacu, S., 289
Nadler, M., 119–121
Nagalakshmi, U., 272, 298–299, 304
Nagamune, T., 64–65
Nagarajavel, V., 164–165
Nagy, G., 253–254
Nair, N. U., 317–318
Narlikar, G. J., 60–62, 65–69
Nekrutenko, A., 134–135
Ngau, W. C., 298–299
Nichols, P. W., 186
Nielsen, C. B., 272–273
Nieuwland, M., 91–92
Ning, Z., 133–134
Nishida, K., 272–273
Noga, S. J., 253
Noordermeer, D., 91–92
Nora, E. P., 91–92, 107–109
Nordman, E., 307
North, J. A., 30–31, 41–42
Northrup, D., 298
Nourani, A., 171
Novick, P., 216
Nugent-Glandorf, L., 9–10

O
Ogletree, D. F., 60
Oh, J., 306–307
Olivares, E. C., 188
Ombodi, T., 253–254
Ong, C. T., 91
Orozco, M., 302
Orth, P., 13

Oshlack, A., 306–307
Ou, W. B., 320–321
Oudet, P., 214–215, 217–218, 223
Owen-Hughes, T., 4–5, 60–62
Ozsolak, F., 273

P
Padgett, R. A., 272–273
Pagie, L., 91–92
Palm, C. J., 272
Palmer, J., 60–62
Palstra, R. J., 90, 115–116
Pan, Y. F., 91
Pan, Z., 272–273
Papamichos-Chronakis, M., 4–5
Pardo, C. E., 186, 190, 197–199, 201–202
Park, S. Y., 4–5
Park, Y. D., 320–321
Park, Y. J., 30
Parkhomchuk, D., 290f
Parnell, T. J., 229–230
Paszkiewicz, K., 147–148, 166, 170–171
Patel, S. S., 50–53
Patterton, H. G., 214–215
Peckham, H., 147, 206–207
Pelechano, V., 273, 290f
Pelizzola, M., 186
Pena-Castillo, L., 229–230
Peng, W., 302–304
Perkins, T. T., 9–10
Perocchi, F., 272
Pertea, G., 306–307
Pession, A., 119–121
Peterson, C. L., 4–5, 20, 23, 30–35, 32f, 36–37, 42, 46, 48, 54, 55f, 60–62, 206–211, 227
Petesch, S. J., 316
Pettersson, E., 276
Pfaffle, P., 214–215
Philipsen, S., 90
Pieters, R., 91–92
Pinskaya, M., 4–5, 229–230, 234
Plath, K., 298
Ploegh, H. L., 64–65
Podhraski, V., 4–5

Poirier, M. G., 41–42
Pombo, A., 115–116
Pomerantz, M. M., 115–116
Pondugula, S., 186, 190
Pop, M., 133–134, 289, 302
Popp, M. W., 64–65
Porreca, G. J., 138
Porter, W., 186
Praly, E., 60–62
Pringle, T. H., 103–104
Proudfoot, N. J., 272–273
Pugh, B. F., 201, 206–207, 207*f*, 234, 272
Pusch, O., 253
Puthenveetil, S., 64–65
Pyle, A. M., 5

Q

Qi, J., 147, 166, 233–234, 272–273
Qi, Z., 16
Qiu, X. W., 116
Qu, C., 298
Quail, M. A., 178–179, 180
Quinones, M., 178–179

R

Rada-Iglesias, A., 298
Radman-Livaja, M., 206–207
Ragoczy, T., 91, 94–96, 116
Raha, D., 272, 298–299, 304
Ranade, S., 147, 206–207
Rando, O. J., 206–207, 234
Raney, B. J., 198*f*
Ransom, M., 30
Rashid, N. U., 303
Rashtchian, A., 195
Rasnik, I., 69, 73–74
Rattner, B. P., 30, 147, 206–207, 207*f*, 214–215
Rau, D. C., 320–321
Ray-Gallet, D., 30
Rearick, T. M., 129
Rechsteiner, T. J., 36–37, 66–69, 217–218
Regula, J. T., 211
Reinberg, D., 30
Ren, B., 302–304
Revyakin, A., 64–65

Reyes, A., 291
Reynolds, B. A., 197–199
Rhead, B., 198*f*
Rhodes, D., 25*f*, 214–215, 217–218
Riccio, A., 272–273
Rice, S., 65–66
Richard, H., 304
Richmond, R. K., 25*f*, 30, 52*f*, 53, 146, 217–218
Richmond, T. A., 91, 114–115, 116, 118–119, 298
Richmond, T. J., 25*f*, 30, 36–37, 52*f*, 53, 66–69, 146, 217–218, 233–234, 316–317, 325
Riemer, C., 134–135
Riggs, B. M., 186, 199, 201–202
Rijkers, E. J., 91–92
Rippe, K., 90, 100–101, 114–115, 116, 319–320
Roach, H. I., 192
Robert, F., 171, 302–304
Roberts, M. S., 164–165, 166
Robinson, J. T., 109
Robinson, K. M., 207–211
Robinson, M. D., 196, 306–307
Robinson, P. J., 214–215
Robyr, D., 91–92
Rogers, J., 216
Roh, T. Y., 206–207, 298–299, 302–304
Rose, M. D., 216
Rosenfeld, J. A., 302–304
Roskin, K. M., 103–104
Ross-Innes, C. S., 115–116
Roth, S. Y., 165–166
Rothberg, J. M., 129
Rougemont, J., 91–92
Roure, V., 91–92
Rousseau, M., 114–115, 118–122, 125, 137
Routh, A., 25*f*
Roy, R., 62
Rozen, S., 100–101
Ruan, J., 133–134, 158, 159–160
Ruby, J. G., 273
Rufiange, A., 171
Russell, R., 60
Ryan, N. K., 90

S

Sabo, P. J., 91–92, 198f
Sabri, A., 186
Sadeh, R., 302
Saenger, W., 13
Safer, D., 65–66
Saha, A., 5, 9–10, 18–20, 23–24, 60–62
Sakai, A., 170–171
Salzberg, S. L., 133–134, 289, 302
Sambrook, J., 226, 227
Sandaltzopoulos, R., 4–5, 60–62
Sanderson, Ch 17
Sanyal, A., 114–115, 118–119, 121–122, 125, 134–137, 138
Sargent, D. F., 25f, 30, 52f, 53, 146, 217–218
Satchwell, S. C., 233–234
Sauvageau, G., 119–121
Scally, A., 178–179, 180
Schaeffer, L., 298–299, 304
Schafer, D. A., 35
Schalch, T., 25f
Schein, J., 110
Scherf, M., 304
Schmidt, C. F., 6–9
Schmidt, D., 115–116
Schnappinger, D., 13
Schneider, B. L., 252
Schnitzler, G. R., 217–218
Schofield, P., 4–5, 229–230, 234
Schones, D. E., 206–207, 298–299, 302–304
Schor, I. E., 272
Schubeler, D., 30, 206–207
Schuettengruber, B., 91–92
Schug, J., 159
Schultz, J., 129
Schultz, M. C., 207–211
Schultz, P. G., 64–65
Schulz, M. H., 304
Schuster, D. M., 195
Schuster, S. C., 147, 166, 233–234
Schutz, K., 116
Schwalie, P. C., 115–116
Schwartz, S., 100–101, 316
Segal, E., 147, 165–166, 206–207, 233–234
Sekiya, S., 146
Selvin, P. R., 60

Selzer, R. R., 91, 114–115, 116, 118–119
Sen, S., 4–5
Seo, C. H., 306–307
Sevastopoulos, E., 65–66
Sexton, T., 90
Shaevitz, J. W., 9–10
Shaffer, C. D., 211
Shah, P., 134–135
Shah, S., 64–65
Shahian, T., 66–67
Sharma, S., 186
Sharon, E., 147–148, 165–166
Sheetz, M. P., 35, 64–65
Sheinin, M. Y., 30–31, 56
Shen, C. H., 149–150, 164–165
Shen, X. T., 4–5
Sheng, J., 303
Shenker, S., 114–115, 118–121
Shepard, P. J., 273
Shi, Y., 273
Shiloach, J., 60–62
Shimko, J. C., 30–31
Shin, H., 298, 302
Shou, C., 272, 298–299, 304
Shundrovsky, A., 30–35, 32f, 36–37, 39–41, 42–43, 46–49, 47f, 50–53, 51f, 52f, 54, 55f, 60–62
Shure, M., 214–215, 223
Sidorova, N. Y., 320–321
Sidow, A., 206–207, 302
Siebzehnrubl, F. A., 197–199
Siegmund, K. D., 186
Sim, R., 30–31
Simon, I., 302–304
Simon, M., 30–31
Simon, R. H., 211–214
Simonis, M., 90, 91–92, 101, 107–109, 116
Simpson, R. T., 165–166, 186
Singer, M. A., 298
Singh, V., 4–5, 229–230, 234
Sinha, M., 4–5
Sirinakis, G., 5, 9–10, 13, 14–15, 14f, 19f, 60–62
Sjolinder, M., 91, 116
Skaletsky, H., 100–101
Slack, M. D., 147, 206–207
Smale, S. T., 146

Smith, B. Y., 50–53
Smith, C. L., 4–5, 9–10, 17, 18–20, 23–24, 30–35, 32f, 36–37, 42, 43, 46, 48, 54, 55f, 60–62, 206–207, 302
Smith, D. E., 30–31
Smith, F., 178–179, 180
Smith, S. B., 5–6, 9–10, 18–20, 23–24, 30–31, 60–62
Smith, S. S., 192–193
Smith, T. F., 133–134
Smolle, M., 4–5, 229–230, 234
Smyth, G. K., 306–307
Snyder, M., 30, 147, 206–207, 207f, 214–215, 272
Soler, E., 91–92
Sondhi, D., 64–65
Song, J. S., 302
Soshnikova, N., 91–92
Soucek, T., 253
Spakowitz, A. J., 30–31, 316
Speed, T. P., 196
Spies, N., 272–273
Spilianakis, C. G., 90, 115–116
Spittle, K. E., 188
Splinter, E., 90, 91–92, 101, 107–109, 115–116
Spooner, E., 64–65
Stadhouders, R., 91–92
Stalker, J., 216
Statham, A. L., 196
Stein, A., 217–218
Steine, E. J., 298
Stephens, P. J., 178–179, 180
Stockdale, C., 5, 60–62
Stone, M. D., 60–62
Strahl, B. D., 146
Straight, A. F., 316
Stralfors, A., 206–207
Stranneheim, H., 276
Strathern, J. N., 317–318
Strattan, J. S., 4–5
Straub, T., 206–207
Struhl, K., 233–234, 272–273
Stryer, L., 60
Stuart, J., 147, 206–207
Stubbs, A., 91–92
Sturzl, S., 206–211

Su, X. Z., 178–179
Sugnet, C. W., 103–104
Sultan, M., 304
Sun, W., 303
Sung, M. H., 91–92
Surtees, J. A., 30–33
Suzuki, H., 119–121
Svaren, J., 163f
Svejstrup, J. Q., 206–211, 227
Swigut, T., 298
Szczelkun, M. D., 5

T

Tabbaa, O. P., 41–42
Tabuchi, T. M., 116
Takai, D., 186
Tan, S., 316–317, 325
Tanaka, T., 64–65
Tanay, A., 186, 199
Tang, F., 307
Tang, Q., 298, 304
Tavoosidana, G., 91, 116
Taylor, J., 134–135
Team, R. D. C., 107–109
Teif, V. B., 319–320
Tekkedil, M. M., 273, 290f
Telling, A., 91, 94–96
Terasima, T., 253
Teunissen, H., 91–92
Teves, S. S., 170–171
Thastrom, A., 36–37
Thevenet, L., 91–92
Thierry-Mieg, D., 272–273
Thierry-Mieg, J., 272–273
Thiriet, C., 171
Thoma, N. H., 30, 206–207
Thomas, J. H., 216
Thongjuea, S., 91–92
Thorsteinsdottir, U., 119–121
Thorvaldsdottir, H., 109
Thurman, R., 198f
Tian, B., 272–273
Tillo, D., 30, 147, 206–207, 207f, 214–215, 229–230, 233–234
Ting, A. Y., 64–65
Tiwari, V. K., 30, 206–207
Tjong, H., 116

Toedling, J., 272
Tolhuis, B., 90, 91–92, 115–116
Tolmach, L. J., 253
Tomsho, L. P., 147, 166, 233–234, 272–273
Tonelli, R., 119–121
Tonthat, T., 147, 206–207
Tonti-Filippini, J., 186
Toro, E., 138
Toth, K., 38–39
Towbin, B. D., 115–116
Trapnell, C., 133–134, 289, 302, 306–307
Travers, A. A., 233–234
Travers, K. J., 188
Trencsenyi, G., 253–254
Tribus, M., 4–5
Tsai, Y. C., 186
Tse-Dinh, Y. C., 60–62
Tsien, R. Y., 64–65
Tsui, K., 229–230
Tsukiji, S., 64–65
Tsukiyama, T., 60–62, 164–165, 234
Tuteja, G., 159
Tyler, J. K., 30

U

Ujvarosi, K., 253–254
Umbarger, M. A., 138

V

Valouev, A., 147, 206–207, 302
van Bakel, H., 229–230
van Baren, M. J., 306–307
van Berkum, N. L., 91, 94–96, 116, 122, 134–137
van de Werken, H. J., 91–92, 107–109
Van Etten, J. L., 186
van Holde, K. E., 214–215, 319–320
van Nimwegen, E., 119–121
van Noort, J., 25f
van Steensel, B., 91, 107–109
Vedam-Mai, V., 197–199
Venters, B. J., 206–207, 272–273
Verstegen, M., 91–92
Verzi, M. P., 115–116
Vinograd, J., 214–215, 223
Viprakasit, V., 91–92

Viswanathan, R., 5, 9–10, 13, 14–15, 14f, 19f, 60–62
von Hippel, P. H., 100–101
von Holt, C., 214–215
Vonkuster, G., 134–135
Voss, T. C., 91–92
Vu, T. H., 116

W

Wacker, D. A., 53
Waern, K., 272, 298–299, 304
Wagner, L., 100–101
Wakefield, M. J., 306–307
Wal, M., 206–207, 207f, 234
Wandzioch, E., 146
Wang, E., 252
Wang, J. P., 317, 324
Wang, L., 306–307
Wang, M. D., 23, 30–35, 32f, 36–37, 39–41, 42–44, 46–49, 47f, 50–53, 51f, 52f, 54, 55f, 56, 60–62
Wang, Q., 298
Wang, S., 91, 116
Wang, X. Q., 119–121, 306–307
Wang, Y., 307
Wang, Z., 206–207, 272, 298–299, 302–304
Ward, E., 206–207, 207f, 234
Watanabe, S., 4–5, 206–211, 227
Waterman, M. S., 133–134
Watts, J., 146
Weber, C. M., 170–171
Webster, D. R., 188
Wei, C. L., 303
Wei, G., 298, 302–304
Wei, L., 302–304
Wei, W., 272
Wei, X., 303
Weiss, S., 60
Weissman, J. S., 316
Welstead, G. G., 298
Weng, Z., 273
West, A., 216
White, C. L., 36–37, 66–69, 233–234
White, P., 159
Whitehouse, I., 5, 234

Widom, J., 19–20, 33–35, 36–37, 62–63, 147, 165–166, 206–207, 217–218, 233–234, 316, 317, 324
Wilkening, S., 273, 290*f*
Willemsen, R., 91–92, 101, 107–109, 116
Williams, B. A., 298–299, 304, 306–307
Williams, L. M., 90, 91, 94–96, 116
Winckler, W., 109
Wippo, C. J., 206–211, 207*f*, 214–215, 217–218, 227, 234
Witten, D. M., 273
Wittmeyer, J., 5, 20
Wold, B., 298–299, 304
Wolff, E. M., 186
Wood, W. H. 3rd., 302–304
Woodcock, C. L., 20
Woodside, M. T., 5–6
Workman, J. L., 272
Wright, M. A., 138
Wu, C., 4–5, 60–62
Wu, L. F., 147, 206–207
Wu, T. D., 289
Wu, W. H., 4–5
Wuller, J. M., 211
Würtele, H., 116
Wyns, K., 30–35, 36–37, 42, 48–49, 50
Wyrick, J. J., 302–304
Wysocka, J., 298
Wysoker, A., 158, 159–160

X

Xi, L., 317, 324
Xie, K. T., 64–65
Xie, X., 298
Xu, H., 91, 303
Xu, J., 146
Xu, M., 186
Xu, N., 307
Xu, Z., 272

Y

Yager, T. D., 319–320
Yamamoto, T., 64–65
Yang, J. G., 60–62, 65–66
Yang, J. O., 306–307
Yang, X., 186
Yao, Z., 302–304
Ybarra, R., 298
Ye, C., 303
Yeh, R. C., 23, 30–31
Yin, H., 41, 43–44
Yoon, O. K., 272–273
Young, T. J., 252
Yuan, G., 147
Yuan, G. C., 206–207

Z

Zahler, A. M., 103–104
Zang, C., 302–304
Zanton, S. J., 147, 166, 206–207, 233–234
Zaret, K. S., 146
Zeitlinger, J., 298
Zhan, Y., 116, 121–122, 125, 137
Zhang, X., 306–307
Zhang, Y. L., 5, 9–10, 13, 14–15, 14*f*, 18–20, 19*f*, 23–24, 30–31, 60–62, 147, 206–207, 207*f*, 214–215, 272–273, 302
Zhang, Z., 100–101, 206–207, 207*f*, 234, 249
Zhao, H., 306–307, 318
Zhao, K., 178–179, 298–299, 302, 303, 304
Zhao, Z., 91, 116
Zheng, M., 298
Zheng, W., 306–307
Zhou, H. M., 320–321
Zhou, Y., 273
Zhu, J., 302–304
Zhu, S., 273
Zhu, Y., 91–92, 107–109
Zhuang, X. W., 60–62
Zofall, M., 60–62
Zuber, J., 252
Zweig, A. S., 198*f*

SUBJECT INDEX

Note: Page numbers followed by "*f*" indicate figures, and "*t*" indicate tables.

A

Adenosine triphosphate (ATP), 234
Aha. *See* Azidohomoalanine (Aha)
Amplicon sequencing
  amplification
    conditions, 195
    denaturation, 195
    parameters, 195
  bisulfite treatment, 194–195
  DNA strands fragment, 192–193
  loci, 192
  primer design
    3′ end, 193
    5′ end, 193
    thymine and cytosines, 193, 194
Atomic force microscopy (AFM) imaging,
  nucleosomal array, 23
ATP. *See* Adenosine triphosphate (ATP)
ATP-dependent chromatin remodeling
  bare DNA
    single molecule experiments (*see* Single
      molecule experiments)
    substrate, 13
    tethered motor assay, 13, 14*f*
    tethered remodeler system, 14–15
  data analysis, 25–26
  description, 4–5
  dual-trap optical tweezers, 9–10
  flow cell assembly
    central channels and protein injection
      tube, 11
    channels, 11
    liquid, 11–12
  high-resolution optical tweezers, 6–9, 7*f*
  microfluidic system, 6–9, 8*f*
  molecular mechanisms, 5
  nucleosome-dependent remodeler
    translocation
    AFM imaging, 23
    DNA labeling, 22
    experimental design, 18–20
    FEC, 23

    NPSs, 20–22
    reconstitution, 22–23
    SWI/SNF and RSC, 20
    translocation, 23–24
  optical traps, 5–6
  power spectrum density, 9–10, 10*f*
  tweezer calibration, 12–13
Azidohomoalanine (Aha)
  H2A/H2B dimers, 171
  labeling and biotin coupling
    incorporation, cellular proteins,
      175, 176*f*
    nuclei preparation, 173–175
    western blot analysis, 175

B

Bioinformatic analysis
  alignment, 289
  biological analyses, 291
  demultiplexing and recovering
    polyadenylated, 289
  filtering, 289–290
  transcriptome and expression analysis,
    290–291, 290*f*
Burrows-Wheeler Aligner (BWA), 133–134
BWA. *See* Burrows-Wheeler Aligner
  (BWA)

C

Centrifugal elutriation
  asynchronous populations, 261, 262*f*
  buffers and reagents, 254
  cell culture conditions, 254
  collecting and processing fractions, 260
  disposables, 254
  FACS, 260–261
  loading cell, 256–260, 259*f*, 262*f*
  materials, 254
  preparation, 256, 257*f*
  protocol, 253–254
  rotor speed and flow rate nomogram,
    256, 257*f*

Centrifugal elutriation (*Continued*)
  system set up, 254–255, 255*f*
ChIP. *See* Chromatin immunoprecipitation (ChIP)
Chromatin immunoprecipitation (ChIP)
  adaptors and primers, 242
  aliquot stocks, 242
  antibody, 242
  antibody-histone attachment, 242–243
  buffers, 242
  description, 252
  enzymes, 242
  FA high-salt wash buffer, 241
  magnetic protein A beads, 240–242
  mitotic cells, 261
  procedure, 243
  sequencing, 298–299, 302–304
  wash series, 243–244
Chromatin organization
  5C technology (*see* Chromosome conformation capture carbon copy (5C) technology)
  description, 114–115
  genome architecture, 115–116
  Hox genes, 119–121
  ion torrent sequencing protocol, 129–131
Chromosome conformation capture carbon copy (5C) technology
  amplification and purification
    composition, 126, 126*t*
    PCR conditions, 126–127, 127*t*
  analysis formats
    HoxA gene cluster, 135–137, 136*f*
    IFL, 135
    my5C, 135–137
  3C libraries generation, 121–122
  control, gene desert region, 137
  data mapping
    BWA and SSHA algorithms, 133–134
    transformation pipeline, 134–135
  description, 131
  Ion Torrent, 131
  libraries
    array hybridization, 119
    detection method, 118–119
    microarrays, 118–119

  normalization, 137–138
  optimal number, 128–129
  preparation
    annealing, 3C libraries, 125
    description, 124
    ligating, 3C junctions, 125
    primers, diluting, 124
    primer design, 122–123
  sequencing, 128
  sequencing quality control
    EmPCR, 131–132
    polyclonal reads, 133
  technology
    3C library, 116–118
    description, 116
    primers, 116–118
Chromosome conformation capture (3C) technology
  availability, 91
  and 5C, 122
  development, 90
  HoxA cluster, 119–121
  libraries, 121–122
  products, 116–118
Circularized chromosome conformation capture (4C) technology
  3C experiments, 90
  cis, data analysis
    FDR, 109
    plotting, 107–109, 108*f*
    running window approach, 107–109
  cross-link
    and cell lysis, 94–95
    ligation junctions, 97–98
    RE digestion, 96–97
    reversal, 98
  description, 90
  DNA circles, 99–100
  genome, mapping, 103–104
  PCR DNA polymerase, 101–102
  PCR products, 103
  primer design, 100–101
  principles
    chromatin, 92, 93*f*
    diploid cells, 92–93
    viewpoint fragment and captures, 92–93

Subject Index

statistics and quality control
  cross-linking hampers, 105
  estimation, 104–105
  PCR products, 105
  read distribution, 105–107
  self-ligated and nondigested, 105, 106f
  trans, 110
Cleavage model
  nucleosome centers, 326, 328f
  template-weighted score pattern, 325–326, 327f

D

DNA methyltransferase accessibility protocol
  amplicons bypasses, 186–188
  amplicon sequencing (see Amplicon sequencing)
  chromatin structure, 186
  MAPit (see Methylation accessibility protocol for individual templates (MAPit))
  materials
    reagents, 188–189
    solutions, 189
    supplies, 190
  sequence analysis
    alignment, 196
    CG methylation, 200f, 202
    DNase I footprinting, 201
    duplication, 197
    graphing percent methylation, 196
    locus and barcode, 195–196
    methylation maps, 201–202
    MethylTracker, 199–201
    MethylViewer, 197–199
    PCR recombination, 197
    percent conversion, 196
  SMRT technology, 188
  workflow, 186, 187f
DNA unzipping
  accuracy and precision, 46–48
  advantages, 31–32
  configurations, 30–31, 32f
  data acquisition, 42–43
  data processing
    conversion, 44–45

  curve alignment, 44
  elastic parameter determination, 43
  trap height determination, 43–44
  description, 30
  experimental sample chambers
    blocking agent, 38–39
    concentration, 39
    description, 38
  force clamp, 41–42
  formation, 37
  loading rate clamp, 41
  nucleosome reconstitution, 36–37, 48–56
  optical trapping apparatus (see Optical trapping apparatus)
  optical trapping system, 41
  single-molecule techniques, 30–31
  template design
    anchor segment, 33
    anchor segment preparation, 33–35
    construction, 33, 34f
    hairpin-capped segment preparation, 35–36
    protocol, 33–35
    segments, 33–35
Double-stranded linkers preparation, 277t, 293–294
Drosophila tissue culture cells
  centrifugal elutriation, 253–261
  ChIP, 252
  FACS, 261–268
  mitotic cells, 253
  separating process, 252–253

F

FACS. See Fluorescent-activated cell sorting (FACS)
False discovery rate (FDR), 109
FDR. See False discovery rate (FDR)
Fluorescence resonance energy transfer (FRET)
  chromatin remodeling enzymes, 60–62
  data analysis
    CCD camera, 77
    description, 77
    efficiencies calculation, 78
    histograms, 78–79
    IDL scripts, 77–78

Fluorescence resonance energy transfer
(FRET) (Continued)
single-molecule time traces, 79–83
description, 60
distance dependence, 60, 61f
fluorescently labeled sample
(see Fluorescently labeled sample)
optical setup
excitation laser intensity, 72
objective-type microscopy, 72
prism-type microscopy, 71–72
TIR, 71
PEG (see Polyethylene glycol (PEG))
sample immobilization and data
acquisition
data collection, histograms, 75
imaging buffer and reagents, 73–74
single-molecule traces, 75–76
solutions/reagents, 74–75
Fluorescent-activated cell sorting (FACS)
buffers and reagents, 263
colchicine treatment
ChIP, 268
control cells preparation, 267–268
detergent, 266
DNA content, 263–264, 265f
formaldehyde, 263
mitotic index, 264, 265f
pre- and post-sorting procedures, 267
staining procedure, 264–266
titration, 266
disposables, 262
materials, 261–262
protocol, 261
Fluorescently labeled sample
chromatin remodeling, 62–63
dye pairs, 62
histone octamer, 66–67
nucleosomes reconstitution and
purification, 67–68
preparation and purification
dye-labeled DNA, 64
nucleic acid molecules, 63
PCR, 63–64
site-specific labeling, proteins
Cys-lite, 65–66
description, 64–65

Force-extension curve (FEC), single
nucleosomal array, 23, 25f

G

GAPDH. See Glyceraldehyde 3-phosphate
dehydrogenase (GAPDH)
Genome-wide mapping, nucleosome
positions
adaptor ligation, 246
A-tailing, 245–246
cell lysis, MNase digestion
agarose, 238
aliquot stocks, 238
bead-beating lysis, 238–239
chromatin, 236–238
2× proteinase K buffer, 237
DNA extraction, 240
enzyme stocks, 237
FA lysis buffer, 237
mechanical force, 236–238
methodology, 239
NP-S buffer, 237
reversing cross-links, 239–240
TE buffer, 237
zirconia/silica beads, 238
cellular process, 234
ChIP and library preparation, 240–244
chromatin remodeling enzymes, 234
description, 233–234
gel purification, 248–249
harvesting, 235–236
kinase reaction, 245
ligation-mediated PCR, 247–248
Phi29, 246–247
qiagen cleanup, 249
Genome-wide polyadenylation site mapping
bioinformatic analysis, 289–291
cDNA, purification, 282–283
dA overhang, 285–286
double-stranded linkers, 293–294
end repair, 285
ligation, 286
mRNA sequencing methods, 273
PCR, 286–287
protocol, 273–274, 274f
purification, double-stranded cDNA, 283
quality control

cloning and Sanger sequencing, 292
in vitro transcripts, 292–293
library size, 291f, 292
recommendations, 275–276
reverse transcription, 277t, 282
sample preparation
  DNA contamination, 280–281
  extraction, total RNA, 276–280
  fragmentation, 281
  purification, 281–282
second-strand synthesis, 283
size selection, 288
$3'$ terminal cDNA fragments
  binding, streptavidin beads, 284–285
  biotinylated sequencing primer, 284
  streptavidin beads, 284
transcription, 272–273
Genomic plasmid library
chromatin, 214–215
cultivation and plasmid preparation, 217
electrocompetent E. coli cells, 216
electroporation, 216
prokaryotic and eukaryotic DNA, 214–215
Glyceraldehyde 3-phosphate dehydrogenase (GAPDH), 192

H

Histones
  Drosophila, 211
  hydroxylapatite, 211–214
  modifications, 302–304
  purification, 211–214
  SDS-PAGE, 215f
Hydroxyl radical cleavage
  locations, 325, 325f
  nucleosomal DNA, 321

I

IFL. See Interaction frequency list (IFL)
Interaction frequency list (IFL), 134–135
In vitro and in vivo, nucleosome positioning
  biochemical isolation, 229–230
  flowchart, 206–207, 207f
  genomic plasmid library
    chromatin, 214–215

cultivation and plasmid preparation, 217
electrocompetent E. coli cells, 216
electroporation, 216
eukaryotic and prokaryotic DNA, 214–215
histones, 211–214
incubation of salt gradient dialysis, 227–228
MNase digestion, 228–229
molecular mechanisms, 206–207
plasmid-based genomic library, 206–207, 207f
salt gradient dialysis (see Salt gradient dialysis)
yeast WCE, 207–211
In vitro transcripts (IVTs)
internal controls, 292–293
stock solution, 293
Ion torrent sequencing protocol
chips, 130–131
HoxA cluster, 130–131
ISPs, 129
PGM^{TM}, 129
IVTs. See In vitro transcripts (IVTs)

M

MAPit. See Methylation accessibility protocol for individual templates (MAPit)
Methylation accessibility protocol for individual templates (MAPit)
  GAPDH promoter, 192
  NSCs, 190
  nuclei isolation, 190–191
  probe, M.CviPI, 191
  qMSRE, 192
MethylTracker
  FASTA format, 199–201
  images, 199, 200f
MethylViewer
  GCG sites, 199
  human NSCs, 197–199, 198f
  "lollipop" diagram, 197
Micrococcal nuclease (MNase), 298–299, 302–304, 316
  agarose, 238
  aliquot stocks, 238

Micrococcal nuclease (MNase) (*Continued*)
bead-beating lysis, 238–239
chromatin, 236–238
digestion
anti-H3 immunoprecipitation, 228–229
concentration, 229t
products, 166
yeast nuclei, 149–150
DNA extraction, 240, 241f
enzyme stocks, 237
FA lysis buffer, 237
mechanical force, 236–238
methodology, 239
NP-S buffer, 237
2× proteinase K buffer, 237
reversing cross-links, 239–240
sequence bias, 165–166
TE buffer, 237
titration, 151
zirconia/silica beads, 238
MNase. *See* Micrococcal nuclease (MNase)

N

NCP. *See* Nucleosome center positioning (NCP)
Neural stem cells (NSCs), 190, 197–199, 198f, 200f
Next-generation sequencing technology
"A" overhangs to 3' ends, 300–301
data analysis, 302
DNA purification and end-repair, 300
histone modifications, 302–304
linker ligation, 301
methylation, 298
mononucleosomes, 299–300
PCR amplification, 301–302
RNA-seq (*see* RNA-sequencing)
transcriptional regulation, 298
NSCs. *See* Neural stem cells (NSCs)
Nucleosome
description, 170
histones
Aha-labeled cells, 171
Aha labeling and biotin coupling, 173–175

chromatin fragmentation and extraction, 175–176
DNA isolation, 177–178
methionine-free growth medium, 173
solutions and materials, 172
streptavidin affinity capture, 176–177
Solexa library, 170–171
Solexa library preparation (*see* Solexa library preparation)
Nucleosome center positioning (NCP)
definition, 329
score, 329
Nucleosome occupancy and positions. *See* Next-generation sequencing technology
Nucleosome occupancy maps
description, 161
sequences, 161
Web-viewer program, 164
yeast, 161–164, 163f
Nucleosome positions. *See* Genome-wide mapping, nucleosome positions
Nucleosome position sequences (NPSs), 20–22
Nucleosomes mapping
chemical cleavage
ethidium bromide-stained agarose gel, 318, 319f
"no mapping control", 318–319
chemically mapped DNA fragments, 323–324
chemical mapping data
base composition, 329–330
cleavage model (*see* Cleavage model)
description, 324
four-template model, 329–330, 330f
NCP score, 329
occupancy scores, 331
sequencing depth, 332
single *vs.* paired-end sequencing, 324
site identification, 325
template-weighted score approach, 327–328, 327f
use, 324
chemical mapping method, 316–317
description, 316
DNA fragments, 321–323

H4S47C growth and permeabilization, 319–320
hydroxyl radical cleavage, 321
labeling, 320–321
MNase, 316
salt and detergent, 321–323
S. cerevisiae strain, 317–318
Nucleosomes position map
  analysis, 164–165
  approach, 164
  complication, construction, 164
  description, 164
Nucleosome, unzipping
  histone–DNA interactions
    constant force, 50–53, 52f
    interaction map, 53–54
    RNA polymerase, 54
  reconstitution, 36–37
  remodeling, 54–56
  signature
    arbitrary sequence, 50, 51f
    forward and reverse directions, 48–49
    tetrasome, 50
Nucleosomes, yeast
  bioinformatic analysis
    data files, 159–160
    length distribution histograms, 160
    occupancy map, 161–164
    position map, 164–165
  core particle DNA
    adapter-ligated nucleosomal, 157–158
    modification, 154–155
    mononucleosomal, 152–153
    paired-end adapters, 155–157
    quality control, 153–154
  core particles
    aim, 148
    gel-purification, 150–152
    nuclei, 149–150
  digestion products, 166
  eukaryotic DNA, 146
  formaldehyde, 166
  hybridization method, 147
  Illumina sequencing, 147–148
  MNase, 147, 165–166
  occupancy map, 148
  pair end sequencing (*see* Pair end sequencing)
  position, 146
  types, 147

O

Optical trapping apparatus
  laser intensity, 39–41
  layout, 39–41, 40f

P

Paired-end adapters
  barcoded, 155
  correct band, 156
  gel piece, 157
  ligation, 155–156
  protocol, 155
Pair end sequencing
  description, 158
  genome, 159
  HiSeq lane, 158–159
PCR. *See* Polymerase chain reaction (PCR)
PEG. *See* Polyethylene glycol (PEG)
Polyadenylation site (PAS). *See* Genome-wide polyadenylation site mapping
Polyethylene glycol (PEG)
  description, 69
  quartz slides, 69, 70f
  sample chambers, 71
  surface coating, 69–71
Polymerase chain reaction (PCR), 188, 195, 197, 286–287

Q

qMSRE. *See* Quantitative methyl-sensitive restriction endonuclease digest (qMSRE)
Quantitative methyl-sensitive restriction endonuclease digest (qMSRE), 192

R

RE. *See* Restriction enzyme (RE)
Restriction enzyme (RE)
  cross-linked chromatin, digestion, 96–97
  digestion and ligation, 99–100
RNA-sequencing

RNA-sequencing (*Continued*)
  amplification, double-strand cDNA, 310–311
  cDNA synthesis
    first-strand, 305, 308
    second-strand, 306, 308–310
  data analysis, 306–307, 311
  fragment, double-strand cDNA, 306
  mRNA purification, 304–305
  primers, 307
  sequencing libraries, 306

S

Salt gradient dialysis (SGD)
  description, 221
  histone–DNA interactions, 217–218
  incubation, 227–228
  mini chamber, 218–220, 219f
  pump and beakers, 218, 219f
  samples, 215f, 220
  titrating
    chloroquine, 224–226
    chromatin, 224
    MNase ladder assay, 226–227
    plasmid, *E. coli*, 224–226
    plasmid supercoils, 223
    supercoil assay, 221, 222f
SGD. *See* Salt gradient dialysis (SGD)
Single molecule experiments
  bead preparation, 16
  buffer, 15
  remodeler solution preparation, 15
  remodeler translocation, bare DNA, 18
  single DNA, 16–18
Single molecule real time (SMRT) sequencing, 188

Single-molecule techniques, 30–31
Single-molecule time traces, FRET
  kinetic parameters, 81–82
  Markov process, 82–83
  Matlab script, 81–82
  mononucleosomes, 79–81, 80f
  phases, 82
  photophysical properties, 81
SMRT sequencing. *See* Single molecule real time (SMRT) sequencing
Solexa library preparation
  adapter ligation and Ampure bead purification, 182
  A-tailing, 181–182
  end repair, 180–181
  Illumina protocol, 178–179
  paired-end adapter and primers, 180
  PCR amplification and final purification, 182–184
  phenol extraction and column purification, 181
  solutions and materials, 179–180

T

TIR. *See* Total internal reflection (TIR)
TMAC. *See* Trimethyl ammonium chloride (TMAC)
Total internal reflection (TIR), 71
Trimethyl ammonium chloride (TMAC), 195

U

Unzipping accuracy technique, 46–48

Y

Yeast whole cell extract (WCE), 207–211

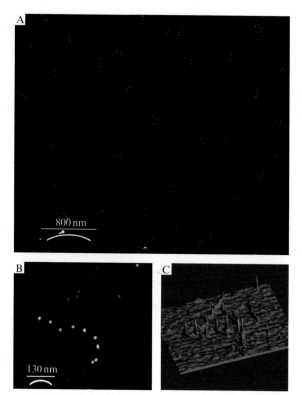

Yongli Zhang et al., Figure 1.8 AFM images of the reconstituted nucleosomal arrays. (A) AFM image of the nucleosome arrays on the plasmid DNA containing nine tandem repeats of NPSs (pUC-N9). Nucleosomes (bright spots) are mainly formed on NPSs located at half of the DNA molecule, with the other half being plasmid DNA free of nucleosomes. (B) Close-up view of a single nucleosomal array containing nine uniformly spaced nucleosomes. (C) The 3D image corresponding to the image in (B). The average height of nucleosomes is 3.1 nm.

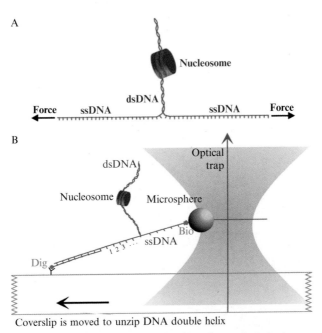

Ming Li and Michelle D. Wang, Figure 2.1 Experimental unzipping configuration. (Adapted from Hall et al., 2009 and Shundrovsky et al., 2006, with permissions from the publishers.) (A) A simplified cartoon of the unzipping configuration. A DNA double helix is mechanically unzipped in the presence of DNA-binding proteins, such as a nucleosome, by the application of opposing forces on the two strands. (B) A typical experimental configuration for unzipping. An optical trap is used to apply a force necessary to unzip through the DNA as the coverslip is moved away from the trapped microsphere.

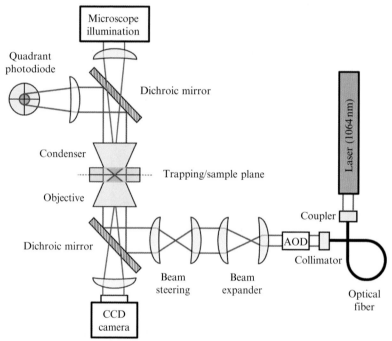

Ming Li and Michelle D. Wang, Figure 2.3 Layout of the optical trapping apparatus. See text for a detailed description of the setup.

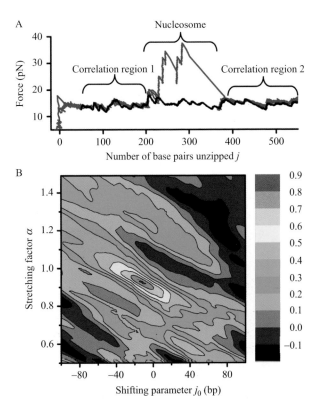

Ming Li and Michelle D. Wang, Figure 2.4 Unzipping curve alignment. (A) An example of force versus number of base pairs unzipped plot for a nucleosome unzipping curve (red) after alignment with the corresponding theoretical curve for naked DNA of the same sequence (black). Unzipping was carried out at a loading rate of 8 pN/s. Regions 1 and 2 flanking the nucleosome were used for correlation. Note also that the initial rise of force is located at $j = 0$ bp, corresponding to stretching of the anchoring segment before strand separation. (B) A two-dimensional intensity graph of the generalized correlation function $R(a, j_0)$ for the trace shown in A. The peak $R = 0.80$ is located at stretching factor $a = 0.93$ and shifting parameter $j = -14$ bp.

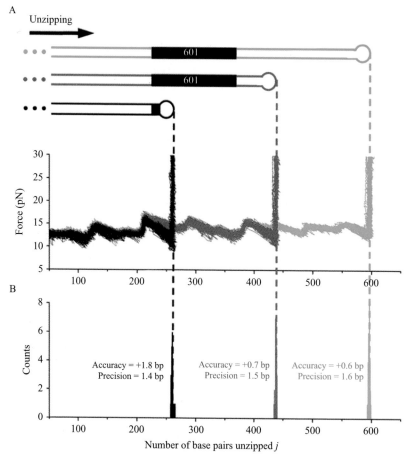

Ming Li and Michelle D. Wang, Figure 2.5 Characterization of the accuracy, and precision, of the unzipping method. (Adapted from Hall et al., 2009, with permission from the publisher.) (A) Three hairpin-capped unzipping templates were unzipped using a loading rate clamp (8 pN/s): 258 bp (black, 21 traces), 437 bp (red, 27 traces), and 595 bp (green, 33 traces). (B) For each template, a histogram was generated from the data points in the vertically rising section only. The measured hairpin location of each template was taken as the mean of the histogram. The accuracy was determined by the difference between the mean of the histogram and the expected value (dashed vertical line). The precision was determined by the standard deviation of the histogram.

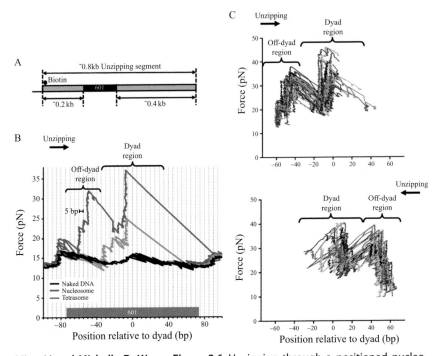

Ming Li and Michelle D. Wang, Figure 2.6 Unzipping through a positioned nucleosome using a loading rate clamp at 8 pN/s. (Adapted from Dechassa et al., 2011, with permission from the publisher.) (A) A sketch of the forward nucleosome unzipping segment. (B) Representative force unzipping signatures of naked DNA (black), DNA containing a nucleosome (red), and DNA containing a tetrasome (green). Both the nucleosome and the tetrasome were assembled onto an unzipping segment containing the 601 positioning element. The arrow indicates the unzipping direction. Two distinct regions of interactions, as well as a 5-bp periodicity within each region, were observed for the nucleosome. The tetrasome signature exhibits only a single region of interactions, which substantially overlaps the dyad region identified in the nucleosome. (C) Multiple traces of unzipping through a nucleosome from both forward (upper panel, 31 traces) and reverse (lower panel, 28 traces) directions. Each color represents data obtained from a single nucleosomal DNA molecule. Distinct regions of interactions and a 5-bp periodicity within each region are highly reproducible.

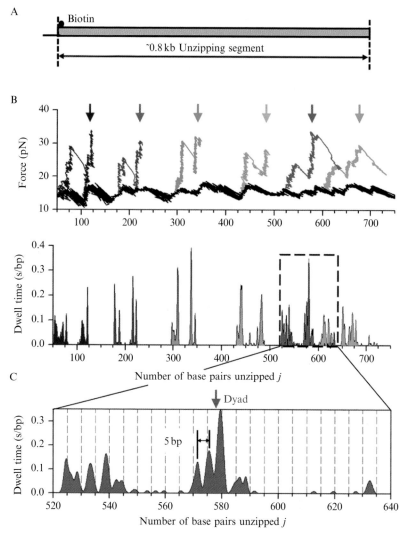

Ming Li and Michelle D. Wang, Figure 2.7 Unzipping through a nucleosomes on an arbitrary sequence using a loading rate clamp at 8 pN/s. (Adapted from Hall et al., 2009, with permission from the publisher.) (A) A sketch of the unzipping segment. (B) Force unzipping signature of a nucleosome at different locations on a DNA template lacking known strong positioning elements. Each color was obtained from a single nucleosome unzipping trace, with the unzipping force shown in the top panel and the corresponding dwell time histogram shown in the bottom panel. The unzipping signature of a naked DNA molecule of the same sequence is also shown (black), as a reference. Vertical arrows indicate the observed dyad locations of these nucleosomes. (C) Close-up of the dwell time histogram for a specific unzipping trace (red) to emphasize the 5-bp periodicity observed in each interaction region of the unzipping signature.

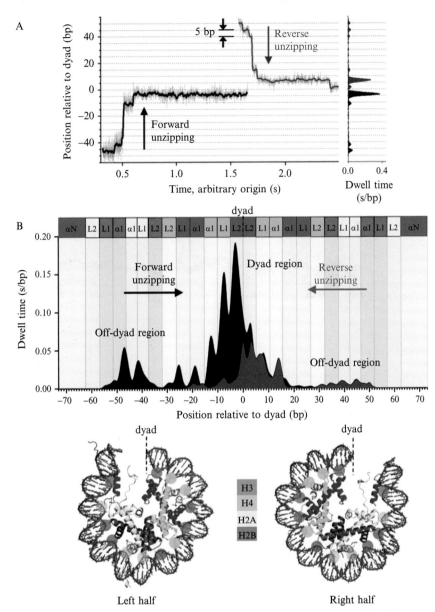

Ming Li and Michelle D. Wang, Figure 2.8 Unzipping through a positioned nucleosome using a force clamp at 28 pN. (Adapted from Hall et al., 2009, with permission from the publisher.) (A) Representative traces of forward (black) and reverse (red) unzipping through a nucleosome under a constant applied force ($\sim$ 28 pN). The unzipping fork paused at specific locations when passing through a nucleosome, which are evident from both the traces (left) and their corresponding dwell time histograms (right). (B) A histone–DNA interaction map is constructed by using a total of 27 traces from the forward direction and 30 traces from the reverse direction. Each peak corresponds to an individual histone–DNA interaction and the heights are indicative of their relative strengths. Three regions of strong interactions are indicated: one located at the dyad and two located off-dyad. The bottom panel is the crystal structure of the nucleosome core particle (Luger et al., 1997), where dots indicate individual histone binding motifs that are expected to interact with DNA. The two halves of the nucleosome are shown separately for clarity. On the top panel, these predicted interactions are shown as colored boxes.

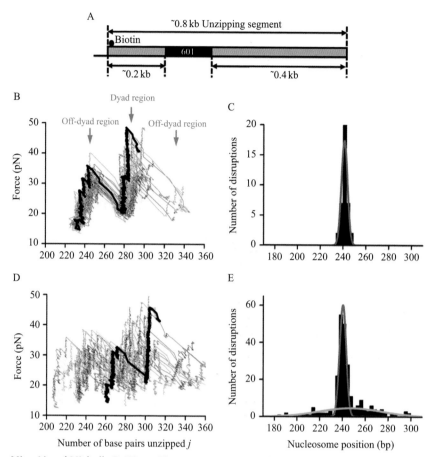

Ming Li and Michelle D. Wang, Figure 2.9 Unzipping through nucleosomes before and after remodeling. (Adapted from Shundrovsky et al., 2006, with permission from the publisher.) (A) A sketch of the nucleosome unzipping segment. (B) Unzipping signatures for 30 unremodeled data curves. A single representative curve is highlighted in black. (C) Histogram of unremodeled nucleosome positions on the DNA. A nucleosome position is defined as the mean position of interaction of the first off-dyad region. Data (black) and their Gaussian fit (red) are shown. The distribution is centered at 241 bp, on this particular template, with a standard deviation (SD) of 2.6 bp. (D) Force unzipping signature for 30 data curves obtained after SWI/SNF remodeling. A single representative curve is highlighted in black. (E) Histogram of remodeled nucleosome positions on the DNA after remodeling reaction times <1 min. Data (black) and their fit to two Gaussians (red and green) are shown. The red fit curve represents the nucleosome population that remained at the original 601 position (unremodeled; center=240 bp, SD=2.8 bp), while the green curve corresponds to those moved by the action of yeast SWI/SNF (remodeled; center=247 bp, SD=28 bp).

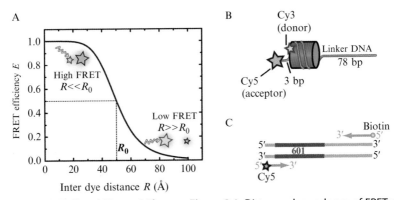

Sebastian Deindl and Xiaowei Zhuang, Figure 3.1 Distance dependence of FRET efficiency and example construct design for single-molecule FRET imaging of nucleosome remodeling. (A) The FRET efficiency as a function of the distance between donor and acceptor dye molecules. The donor and acceptor fluorophores are represented as green and red stars, respectively. At the Förster distance, R_0, the donor fluorophore transfers 50% of its energy to the acceptor fluorophore. (B) Example nucleosome with a single donor dye on the proximal H2A subunit. The histone octamer and the nucleosomal DNA are shown in blue and brown, respectively. The Cy5 (acceptor) and Cy3 (donor) fluorophores are depicted as red and green stars, respectively. (C) The Cy5 dye (red star) and the biotin moiety (orange dot) are included in the PCR primers used to generate the nucleosomal DNA.

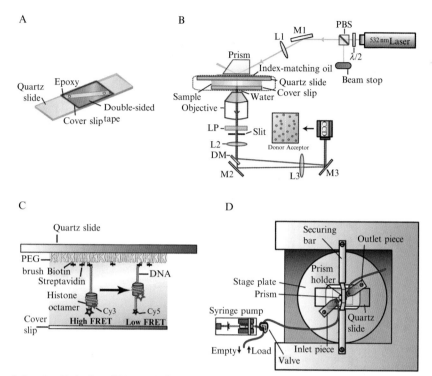

Sebastian Deindl and Xiaowei Zhuang, **Figure 3.2** Sample chamber and optical setup for single-molecule FRET imaging with buffer exchange during data acquisition. Individual components are not drawn to scale. (A) Components used to assemble a flow chamber on a quartz slide. (B) Schematic of the optical setup for single-molecule FRET imaging using the prism-type TIR geometry. The excitation laser beam is focused into a Pellin-Broca fused silica prism with a shallow incident angle (<23°) such that an evanescent field is created at the quartz–water interface on the quartz slide. Fluorescence emission from donor and acceptor dyes is collected by the objective, and scattered excitation laser light is removed by a long pass filter. Donor and acceptor emission is separated by a dichroic mirror. The dichroic mirror and an additional mirror are used to slightly offset the paths of donor and acceptor fluorescence such that they can be imaged onto two halves of the CCD camera. The image is reduced by a vertical slit in the imaging plane to fit onto half of the CCD chip area. $\lambda/2$, half waveplate; PBS, polarizing beam splitter; M1–M3, mirrors; L1–L3, lenses; LP, long pass filter; DM, dichroic mirror. (C) Larger view of the boxed region (dotted line) in (B). The slide is covered with a PEG brush to minimize nonspecific sticking of proteins to the quartz glass, and the sample is immobilized via a biotin–streptavidin linkage. The flow channel is formed between the quartz slide and the cover slip. (D) Top view of the microscope stage. A stage plate holds the sample cell. The prism is attached to the prism holder and placed on top of the quartz slide using a securing bar. Flow in the sample chamber is introduced with a motorized syringe pump that is connected to the inlet piece of the flow channel.

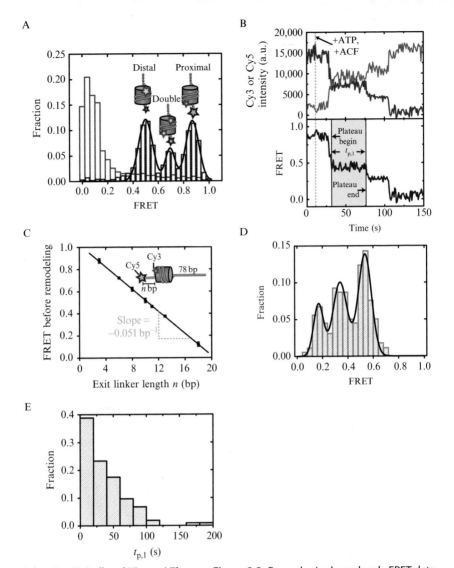

Sebastian Deindl and Xiaowei Zhuang, Figure 3.3 Example single-molecule FRET data of mononucleosomes. (A) FRET histogram constructed from many nucleosomes before (blue bars) and after (red bars) ACF-catalyzed remodeling. The three initial peaks (blue bars) arise from the three distinct donor dye labeling configurations (single proximal donor, single distal donor, proximal and distal donor present as described in the text). The black line represents a fit with three Gaussians. Upon remodeling with ACF, the FRET values are shifted toward very low values (red bars). (B) Donor fluorescence (green), acceptor fluorescence (red), and FRET (blue) traces depicting ACF-induced remodeling of an individual nucleosome with a single proximal donor fluorophore. The first intermediate FRET plateau is shaded in yellow. The duration of the first translocation pause is labeled as $t_{p,1}$, and begin and end time points are indicated for the first intermediate FRET plateau. (C) Initial FRET value (before remodeling) as a function of the exit linker DNA length (n bp). The black line depicts a linear fit to the data with a slope of -0.051 ± 0.002. The deviation from the expected nonlinear dependence is caused by the flexible linkers connecting the dyes to the nucleosome or the DNA. For this ruler, data from nucleosomes with a single Cy3 dye on the proximal H2A subunit are used. (D) FRET distribution of the pauses. (E) Distribution of $t_{p,1}$, the duration of the first translocation pause.

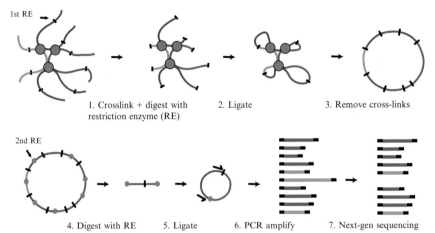

Harmen J.G. van de Werken et al., Figure 4.1 4C technology. Outline of the strategy is provided. After cross-linking by formaldehyde and digestion with the first restriction enzyme (1st RE), "hairballs" of cross-linked DNA are created (1). Chromatin is diluted and religated to fuse the ends of DNA fragments present in the same "hairball" (2). The ultimate outcome of this ligation event is large DNA circles encompassing multiple restriction fragments. Cross-links are removed by heating (3); DNA is digested by a second RE (usually a four cutter) (4) and religated under diluted conditions to create small DNA circles, most of which carry a primary ligation junction. Inverse PCR primers specific for the fragment of interest ("viewpoint," "bait") and carrying 5′ adapter overhangs for next-gen sequencing allow amplification of all its captured sequences followed by high-throughput sequencing.

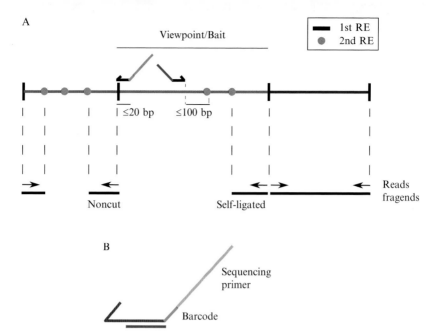

Harmen J.G. van de Werken et al., Figure 4.2 Details of the 4C PCR strategy. (A) First (read) primer is designed on top of the primary restriction site; second primer is designed within 100 base pairs of neighboring secondary restriction site on the viewpoint fragment (red). Green and blue: neighboring restriction fragments. Black horizontal bars indicate the amplifiable genomic parts in 4C (so-called fragends). Horizontal arrows show start and direction of sequencing reads. (B) Details of first primer, designed on top of a *Hin*dIII restriction site. Barcode (2–3 nucleotides) allows pooling of multiple 4C experiments from the same viewpoint in a single sequencing Illumina lane.

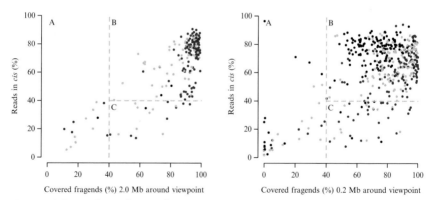

Harmen J.G. van de Werken et al., Figure 4.4 4C quality assessment. Percentage of total number of reads mapped in *cis* (i.e., to the chromosome containing the viewpoint) is plotted versus local coverage around the viewpoint for 4C experiments performed with a six (left, $n=225$) and four (right, $n=467$) cutter as primary restriction enzyme. Colors refer to the total number of reads obtained per experiment, being <300,000 (black), 300,000–1,000,000 (gray), 1–3 M (blue), and >3 M (red). In a high-quality 4C experiment, the great majority of reads map in *cis* and nearly all fragends 1 Mb (left, six cutter) and 0.1 Mb (right, four cutter) on either side of the viewpoint are read more than once. Note that reads from self-ligation and nondigested products were removed from the analysis.

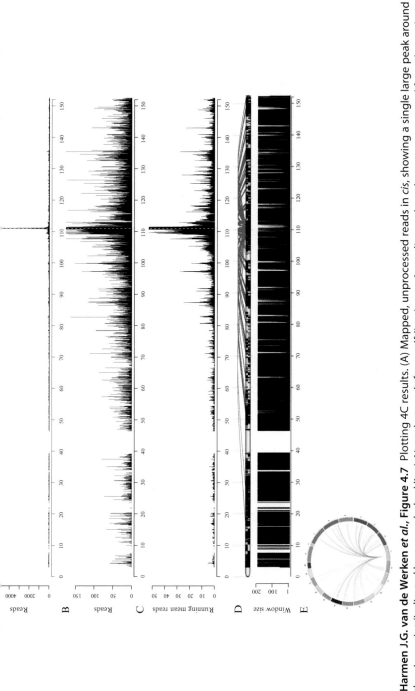

Harmen J.G. van de Werken et al., Figure 4.7 Plotting 4C results. (A) Mapped, unprocessed reads in *cis*, showing a single large peak around the viewpoint (indicated by a gray dashed line). Note that reads from self-ligation and nondigested products were removed from the analysis. (B) Same, but with y-axis scaled to show the many other reads in *cis*. (C) Same data processed by a running mean sliding window approach (31 fragends window) to allow better viewing of specific chromosomal contacts. (D) High-level analysis using a domainogram to show significance scores of contacts across a range of chromosomal window sizes. On top an arachnogram shows the contacts for a given window size. (E) Circo plot visualizing *trans* interactions across all other chromosomes.

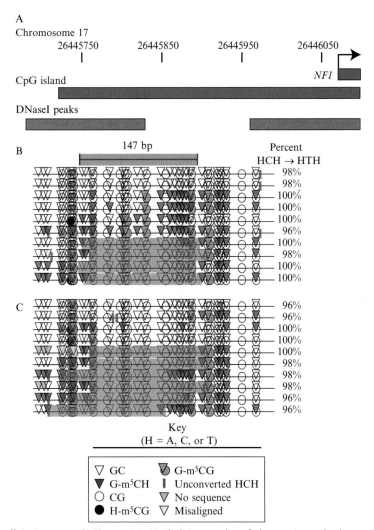

Russell P. Darst et al., Figure 8.2 MethylViewer plot of chromatin at the human *NF1* promoter in NSCs. (A) The 303-bp amplicon was on the transcribed strand ∼200 bp 5′ of the *NF1* transcription start site (bent arrow) and gene body (dark blue bar), overlapping both a CG island (annotated by UCSC Genome Browser; http://genome.ucsc.edu/; Dreszer et al., 2012; Fujita et al., 2011), and sites hypersensitive to DNase I in several cell lines (Sabo et al., 2006). Nucleotide coordinates on chromosome 17 are indicated at the top. (B) MethylViewer plot (to scale with A) of 10 molecules randomly chosen from over 1000 sequenced with forward primer. Of these, four featured a protected footprint on the nucleosome scale, marked by the light blue shading. Note that the top two molecules appear to be duplicates, having identical unconverted HCH site and methylation status of all sites. Nucleosome core particle length is indicated by bar with light blue shading. (C) Plot of 10 randomly chosen from over 6000 sequenced with the reverse primer.

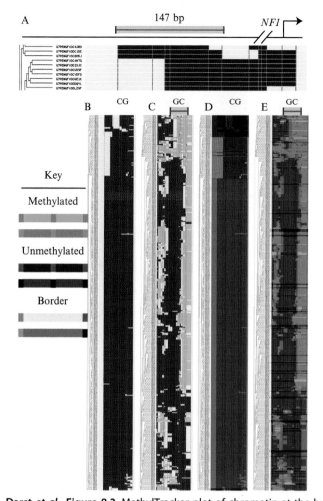

Russell P. Darst et al., Figure 8.3 MethylTracker plot of chromatin at the human *NF1* promoter in NSCs. (A) Diagram of the same amplicon as in Fig. 8.2 and a close-up view of the MethylTracker plot of CG methylation with these settings: two sites to make a patch, one to break, and exclude GCG. Each row represents one sequence read. See key below: methylated and unmethylated sites are indicated by dark green and bright red vertical bars, respectively. Green patches span successive methylated sites and dark red patches span successive unmethylated sites. Yellow patches mark the borders between sites of different methylation status, as well as sequence not bordered on both sides by sites, that is, at each end. Blue bar indicates 147-bp scale. (B) The complete CG methylation plot for 500 randomly chosen molecules. (C) The GC methylation plot for the same molecules. Blue bar at the top indicates 147-bp nucleosome core particle-length scale. Note variable nucleosome positioning among the molecules inferred from ~150 bp or longer spans of protection against GC methylation (dark red patches). As CG and GC methylation were clustered separately, the order of molecules differs from that in (B). (D) Molecules were divided into three classes by CG methylation structure, tagged red, green, or blue, and clustered for CG methylation again. Dark shading now marks unmethylated spans, spans of bright color mark successively methylated sites, and intermediate shading represents the borders. (E) The GC methylation plot for the molecules barcoded by CG structure. Dispersed distribution of molecules from each of the three CG clusters in (D) over the range of chromatin structures clustered in (C) based on GC methylation shows that CG methylation does not correlate with nucleosome positioning at the locus.

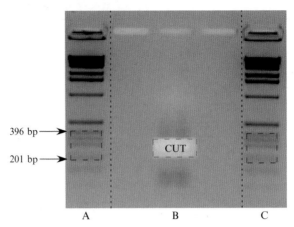

Megha Wal and B. Franklin Pugh, Figure 10.2 Agarose gel purification. Regions A and C indicate 1-kb ladder. Region B: red box indicates the excised gel portion.

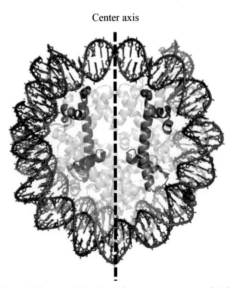

Kristin R. Brogaard et al., Figure 14.1 A nucleosome structure highlighting histone H4 and residue serine 47 (red). Serine 47 is mutated to a cysteine and is the site where the sulfhydryl-reactive label covalently binds. Adapted from Brogaard et al., 2012.

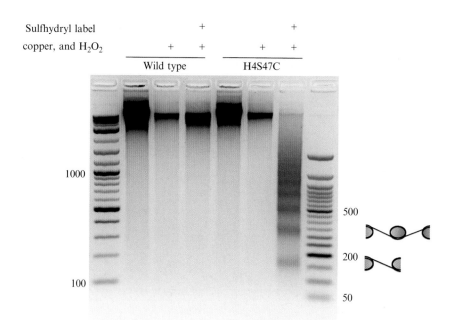

Kristin R. Brogaard et al., Figure 14.2 Ethidium bromide-stained agarose gel showing the chemical mapping results in a DNA banding pattern. Mapping (and observable DNA cleavage) only occurs when the reaction includes (indicated by "+") the sulfhydryl-reactive label, copper, H_2O_2, and the H4S47C mutant yeast. The cartoons adjacent to the agarose gel illustrate that the banding pattern is produced from mapping successive nucleosomes' centers. Adapted from Brogaard et al., 2012.

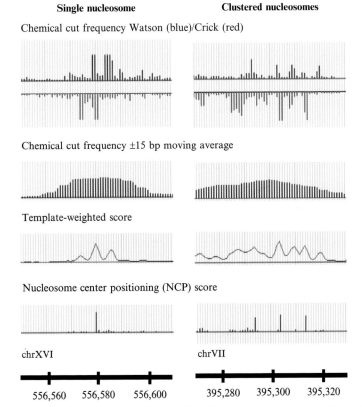

Kristin R. Brogaard et al., Figure 14.5 A well-positioned nucleosome results in a three-peak template-weighted score pattern with the middle highest peak corresponding to the nucleosome center position. A single, well-positioned nucleosome also results in a dominant NCP score peak indicating the nucleosome center. Clustered nucleosomes produce many peaks in the template-weighted score, causing difficulty in identification of true nucleosome centers. The deconvolution algorithm is able to identify three clear major NCP score peaks, defining three overlapping nucleosomes, separated by 10 bp. Adapted from Brogaard et al., 2012.